More than 1 billion people live with inadequate access to safe drinking water, with dramatic consequences for lives, livelihoods and development. Transparency International's *Global Corruption Report 2008* demonstrates in its thematic section that corruption is a cause and catalyst for this water crisis, which is likely to be further exacerbated by climate change. Corruption affects all aspects of the water sector, from water resources management to drinking water services, irrigation and hydropower. In this timely report, scholars and professionals document the impact of corruption in the sector, with case studies from all around the world offering practical suggestions for reform.

The second part of the *Global Corruption Report 2008* provides a snapshot of corruption-related developments in thirty-five countries from all world regions. The third part presents summaries of corruption-related research, highlighting innovative methodologies and new empirical findings that help our understanding of the dynamics of corruption and in devising more effective anti-corruption strategies.

Transparency International (TI) is the civil society organisation leading the global fight against corruption. Through more than ninety chapters worldwide and an international secretariat in Berlin, Germany, TI raises awareness of the damaging effects of corruption, and works with partners in government, business and civil society to develop and implement effective measures to tackle it. For more information, go to www.transparency.org.

Global Corruption Report 2008

Corruption in the Water Sector

CAMBRIDGE UNIVERSITY PRESS
Cambridge, New York, Melbourne, Madrid, Cape Town, Singapore, São Paulo, Delhi

Cambridge University Press
The Edinburgh Building, Cambridge CB2 8RU, UK

Published in the United States of America by Cambridge University Press, New York

www.cambridge.org
Information on this title: www.cambridge.org/9780521727952

First published 2008

Printed in the United Kingdom at the University Press, Cambridge

A catalogue record for this publication is available from the British Library

ISBN 978-0-521-72795-2
ISSN: 1749-3161

Edited by Dieter Zinnbauer and Rebecca Dobson

Contributing editors: Krina Despota, Craig Fagan, Michael Griffin, Robin Hodess and Mark Worth

Contents

Illustrations

Figures

Tables

Boxes

Contributors

- Enriqueta Abad – Transparency International
- Sonny Africa – IBON Foundation
- Federico Arenoso – Poder Ciudadano (TI Argentina)
- Martín Astarita – Poder Ciudadano (TI Argentina)
- Sona Ayvazyan – Center for Regional Development (TI Armenia)
- Alma Rocío Balcázar – Transparencia por Colombia (TI Colombia)
- Patrick Barron – World Bank
- Paramjit S. Bawa – TI India
- Marianne Bertrand – University of Chicago
- Peter Bosshard – International Rivers
- John Butterworth – International Water and Sanitation Centre
- Louis Bwalya – TI Zambia
- Vanja Calovic – (The Network for Affirmation of the NGO Sector – MANS)
- Sue Cavill – Loughborough University
- Camrin Christensen – TI Georgia
- Bernard Collignon – Hydroconseil
- Iulia Cospanaru – TI Romania
- Frosse Dabit – Transparency Palestine (AMAN)
- Yusuf Umaru Dalhatu – National Accountability Group (TI local partner, Sierra Leone)
- Ramesh Nath Dhungel – TI Nepal
- Simeon Djankov – World Bank
- Raymond Dou'a – TI Cameroon
- Drewery Dyke – Amnesty International
- Dolores Español – TI Philippines
- Carlos Filártiga – TI Paraguay
- Ray Fisman – Columbia University
- Marta Foresti – Overseas Development Institute
- Verena Fritz – Overseas Development Institute
- Zanda Garanca – TI Latvia
- Lawrence Gikaru – Consultant, Kenya
- Syed Adil Gilani – TI Pakistan
- Lawrence Haas – formerly of the World Commission on Dams
- Rema Hanna – New York University
- Gørild M. Heggelund – Fridtjof Nansen Institute
- Nathaniel Heller – Global Integrity
- Nicholas Hildyard – Corner House
- Goran Hyden – University of Florida
- Lisa Karanja – TI Kenya
- Tamuna Karosanidze – TI Georgia

- Anung Karyadi – TI Indonesia
- Charles Kenny – World Bank
- Idrissa Alichina Kourgueni – Association Nigérienne de Lutte contre la Corruption (TI Niger)
- Huguette Labelle – Transparency International
- Johann Graf Lambsdorff – University of Passau
- Aileen Laus – TI Philippines
- Emmanuelle Lavallée – DIAL, Paris
- Virginia Lencina – Poder Ciudadano (TI Argentina)
- Roberto Lenton – Global Water Partnership
- Kristen Lewis – Consultant
- Per Ljung – PM Global Infrastructure
- Reinier Lock – International Private Water Association
- Matthew Loftis – TI Romania
- Goodwell Lungu –TI Zambia
- Wangari Maathai – Green Belt Movement
- Tanvir Mahmud – TI Bangladesh
- Grit Martinez – Ecologic, Institute for International and European Environmental Policy
- Olga Mashtaler – NGO 'Anticorruption Committee' (TI national contact group, Ukraine)
- Kennedy Masime – Centre for Governance and Development, Kenya
- Kenneth Mease – University of Florida
- Edward Miguel – University of California, Berkeley
- William Mishler – University of Arizona
- Jack Moss – Aquafed
- Sendhil Mullainathan – Harvard University
- Andreea Nastase – TI Romania
- Doron Navot – Hebrew University and the Israel Democracy Institute
- Venkatesh Nayak – Commonwealth Human Rights Initiative
- Semou Ndiaye – Forum Civil (TI Senegal), Université Cheikh Anta Diop de Dakar
- Natalie P. W. Ng – TI Malaysia
- Maurice Nguefack – TI Cameroon
- Juanita Olaya – Transparency International
- Donal T. O'Leary – Transparency International
- Benjamin Olken – Harvard University
- Minoru O'uchi – TI Japan
- Fred Owegi – Kenya Institute for Public Policy Research and Analysis
- Silke Pfeiffer – Transparency International
- Janelle Plummer – Consultant
- Lucila Polzinetti – Poder Ciudadano (TI Argentina)
- Sarah Repucci – Transparency International
- Juanita Riaño – Transparency International
- Frank B. Rijsberman – Google.org
- Jean-Daniel Rinaudo – French Geological Survey
- Byron López Rivera – Grupo Civico Ética y Transparencia (TI Nicaragua)
- Segundo Romero – TI Philippines
- Richard Rose – University of Aberdeen
- Gastón Rosenberg – Poder Ciudadano (TI Argentina)

- Dagmar Schröder-Huse – TI Germany
- Thayer Scudder – California Institute of Technology
- Pablo Secchi – Poder Ciudadano (TI Argentina)
- Mitchell Seligson – Vanderbilt University
- Kathy Shandling – International Private Water Association
- Kathleen Shordt – International Water and Sanitation Centre
- Emilia Sičáková-Beblava – TI Slovakia
- Hubert Sickinger – TI Austria
- Muhammad Sohail – Loughborough University
- Felipe de Solar – Corporación Chile Transparente (TI Chile)
- Bruno W. Speck – State University of Campinas
- Liga Stafecka – TI Latvia
- Patrik Stålgren – Göteborg University
- Varina Suleiman – Poder Ciudadano (TI Argentina)
- TI Papua New Guinea (Inc.)
- TI Switzerland
- TI UK
- TI USA
- Transparencia Mexicana (TI Mexico)
- Toru Umeda – TI Japan
- Manuel Villoria – TI Spain
- Jonathan Werve – Global Integrity
- Keiichi Yamazahi – TI Japan
- Kavwanga Yambayamba – TI Zambia
- Anna Yarovaya – NGO 'Anticorruption Committee' (TI national contact group, Ukraine)
- Richard Y. W. Yeoh – TI Malaysia
- Iftekhar Zaman – TI Bangladesh
- Dominique Zéphyr – Vanderbilt University

Preface

Huguette Labelle, Chair of Transparency International

Transparency International's flagship publication, the *Global Corruption Report*, sets out to explore how corruption corrodes the foundations of our societies and to suggest what we can do to reverse this course. In 2008 the report tackles the crucial issue of the water sector, examining how the failure to govern this essential life resource more transparently and accountably has an enormous price – both today and for future generations.

Now in its seventh edition, the *Global Corruption Report* has powerfully documented how corruption hinders democratic self-determination and thwarts the course of justice. It has provided proof positive that corruption undermines liberty, prosperity and individual empowerment. Drawing on the expertise of the TI movement, particularly that of our national chapters around the world, the report provides a unique perspective on the global state of corruption – and on the many efforts to combat it.

The special focus section of this year's report, *corruption in the water sector*, shows that in perhaps no other area does corruption so directly and profoundly affect the lives and livelihoods of billions of people as in the provision of water. Water is a natural resource, a commodity and the foundation of life on our planet. That is why we made it the focal topic for this year's report.

It is difficult to overstate the importance of water for health and secure livelihoods, for economic development, environmental integrity and social cohesion. As the United Nations (UN) Millennium Report in 2000 concludes: 'No single measure would do more to reduce disease and save lives in the developing world than bringing safe water and adequate sanitation to all.' It is also difficult to overstate the scope and consequences of the current global water crisis, one that leaves more than 1 billion people without access to safe drinking water. At the same time, growing water shortages – exacerbated by corruption – threaten development and political stability.

Let's remind ourselves about what we are capable of achieving in the water sector – and how far we still have to go to claim success. No other sector pits our boldest achievements in human progress so starkly against our most abject failures in delivering development to all. The introduction of public water and sanitation systems ushered in dramatic improvements in a very short time frame – a mere hundred years ago, child mortality in urban centres in Europe due to water contamination was as high as today in sub-Saharan Africa. Yet more than 2.6 billion people still do not have access to sanitation systems that are so crucial for human health.

Experts concur that the water crisis is a crisis of water governance. Corruption is certainly not its only cause, but, as the *Global Corruption Report 2008* shows, it is a major factor and a catalyst in this crisis. Contributions to this report document how corruption pervades all aspects of the water sector, how it inflates costs for drinking water in India, Kenya and elsewhere, how it is detrimental for irrigation in Pakistan or large dams in Latin America and how it abets large-scale water pollution in China. Corruption creeps into water management in many industrialised countries and makes the global adaptation to climate change even more difficult. Women and the poor are most often the main victims of corruption in water governance, unduly punishing the weakest in societies.

The sheer scope of corruption in water governance also bears a grain of hope. It points towards a unique opportunity to forge a powerful coalition for change. Fighting corruption in water is in the common interest of people who are concerned about poverty, food security and economic development, about sustainable environments and climate change, about health and gender equality and about social cohesion. The international community has made enormous commitments to improving the lives of the poor via the Millennium Development Goals, which include a commitment to safe and secure access to water. It is now for this same community, and the many stakeholders engaged in the water sector, to make sure that corruption does not prevent the achievement of this goal.

Transparency International will work to expand and invigorate the global coalition against corruption to include the many stakeholders involved in the water sector. Our alliance with the Water Integrity Network, an international coalition of water experts, field workers, academics and activists dedicated to tackling corruption in the sector, offers TI an excellent opportunity to pursue enhanced anti-corruption efforts in water. As the first report of its kind examining corruption in water, the *Global Corruption Report 2008* delivers a compelling invitation to join this important and rewarding fight. We owe it to our societies to remove the scourge of corruption and make this life resource work in favour of better and more sustainable human development.

Foreword
Water in the community: why integrity matters

Hon. Prof. Wangari Maathai[1]

Water is the driving force of all nature. It is essential for the workings of our ecological systems. It is essential for our health and the health of our communities. It features prominently in our spiritual life. It binds us together through shared waterways and shared water sources. It shapes our relationship with nature, politics and economies.

Managing water wisely is as paramount to our common future as it is difficult to achieve. Different visions, values and interests compete for shaping water governance. But one fact is clear: the global water crisis that destroys sources of water and waterways, and leaves a large portion of the world without access to safe drinking water, that destroys lives and livelihoods all over the world and that continues to create ecological disasters at an epic and escalating scale is a crisis of our own doing.

It is a crisis of governance: man-made, with ignorance, greed and corruption at its core. But the worst of them all is corruption.

Corruption means power unbound. It gives the powerful the means to work against and around rules that communities set themselves. This makes corruption in water particularly pernicious. It allows the powerful to break the rules that preserve habitats and ecosystems, to plunder and pollute the water sources that entire world regions depend upon and to steal the money that is meant to get water to the poor. Corruption shuts smallholders out of irrigation systems, displaces communities with impunity during dam construction, disrespects carefully crafted arrangements for water-sharing across borders, and permits the poor and ignorant to carry out activities that undermine the environment and their livelihoods, all with grave consequences for environmental sustainability, social cohesion and political stability. Perhaps most destructive of all, the force of corruption threatens to create a situation in which the rules continue to be gamed in favour of the powerful and efforts for reform are thwarted.

Tackling corruption in water is therefore a prerequisite for tackling the global water crisis. With the stakes so high, Transparency International's *Global Corruption Report 2008* could not come at a better moment. The report helps us to better understand the many different forms that corruption takes and it describes in detail the effects it has wrought. But, most importantly, it does not end on a gloomy note; it also describes some very practical initiatives that can be taken to combat corruption in water.

1 Hon. Prof. Wangari Maathai is the 2004 Nobel Peace Prize Laureate and founder of the Green Belt Movement.

Nowhere are the global water crises and the havoc that corruption inflicts on the sector more shockingly on display than in Africa, where a rich and powerful elite oversee a rich region inhabited by an impoverished and disempowered population. But Africa is not alone. With case studies from around the world the report clearly demonstrates that corruption in water is a global phenomenon. It is global in two senses. Not only does it occur in all regions of the world, confirming that industrialised countries are not immune, but also tackling it is a global responsibility and in the interest of all stakeholders, communities, policy-makers, business, civil society and donors.

With my own experience as an activist, I sincerely believe that the analysis presented in this report provides a strong impetus to bind together more firmly governments, corporations and civil society activists striving for environmental justice, poverty alleviation and good governance for a strong coalition to fight corruption in the water sector.

I have always believed that our treatment of the natural environment reflects the strength of our societies. As the report underscores, everyone can and must do their share. Only by acting together is progress attainable and sustainable. Our world's well-being depends upon it.

Acknowledgements

The *Global Corruption Report 2008* could not have been prepared without the dedicated efforts of many individuals, above all the authors, who have worked with purpose and passion on their contributions.

We are grateful to the Transparency International movement, from national chapters around the world to the TI Secretariat in Berlin, for constant input and enthusiasm. The TI national chapters merit a special word of appreciation for continuing to fill the country report section with a wealth of insights and experiences.

We are particularly indebted to the members of our Editorial Advisory Panel whose advice and expertise helped to develop and refine the report, in particular the thematic section on water: Dogan Altinbilek, Eduardo Bohórquez, Jermyn Brooks, Sarah Burd-Sharps, Piers Cross, Hansjörg Elshorst, Håkan Tropp, Tony Tujan, Surya Nath Upadhyay and Frank Vogl.

The members of TI's Index Advisory Committee offered their input on the research section of the report: Jeremy Baskin, Julius Court, Steven Finkel, Johann Graf Lambsdorff, Daniel Kaufmann, Emmanuelle Lavallée, Richard Rose and Susan Rose-Ackerman.

Thanks to those individuals who graciously agreed to referee the report's contributions and offered thoughtful and thorough responses: David Abouem a Tchoyi, Andrew Aeria, Graham Alabaster, Andrew Allan, Laurence Allan, Jens Andvig, Dominique Arel, Livingston Armytage, Manuhuia Barcham, Linda Beck, Predrag Bejakovic, Bernhard Bodenstorfer, Emilio Cárdenas, José Esteban Castro, Emil Danielyan, Marwa Daoudy, Phyllis Dininio, Gideon Doron, Juris Dreifelds, Anton Earle, Eduardo Flores-Trejo, Elizabeth Fuller, Michelle Gavin, Mamoudou Gazibo, Charles Goredema, Åse Grødeland, Ernest Harsch, Clement Henry, Paul Heywood, Jonathan Hopkin, Jarmo Hukka, Karen Hussmann, Sorin Ionita, Michael Johnston, John-Mary Kauzya, George Kegoro, Michael Kevane, Gopathampi Krishnan, Daniel Kübler, Peter Lambert, Evelyn Lance, Peter Larmour, Nelson Ledsky, Michael Likosky, Joan Lofgren, Xiaobo Lu, Cephas Lumina, Stephen Ma, Richard Messick, Arnauld Miguet, Stephen Morris, Andrew Nickson, Bill O'Neill, Katarina Ott, Michael Palmer, Jan Palmowski, Heiko Pleines, Som Nath Poudel, Miroslav Prokopijevic, Gabriella Quimson, Isha Ray, William Reno, Carlos Buhigas Schubert, Anja Senz, Erik Swyngedouw, Celia Szusterman, Madani Tall, Anthony Turton, Nicolas van de Walle, Shyama Venkateswar, Jeroen Warner, Kai Wegerich, Laurence Whitehead, Melvin Woodhouse, José Zalaquett, Mark Zeitoun and Darren Zook.

We would like to acknowledge the meticulous work of our fact checkers: Cecilia Fantoni, Yolanda Fernandez, Tita Kaisari, Steven Liu, Charlotte Meisner, Friederike Meisner, Ariana Mendoza, Pamela Orgeldinger, Sarah Pellegrin, Veronica Rossini and Katherine Stecher.

Thanks are due to Kathleen Barrett, Joss Heywood, Richard Leakey, Kyela Leakey, Susan LeClercq, Agatha Mumbi, Mijako Nierenkoether, David Nussbaum, Paula O'Malley, Jamie Pittock, David Tickner, Lucy Wanjohi and Petra Wiegmink for their varied and valued input and assistance.

Once again we are grateful to the law firm Covington and Burling for their pro-bono libel advice, specifically Enrique Armijo, Sara Cames, Jason Criss, Tim Jucovy, Raqiyyah Pippins, Eve Pogoriler, Brent Powell, Sumit Shah, Robert Sherman, Jodi Steiger, Lindsey Tonsager and Steve Weiswasser. Finola O'Sullivan, Richard Woodham, Daniel Dunlavey and Mainda Kiwelu at Cambridge University Press provided guidance and professionalism.

Jill Ervine, Diane Mak and Diana Rodriguez contributed invaluably to the early stages of this report and external editor Mark Worth must be thanked for placing many final touches. As ever, we owe a special thanks to our external editor, Michael Griffin, for his quick pen and quicker wit. We are indebted to Robin Hodess for her editorial input and tremendous support throughout the production process.

The *Global Corruption Report 2008* could not have explored corruption in the water sector so successfully without the assistance, expertise and financial support of the Water Integrity Network, which we gratefully acknowledge. We wish to thank especially Jens Berggren, Manoj Nadkarni and Birke Otto at the WIN Secretariat for their unrelenting support and good humour. We also thank the former and current WIN Steering Committee members who provided invaluable input to the report: Franz-Josef Batz, John Butterworth, Piers Cross, Grit Martinez, Jack Moss, Henk van Norden, Donal T. O'Leary, Janelle Plummer, Kathleen Shordt, Patrik Stålgren, Håkan Tropp and Tony Tujan.

The *Global Corruption Report 2008* received special funding from the Water Integrity Network and their funding partners, the Swedish International Development Cooperation Agency, the Dutch Directorate-General for International Cooperation and the German Federal Ministry for Economic Cooperation and Development.

Dieter Zinnbauer, Rebecca Dobson and Krina Despota,
Editors

Executive summary

Transparency International

Corruption in the water sector puts the lives and livelihoods of billions of people at risk. As the *Global Corruption Report 2008* demonstrates, the onset of climate change and the increasing stress on water supply around the world make the fight against corruption in water more urgent than ever. Without increased advocacy to stop corruption in water, there will be high costs to economic and human development, the destruction of vital ecosystems, and the fuelling of social tension or even conflict over this essential resource. This report clearly shows that the corruption challenge needs to be recognised in the many global policy initiatives for environmental sustainability, development and security that relate to water.

As the *Global Corruption Report 2008* reveals, there are several encouraging initiatives from all over the world that demonstrate success in tackling water corruption. This is the pivotal message that more than twenty experts and practitioners emphasise in this report. In addition, the *Global Corruption Report 2008* – which is the first report to assess how corruption affects all aspects of water – reflects on what more can be done to ensure that corruption does not continue to destroy this basic and essential resource, one that is so fundamental to the lives of people all over the planet.

Water and corruption: putting lives, livelihoods and sustainable development at risk

Water is vital and has no substitutes. Yet a water crisis that involves corruption engulfs many regions of the world. Nearly 1.2 billion people in the world do not have guaranteed access to water and more than 2.6 billion are without adequate sanitation, with devastating consequences for development and poverty reduction. In the coming decades the competition for water is expected to become more intense. Due to overuse and pollution, water-based ecosystems are considered the world's most degraded natural resource. Water scarcity already affects local regions on every continent, and by 2025 more than 3 billion people could be living in water-stressed countries.

The human consequences of the water crisis, exacerbated by corruption, are devastating and affect the poor and women most of all. In developing countries, about 80 per cent of health problems can be linked back to inadequate water and sanitation, claiming the lives of nearly 1.8 million children every year and leading to the loss of an estimated 443 million school days for the children who suffer from water-related ailments. In Africa, women and girls often walk more than 10 kilometres to gather water for their families in the dry season, and it is estimated that an amount equivalent to about 5 per cent of gross domestic product (GDP) is lost to illness and death caused by dirty water and poor sanitation there, as well. When clean water is denied, the stakes are very high.

The *Global Corruption Report 2008* argues that the crisis of water is a crisis of water governance, with corruption as one root cause. Corruption in the water sector is widespread and makes water undrinkable, inaccessible and unaffordable. It is evident in the drilling of rural wells in sub-Saharan Africa, the construction of water treatment facilities in Asia's urban areas, the building of hydroelectric dams in Latin America and the daily abuse and misuse of water resources around the world.

The scale and scope of the water and corruption challenge

The *Global Corruption Report 2008* explores corruption in water through four key sub-sectors.

Water resources management (WRM), which involves safeguarding the sustainability and equitable use of a resource that has no substitutes, is shown in this report to be susceptible to capture by powerful elites. Water pollution has often gone unpunished due to bribery, and funds for WRM end up in the pockets of corrupt officials. In China, for example, corruption is reported to thwart the enforcement of environmental regulations and has contributed to a situation in which aquifers in 90 per cent of Chinese cities are polluted and more than 75 per cent of river water flowing through urban areas is considered unsuitable for drinking or fishing.

The need to adapt to climate change makes cleaning up corruption in water resources all the more urgent. Changing water flows and more floods may require massive new investment in water infrastructure and the resettlement of 200 million people globally, and demand more frequent emergency relief efforts. All of the above are particularly vulnerable to corruption, as the *Global Corruption Report 2008* shows.

Where corruption disrupts the equitable sharing of water between countries and communities, it also threatens political stability and regional security. Two in every five people in the world today live in international water basins, and more than fifty countries on five continents have been identified as hotbeds for potential future conflicts over water. Water 'grabs', the irresponsible appropriation or diversion of water without consideration for other users, abetted by corruption, may translate tension into open conflict.

In **drinking water and sanitation services,** the second water sub-sector explored in the *Global Corruption Report 2008*, corruption can be found at every point along the water delivery chain: from policy design and budget allocations to operations and billing systems. Corruption affects both private and public water services and hurts all countries, rich and poor. In wealthier countries, corruption risks are concentrated in the awarding of contracts for building and operating municipal water infrastructure. The stakes are high: this is a market worth an estimated US$210 billion annually in Western Europe, North America and Japan alone.

In developing countries, corruption is estimated to raise the price for connecting a household to a water network by as much as 30 per cent. This inflates the overall costs for achieving the Millennium Development Goals (MDGs) for water and sanitation, cornerstones for remedying the global water crisis, by more than US$48 billion.

Irrigation in agriculture, the third water sub-sector examined in this report, accounts for 70 per cent of water consumption. In turn, irrigated land helps produce 40 per cent of the world's food. Yet irrigation systems can be captured by large users. In Mexico, for example, the largest 20 per cent of farmers reap more than 70 per cent of irrigation subsidies. Moreover, corruption in irrigation exacerbates food insecurity and poverty.

Irrigation systems that are difficult to monitor and require experts for their maintenance offer multiple entry points for corruption, leading to wasted funding and more expensive and uncertain irrigation for small farmers. One particular problem is the regulation of irrigation with groundwater resources. As a result of weak regulation, large users in places such as India or Mexico can drain groundwater supplies with impunity, depriving smallholders of essential resources for their livelihoods. In India, the total corruption burden on irrigation contracts is estimated to exceed 25 per cent of the contract volume, and is allegedly shared between officials and then funnelled upwards through the political system, making it especially hard to break the cycle of collusion.

The fourth water sub-sector to be covered in the *Global Corruption Report 2008* is that of **hydropower**, involving dams. Few other infrastructure projects have a comparable impact on the environment and people. The hydropower sector's massive investment volumes (estimated at US$50–60 billion annually over the coming decades) and highly complex, customised engineering projects can be a breeding ground for corruption in the design, tendering and execution of large-scale dam projects around the world. The impact of corruption is not confined to inflated project costs, however. Large resettlement funds and compensation programmes that accompany dam projects have been found to be very vulnerable to corruption, adding to the corruption risks in the sector.

Corruption in water: a challenge beyond the water sector

The importance of water for human development and environmental sustainability is well established and the global water crisis has assumed a central role in the development and environment debate. The *Global Corruption Report 2008* highlights that corruption in water is a significant factor in this crisis and therefore also a critical issue for global public policy. The impact of corruption in the water sector on lives, livelihoods, food security and international cooperation also underscores the many linkages to global policy concerns.

Corruption in water is a concern not only for the water sector. It also complicates the global challenge to confront climate change, and must be addressed in the building of a governance framework that updates and expands the Kyoto Protocol. Further, corruption in water must feature more prominently in any debate on environmental sustainability. It also matters for a global security agenda that is concerned about the root causes of conflict, extremism and failing states. Finally, corruption needs to be recognised as an obstacle to the global resolve to bring development to all, most prominently articulated in the Millennium Development Goals and related policy initiatives.

Water: a high-risk sector for corruption

The *Global Corruption Report 2008* draws some preliminary conclusions about why water is especially vulnerable to corruption.

- **Water governance spills across agencies.** Water often defies legal and institutional classification, creating a regulatory lacuna and leaving governance dispersed across countries and different agencies with many loopholes to exploit.
- **Water management is viewed as a largely technical issue in most countries.** Managing water is still predominantly approached as an engineering challenge. Consideration for the political and social dimensions of water, including corruption issues and their costs, is limited.
- **Water involves large flows of public money.** Water is more than twice as capital-intensive as other utilities. Large water management, irrigation and dam projects are complex and difficult to standardise, making procurement lucrative and manipulation difficult to detect.
- **Private investment in water is growing in countries already known to have high risks of corruption.** Nine of the ten major growth markets for private sector participation in water and sanitation are in countries with high risks of corruption, posing particular challenges for international investors.
- **Informal providers, often vulnerable to corruption, continue to play a key role in delivering water to the poor.** Informal water providers provide important bridging functions in many developing countries to bring water to the poor. They often operate in a legal grey zone, however, making their operations vulnerable to extortion and bribery.
- **Corruption in water most affects those with the weakest voice.** Corruption in water often affects marginalised communities, the poor or – in the case of its impact on the environment – future generations. These are all stakeholders with a weak voice and limited ability to demand more accountability.
- **Water is scarce, and becoming more so.** Climate change, population growth, changing dietary habits and economic development all exacerbate local water scarcities. The less water there is available, the higher the corruption risks that emerge in control over the water supply.

From diagnosis to action: lessons for fighting corruption in the water sector

The case studies and experiences presented in the *Global Corruption Report 2008* yield a set of four key lessons for fighting corruption in the water sector.

- **Lesson one: prevent corruption in the water sector, as cleaning up after it is difficult and expensive**

  When corruption leads to contaminated drinking water and destroyed ecosystems, the detrimental consequences are often irreversible. When subsidised water gives rise to powerful agricultural industries and lobbies, refocusing subsidies on the poor becomes more difficult.

- **Lesson two: understand the local water context, otherwise reforms will fail**
  One size never fits all in fighting corruption, but this is particularly the case in the water sector, where conditions of supply and demand, existing infrastructure and governance systems vary widely. Understanding local conditions and the specific incentive systems that underpin corruption is a prerequisite for devising effective reforms.
- **Lesson three: cleaning up water corruption should not be at odds with the needs of the poor**
  The costs of corruption in the water sector are disproportionately borne by the poor. Pro-poor anti-corruption efforts should focus on the types of service provision that matter most to them, such as public standpipes or drilling rural wells. Such efforts need to be designed so that they do not undercut peoples' basic livelihoods: for example, a crackdown on informal service providers may eliminate an important way for the poor to secure reliable access to water.
- **Lesson four: build pressure for water reform from above *and* from below**
  Ending corruption in the water sector requires breaking the interlocking interests and relationships that are perpetuating the problem. This is a formidable challenge. Leadership from the top is necessary to create political will and drive institutional reform. Bottom-up approaches are equally important to curbing corruption, by adding checks and balances on those in power that include the monitoring of money flows or benchmarks of utility performance.

Stemming the corruption tide: recommendations for reform

The *Global Corruption Report 2008* presents a number of promising strategies and tools to tackle corruption in water resources management, drinking water and sanitation, irrigation and hydropower. A particular country's dynamics determines the right mix and sequence of anti-corruption reforms, but the following is a summary of the most promising recommendations.

- **Recommendation one: scale up and refine the diagnosis of corruption in water – the momentum and effectiveness of reform depend on it**
  Much work remains to be done on studying the scope and nature of corruption in water. Tools such as corruption impact assessments for different areas of the water sector, public expenditure tracking or poverty and corruption risk-mapping help to shed valuable light on different aspects of the puzzle. These tools need to be refined, adopted widely across the water sector and adapted to specific local contexts to lay the foundations for targeted reform.

- **Recommendation two: strengthen the regulatory oversight of water management and use**
  Government and the public sector continue to play the most prominent role in water governance and should establish effective regulatory oversight, whether for the environment, water and sanitation, agriculture or energy. There are a number of institutional reforms that can make regulatory capture less likely and therefore should be prioritised: capacity building and training for regulatory staff, the provision of adequate resources (human, financial, technical and administrative), the creation of a clear institutional mandate, the

implementation of transparent operating principles and the introduction of a public consultation and appeals process.

- **Recommendation three: ensure fair competition for and accountable implementation of water contracts**
  In many countries, the private sector has embraced basic anti-corruption measures as part of its standard operating procedures, but more must be done for this to have an impact on water. Governments and contractors can enter into integrity pacts (IPs) for public procurement processes. The large investment demand in the water sector means that export credit agencies, commercial banks and the lending wings of international financial institutions can play an important role in fighting corruption and should expand their due diligence requirements to include anti-bribery provisions.
- **Recommendation four: adopt and implement transparency and participation as guiding principles for all water governance**
  Transparency lays the foundation for public oversight and accountability and must come to characterise how water sector business is done by public and private stakeholders alike. Too often, commitments to this principle have not been translated into action. There are, however, some examples of how transparency is being practised in water governance in the *Global Corruption Report 2008* – from opening up project budgets to disclosure of performance indicators. These must be repeated and used as the basis for learning and improvement.

  Increased *participation* has been documented throughout the *Global Corruption Report 2008* as a mechanism for reducing undue influence and capture of the sector. Participation by marginalised groups in water budgeting and policy development can provide a means for adding a pro-poor focus to spending. Community involvement in selecting the site of rural wells and managing irrigation systems helps to make certain that small landholders are not last in line when it comes to accessing water. Civil society participation in auditing, water pollution mapping and performance monitoring of water utilities creates important additional checks and balances. Transparency and participation build the very trust and confidence that accountable water governance demands and civil society plays a critical role in turning information and opportunities for participation into effective public oversight.

Creating momentum for change: a global coalition against corruption in water

Implementing these recommendations requires a strategic vision. The global challenge of corruption in the water sector needs a global response, local expertise and adaptation and buy-in from a wide range of stakeholders. Transparency International, with its network of corruption experts and advocates in more than ninety countries, is well positioned to make a significant contribution. Efforts to bring more transparency to the water sector, for example, can benefit from TI's long-standing research and advocacy on raising the standard of freedom of information and transparency in governance systems around the world. Initiatives for more integrity in corporate participation in the water sector can adopt TI's private sector anti-corruption tools for their purposes and link into TI's extensive work on accountable public procurement. The Water Integrity Network, a fast-growing international coalition of water experts, field workers, academics and activists that worked with TI in the development of this report, is spearheading the fight against corruption in the water sector. The *Global Corruption Report 2008* presents strong reasons why many others should join in and help generate the momentum for sustained reform.

The onset of climate change and increasing stress on water resources means that a critical crossroads has been reached. As the *Global Corruption Report 2008* shows, tackling corruption in the water sector is not only a moral imperative that serves the interests of many, particularly the poor. It is also feasible. The time for action is now.

Part one
Corruption in the water sector

1 Introducing water and corruption

In her lead chapter for the thematic section of the Global Corruption Report 2008, *Janelle Plummer outlines the main parameters of the global water crisis, provides an overview of the different types and dynamics of corruption in the sector and explores their implications. Charles Kenny adds to this overview with calculations that provide a stark reminder of the fatal consequences of corruption in the water sector.*

Water and corruption: a destructive partnership

Janelle Plummer[1]

Water is vital for people, food, energy and the environment. When water is scarce or absent, countries and their citizens suffer incalculable costs – economic, political, social, cultural and environmental. Corruption exacerbates these impacts and amplifies the pivotal challenge of water governance. Urgent action is needed to mobilise all stakeholders to develop practical ways of tackling corrupt practices in the many and varied parts of the water sector. This is the central message of the *Global Corruption Report 2008*.

The global water crisis: a crisis of governance

The story of corruption in the water sector is a story of corruption in resources and services vital for life and development. It is also the story of a sector in crisis. Each year millions of people die of waterborne diseases because access to safe drinking water and adequate sanitation has not been prioritised. In 2004 more than 1 billion people lacked access to safe drinking water and 2 billion did not have access to adequate sanitation – and, despite successes in many regions, the population without access to water services is increasing. Corrupt practices exacerbate these gaps, removing investment that might be used to extend services to the poor, diverting finance from the maintenance of deteriorating infrastructure and taking cash from the pockets of the poor to pay escalated costs and bribes for drinking water.

1 Janelle Plummer was a governance and anti-corruption consultant, currently working for the World Bank. She is currently a governance adviser in the World Bank. This chapter draws on J. Plummer and P. Cross, 'Tackling Corruption in the Water and Sanitation Sector in Africa: Starting the Dialogue', in E. Campos and S. Pradhan (eds.), *The Many Faces of Corruption* (Washington, DC: World Bank, 2007). The opinions expressed are those of the author and do not necessarily reflect those of the World Bank, its executive directors or the countries they represent.

Water scarcity is also a significant and growing problem. The livelihoods of hundreds of millions of people across all regions are threatened from shortages of water for irrigation. Agriculture uses around 70 per cent of the water drawn from rivers and groundwater. High levels of human activity, the pressures of increased water demand and higher populations take their toll.[2] Climate change adds new pressures to the problem. By 2025 more than 3 billion people could be living in water-stressed countries.[3] Over the coming decades crop yields are expected to fall by 25 per cent and global malnutrition may rise by nearly as much if current projections on climate change prove true.[4]

Managing water requires a careful balance of food security, poverty reduction and ecosystem protection. Degraded ecosystems increase the risk of disaster – removing buffers against floods, droughts and other natural hazards. The impact of environmental degradation, inadequate water management and chronic underinvestment are known to us all: the tragedy of Darfur is both a collapse of governance and an emergency of land and water degradation that has escalated to an unprecedented humanitarian disaster.

At the heart of these failures is the crisis of *governance* in water – *a crisis in the use of power and authority over water and how countries manage their water affairs.*[5] And yet, despite the imperatives of water for citizens' livelihoods and a country's growth, water governance has not been prioritised. Institutional dysfunction, poor financial management and low accountability mean that many governments are not able to respond to the crisis, and weak capacity and limited awareness leave citizens and non-governmental organisations (NGOs) in many countries unable to demand change.

Water and corruption: a concern for all

Corruption in and around the development of the water sector is a key dimension of this governance failure. It is evident in the drilling of rural boreholes in sub-Saharan Africa, the operation of treatment facilities in Asia's urban areas, the construction of hydroelectric dams in Latin America and the daily abuse and misuse of water resources entrusted to governments and other decision-makers around the world. Efforts to tackle the multiple aspects of corruption form a critical part of the battle to get water to people who need it. Corruption is both a cause and an effect of weak governance in the sector.

While the impacts of corruption are more extreme in developing countries, the phenomenon of corrupt water is not one limited to low- or middle-income countries. In Europe, North America and Australia, corrupt practices involving or affecting water resources and services are not uncommon. Industrialised countries have their own forms of nepotism in their board-

2 United Nations Development Programme (UNDP), *Human Development Report 2006. Beyond Scarcity: Power, Poverty and the Global Water Crisis* (New York: Palgrave Macmillan, 2006).

3 Ibid.

4 Ibid.

5 Adapted from Department for International Development (DfID), 'Governance, Development and Democratic Politics: DFID's Work in Building more Effective States' (London: DfID, 2007).

rooms and institutions; fraud and embezzlement feature frequently in the press. Even high levels of regulation and oversight have not prevented corruption from playing out where the public and private sector meet – or from being exported abroad, where governance and controls are weaker.

The global push by the international community to remedy the lack of access to water and sanitation for the world's poorest citizens provides an unprecedented opportunity for governments, the private sector and civil society to work in partnership to combat corruption in drinking water and sanitation. To speed progress towards ending poverty, 189 countries committed in 2000 to the United Nations Millennium Declaration.[6] Better water and sanitation services for all people form part of the declaration's eight goals – the MDGs – that world governments have pledged to achieve by 2015.

Since the MDGs are inextricably linked to each other, achieving improvements in water and sanitation produces positive impacts on the other goals – from reducing poverty and hunger, to cutting child and maternal mortality rates and eliminating gender inequalities. Unless primary blockages such as corruption are identified and addressed, it will be impossible to meet the MDG target of halving the number of people without access to safe drinking water and basic sanitation. Too much money is being lost from sector inefficiencies. Based on country and regional estimates compiled by the UN, fifty-five countries will fall short of increasing water access sufficiently, while another seventy-four nations are off track in realising promised improvements in sanitation.[7]

Sub-Saharan Africa is one of the regions where progress is slow and challenges for combating corruption are great. The 2007 *Corruption Perceptions Index* (CPI) compiled by Transparency International finds that nearly a half of the twenty nations that perform worst in the index come from the region.[8] And, according to the latest data, 63 per cent of the region's citizens lack basic sanitation facilities – an insignificant improvement from the 68 per cent recorded in 1990, the baseline year used to track the MDGs' progress towards the 2015 target year.[9] Over the same period the number of people in the region without access to water has actually increased by more than 20 per cent, due to high population growth rates.[10]

Water is an immensely political issue, wide open to manipulation, globally and nationally, and open to capture and conflict among communities and households. These macro and micro dimensions mean that the dialogue over corruption in water must reflect the diversity in practices, and actors, their motivations and levels of impact. It is vital that all countries

6 Subsequently, in 2002, the target for sanitation was adopted. This was a key development, as sanitation is often excluded from consideration.
7 UNDP, 2006.
8 These figures are based on the 2007 results of the Corruption Perceptions Index, available at www.transparency.org/policy_research/surveys_indices/cpi/2007.
9 Data based on 2004 figures provided by the UN Department of Public Information, 'Africa and the Millennium Development Goals, 2007 Update' (New York: UN, 2007).
10 Composite data are misleading, but there is also some debate over the accuracy of country-level data and the internal disparities and horizontal inequalities that are hidden in aggregate statistics.

urgently learn about the corruption taking place in their water sectors, identify the impacts and develop practical and targeted anti-corruption policies and tools.

The nature and scope of corruption

Corruption – the abuse of entrusted power for personal gain – can be found in a vast range of interactions at all levels and in all aspects of the water sector. At present, however, the diagnosis of corruption in the water sector is still developing, and anti-corruption efforts are often marred by narrow views and biased perceptions of what corruption is and where the key risks lie. To overcome these obstacles, a better understanding is needed of what forms corruption takes in the sector, where it is concentrated and what the incentives of stakeholders are. Given the diversity of the water and sanitation, irrigation, water resources management (WRM) and hydropower problem, this represents a major challenge.

Most types of corruption are found in the water sector. When bureaucratic or *petty* corruption occurs, a hierarchy of public servants abuse their power to extract small bribes and favours. A water meter reader offers to reduce a customer's bill in return for payment or a utility official only responds to water service complaints when favours are traded. When *grand* corruption happens, a relatively small cadre of public and private sector actors are involved and the rewards are high. For example, public funds for a rural water network are diverted into the pockets of ministry officials or a large dam construction contract is captured by a group of colluding companies. When *state capture* occurs, the decision-making process and enforcement of water policies are manipulated to favour the interests of a few influential water users or service providers at the expense of the broader public.[11]

A corruption risk map captures the different types of corruption in the water sector, including fraud, embezzlement, bribery, collusion and nepotism. It points towards the differing incentives of actors and various instruments needed to tackle the diverse nature of the corruption problem.

Typically there are three sets of corrupt interactions.

- Corruption in water occurs between *public officials and other public officials*. This includes corrupt practices in resource allocation – such as diverting funds for a water supply network to pay for upgrading a road near a politician's house. It can also involve using bribes to determine the outcome of personnel management decisions – such as payments to individuals for transfers and appointments to lucrative positions. The larger the potential salary, the higher the bribe to get the post.
- It also occurs between *public officials and private actors*, and includes forms of bribery and fraud that occur in relation to licensing, procurement and construction. Collusion or

11 This disaggregation of corruption follows M. Schacter and A. Shah, 'Look before You Leap: Notes for Corruption Fighters', Policy Brief no. 11 (Ottawa: Institute on Governance, 2001).

bid-rigging is typical of tendering processes in developed and developing countries and involves both international and national actors.[12]

- Corrupt practices also occur between *public officials and users/citizens/consumers*. These practices, known as administrative or petty corruption, enable poor and non-poor households, farmers and other users to get water, get it more quickly or get it more cheaply.

The series of corrupt practices in the sector extends from policy capture, to large and small public–private transactions in construction and operations, to interactions at the point of service delivery, which together can be plotted on a water 'value chain'. The framework shown in table 1 highlights these three sets of interactions in terms of the functions of the water sector: a cycle of policy-making and regulation, budgeting and planning, financing, programme design and management, tendering and procurement, construction, operation and maintenance, and monitoring and enforcement functions.

Table 1 Value chain framework: corrupt interactions from policy-making to water delivery

	Public–public	Public–private	Public–consumers/civil society
Policy-making and regulation	• Policy and regulatory capture over management of water resources, competition and monopolies • Inter-ministerial collusion: cover-up over environmental/social impacts of hydropower projects	• Policy capture over WRM decision-making • Bribery for water rights, extortion for permits and processing of permits • Regulatory capture (e.g. waivers to licences, bypassing EIAs, overlooking social impacts) • Kickbacks to cover up pollution	• Bribery to silence public protest over environmental and social impacts
Planning and budgeting	• Distortionary decision-making by politicians (location/type of investments) • Diversion of funds to individuals, other projects inter-ministerial bribery for fund allocation • Corruption in local budget management (fraud, falsification of accounts/documents, village-level collusion)	• Bribery to influence allocation of funding to higher-capital-investment projects (e.g. bulk water supply vs. improving networks or low-cost efficiency solutions)	

(Continued)

12 While it is possible that private–private interactions or NGO–private interactions are also prevalent in the sector (e.g. bribery or fraud between contractors and subcontractors), these interactions are defined as corruption only if the firm/organisation has been entrusted with public office.

Table 1 (continued)

	Public–public	Public–private	Public–consumers/civil society
Donor financing, funding and fiscal transfers	• Donor–government collusion in negotiations to meet spending targets, progress and quality, to influence type of sector investment • Bribery, rent-seeking and kickbacks to ensure fund transfers between MoF and sector ministries	• Donor and national private operator collusion (outside legal trade agreements)	
Management and programme design	• Corruption in personnel management – payments for lucrative positions (e.g. utility directorships, project management posts) – bribes for promotions, transfers, salary perks • Distortionary decision-making (collusion with leaders in selection/approval of plans, schemes) • Corruption in LG and departmental planning and budget management • Bribery to distort water management and canal construction to benefit officials	• Bribery to shift design to increase potential for kickback and fraud	• Influence project decision-making to benefit some users (project-level site selection, equipment, construction) • Bribery to distort water management, canal construction, sequencing to benefit rich or powerful users
Tendering and procurement	• Administrative corruption (fraud, falsification of documents, silence payments) • Inter-department/agency collusion over corrupt procurement, fraudulent construction • Cover-up and silence payments linked to corrupt procurement • Kickbacks in cash or jobs to help politicians secure preferred contractor	• Bribery/kickbacks to influence contract/bid organisation • Kickbacks to win large-scale projects: to secure contracts, to influence negotiations, for information • Corruption in supply procurement/inflated estimates for capital works, supply of chemicals, vehicles, equipment • Corruption in delegating O&M: awarding contracts, overestimating assets, selection, type,	

Table 1 (continued)

	Public–public	Public–private	Public–consumers/civil society
		duration of concessions, exclusivity, tariff/subsidy decisions • Fraudulent documentation, uncertified materials in construction	
Construction	• Cover-up and silence payments linked to corrupt construction	• Bribery and fraud in construction – not building to specification, concealing substandard work, unspecified materials, underpayment of workers – failure to complete works, delays • Fraudulent invoicing – marked-up pricing, over-billing by suppliers	• Corruption in community-based construction (with similar types of practices as for public–private interactions)
Operation and maintenance		• Over-billing by suppliers, theft/diversion of inputs (chemicals) • Avoiding compliance with regulations, specifications, health and safety rules • Falsification of accounts • Bribery for diversion of water for commercial irrigation or industry • Bribes to cover up wastewater discharge and pollution	• Administrative corruption for water (access to water – installing/ concealing illegal connections, avoiding disconnection, illicit supply, using utility vehicles) • Administrative corruption for speed (or preferential treatment) – irrigation canal repairs, new connections
Payment (for services)		• Bribery for excessive extraction by industry • Bribery, collusion in falsified billing in commercial irrigation and industry	• Administrative corruption – repayment/billing for WSS and irrigation water – fraudulent meter reading, avoidance or partial payment, overcharging

Source: Adapted from J. Plummer and P. Cross, 2007.[13]

13 'EIA' stands for 'environmental impact assessment', 'MoF' for 'Ministry of Finance', 'LG' for 'local government', 'O&M' for 'operation and maintenance' and 'WSS' for 'water supply and sanitation'.

Linkages and legality add to the complexity of any map of corruption in water. These interactions reinforce each other and double the impacts. A legal decision to construct a dam may enable officials to capture resources, private contractors to skim profits and officials to use the power of their office to divert the dam's water to powerful landowners for kickbacks. The accumulative cost of this network of interactions is high, with many losers along the way.

Ultimately, however, corruption scenarios play out very differently in different contexts. Political regimes, legal frameworks, the degree of decentralisation, regional disparities, power relations, cultural norms and levels of accountability (e.g. between state and civil society) will influence the patterns and risks. Understanding the channels where corruption can occur helps in its prevention. Mapping makes it possible to identify 'hot spots', in a particular context, where corruption tends to concentrate along the water value chain.

The impact of corruption: putting billions of lives at stake

The impact of corruption can be described in financial, economic, environmental and socio-political terms, and can also involve issues of security.

Putting an exact *financial* cost on corruption is difficult. While a best-case scenario might suggest that 10 per cent is being siphoned off from the sector annually in corrupt practices, a worst-case scenario places the figure at 30 per cent. If estimates are correct that an additional US$11.3 billion is needed each year to achieve the MDGs on water and sanitation, a 30 per cent leakage rate would mean that corruption could raise the costs of this pivotal development initiative by more than US$48 billion over the next decade.[14]

Weak governance and endemic corruption exact a *social* impact that financial calculations can never estimate. The barriers to access fall disproportionately on the poor in all regions. Chronically low levels of access are found among poorer households and, accordingly, many households find ways – often creative ways – of obtaining water informally. They vary the sources from which they obtain water and pay higher prices when they can afford it. The poorest households in countries such as El Salvador, Jamaica and Nicaragua spend more than 10 per cent of their income on water while their cohorts in rich nations such as the United States pay only a third as much.[15] In many situations elevated costs can be attributed to the corrupt transactions between informal providers and utility officials.

But poverty is multidimensional and household costs are not all financial. Whether poor households engage in corrupt transactions or not, they suffer due to the inefficiencies that corruption produces. Where corruption removes or increases the costs of access to water effects can be measured in terms of lost days, human development and lives. Close linkages have been found between access to safe water and infant mortality, girls' education and the prevalence of waterborne disease.[16]

14 World Health Organisation (WHO) and United Nations Children's Fund (UNICEF), 'Water for Life: Making it Happen' (Geneva: WHO Press, 2005).

15 UNDP, 2006.

16 See articles starting on pages 28 and 40.

It is not only with poverty that water problems are strongly associated. Water is also a key driver of *growth*, being an indispensable input to production (in agriculture, industry, energy and transport). Currently, the extremely low levels of hydraulic infrastructure and limited water resources management capacity in the poorest countries undermine attempts to manage variability in water availability.[17] Water reservoir storage capacity (per capita) in countries such as Morocco or India is less than one-tenth of the volume that Australia has in place.[18] In many countries in Africa, highly variable rainfall and the regular droughts that devastate parts of the region all ripple through national economies. In Ethiopia, for example, the lack of hydraulic infrastructure is estimated to cost the Ethiopian economy over one-third of its growth potential.[19] Reports of the disaster in New Orleans in 2005 suggest that it was not only natural, but exacerbated by unsubstantiated, unaccountable decision-making.[20] Corruption reduces the levels of investment in infrastructure, reduces resilience to shocks and undermines growth.

The impact of corruption in water can also be *environmental*. The lack of infrastructure for water management whether man-made (e.g. dams, inter-basin transfers, irrigation, water supply) or natural (e.g. watersheds, lakes, aquifers, wetlands) in developing countries presents a management challenge almost without precedent.[21] The ever-increasing impact of climate change and the lack of human and financial capacity to manage the water legacy result in far greater shock in developing countries, making the poorest countries ever more vulnerable. Corrupt practices that increase pollution, deplete groundwater and increase salinity are evident in many countries and are closely linked to deforestation and desertification across the globe. Stemming the leakage of funds from the sector is vital to address these issues.

The importance of water – on health, poverty, development and the environment – underscores how it is fundamentally linked to questions of *power and security*. Corruption can turn the control of water into a force that aggravates social tensions, political frictions and regional disputes. Tensions over water are frequent within states. Dire water shortages in Egypt triggered widespread public protest and roadblocks in the summer of 2007. The outcry was fuelled by the perception that corruption had caused the water crisis.[22] In Sierra Leone, a director of the Freetown utility was killed in 2007 during a clampdown on firefighters over their illegal resale of water.[23] Inevitably, internal pressures also spill across borders. Over the last fifty years

17 World Bank, 'Managing Water Resources to Maximize Sustainable Growth: A Country Water Resources Assistance Strategy for Ethiopia' (Washington, DC: World Bank, 2006).

18 UNDP, 2006.

19 UNDP, 2006. Ethiopia is ranked 138 out of 180 countries, based on the TI Corruption Perceptions Index.

20 See article starting on page 28.

21 D. Grey and C. Sadoff, 'Water for Growth and Development: A Framework for Analysis', baseline document for the fourth World Water Forum (Washington, DC: World Bank, 2006).

22 *Al-Ahram* (Egypt), 12 July 2007; *Al-Ahram* (Egypt), 2 August 2007; Land Center for Human Rights, 'Water Problems in the Egyptian Countryside: Between Corruption and Lack of Planning', Land & Farmer Series no. 32 (Cairo: Land Center for Human Rights, 2005).

23 Live from Freetown [blog], 2 June 2007, available at www.livefromfreetown.com/2007/06/.

water has been the source of twenty-five international conflicts, such as communal clashes at the Mali–Mauritania border over access to watering holes in 1999.[24] The potential for future disputes is ever present. Water basins that span more than fifty countries on five continents have been identified as hotbeds of conflict.[25] Corruption, particularly grand corruption, is a potential trigger to ignite these latent tensions.

The drivers of corruption

The equation *Corruption = Monopoly + Discretion − Accountability,* developed by Robert Klitgaard,[26] is very useful and relevant for understanding the problems posed for the water sector. It highlights the aggregate effect of monopoly and discretionary power, which are common in water institutions.[27] The water and sanitation sub-sector tends to be highly monopolistic and has many traits such as high capital costs and economies of scale[28] that help to keep it that way. In hydropower, the need for many tailored, non-standard investments serves as a barrier for new entrants to the market and reduces levels of competition. In addition, agencies and officials involved in all different aspects of the water sector have historically seen enormous discretionary power in the planning, design, contracting and implementation of water projects. Their influence is difficult to address because the sector is highly technical and the professionals involved have a clear information advantage.

Other idiosyncrasies of the water sector also suggest a high potential for corruption. Water investment involves a large flow of mostly public money, often with inadequate planning and oversight. In developing countries, funding sources for projects are often uncoordinated and spending and decision-making are non-transparent. And the sector is a costly one – water services assets, for instance, can be three to four times higher than telecommunications and power.[29] Because water policy, planning and budgeting decisions impact on inputs vital for agriculture, industry and property, political interference is significant. The result is a game of winners and losers who often adopt alternative means to gain access to water.

The funding provided by donors to the sector through official development assistance (ODA) creates additional opportunities for corruption to occur. Financing to the water supply and sanitation sector reached almost US$6 billion in 2005.[30] While this represents roughly 5 per cent of all aid flows, secondary spending leads to a multiplier effect for the money coming into the sector. The flows are particularly vulnerable to corruption, high levels of manipula-

24 See International Water Event Database, www.transboundarywaters.orst.edu/data/.

25 S. Postel and A. Wolf, 'Dehydrating Conflict', *Foreign Policy*, no. 126 (2001).

26 R. Klitgaard, *Controlling Corruption* (Berkeley, CA: University of California Press, 1988).

27 A number of anti-corruption advocates including Klitgaard and Susan Rose-Ackerman identify four key factors that engender opportunities for corruption: monopoly power, wide discretion, weak accountability and lack of transparency.

28 A reduction in unit cost achieved by increasing the amount of production.

29 C. Kirkpatrick *et al.*, 'State versus Private Sector Provision of Water Services in Africa: A Statistical, DEA, and Stochastic Cost Frontier Analysis', Paper no. 70 (Manchester: University of Manchester, 2004).

30 See Organisation for Economic Co-operation and Development (OECD), Official Development Statistics Database.

tion and patronage can occur and donors are often under pressure to disburse – be it grant money or loans.

In water and sanitation services it is also the failure of monopolistic state delivery that creates opportunities for petty corruption. A multitude of small-scale providers fill the gap in provision, often functioning in an informal zone that makes them and their clients vulnerable to exploitation. Government institutions are not well structured to deal with these informal water providers or the forms of bribery that develop.[31] Another driver of corruption in the water sector is related to the fact that the demand for accountability is very limited in developing countries. This is particularly true in relation to the service provider/consumer accountability relationship.[32] When civil society is weak and the concept of customer rights undeveloped, the challenge is multiplied.

The existence of state and non-state actors, systems, service levels and institutions creates a highly complex sector. Responsibilities for water affairs can be found in a multitude of different ministries and agencies and at various levels of government. The lack of clarity in the roles and responsibilities of all these stakeholders results in a lack of transparency and accountability and, inevitably, in a severe asymmetry of information between user, provider and policy-maker. The diversity of arrangements for delivering water services adds to the challenge. Utilities, alternative providers, community management and self-supply, whether formal or informal, all exist side by side in the context of different government structures and institutional challenges. These unique characteristics make water a fertile sector for corruption.

In addition, water has many linkages to other sectors that are particularly vulnerable to corruption. As part of the high-risk construction sector,[33] water displays the resource allocation and procurement-related abuses which arise when the public and private sectors meet. As water services and resource management is one of the functions of a country's administrative or civil service, the sector also confronts a different set of obstacles: low capacity, low wages, lack of clear rules and regulations, and dysfunctional institutions. These conditions make it susceptible to the common practices of fraud, bribery, embezzlement and favouritism.

Addressing incentives for change

Preventing corruption from taking root is less costly and complicated than having to tackle the problems once they begin. Effective prevention involves identifying and understanding the incentives at play. Corruption can be driven by need, greed, the opportunity for money or power[34] – or simply the basic need for water.

---

31 See article starting on page 40.

32 C. W. Gray and D. Kaufmann, 'Corruption and Development', *Finance and Development*, vol. 35, no. 1 (1998).

33 See Transparency International, *Global Corruption Report 2005* (London: Pluto Press, 2005).

34 R. Klitgaard *et al.*, *Corrupt Cities: A Practical Guide to Cure and Prevention* (Oakland, CA: Institute for Contemporary Studies, 2000).

Understanding the incentives of individuals, communities and firms requires careful analysis and knowledge of the local context. Incentives are influenced by a range of interconnected factors: social, political, economic and institutional. As corrupt activities unfold, stakeholders are pulled into a complicated web that connects various institutional levels and involves one or more types of corruption. Powerful patronage networks and patron–client relationships shape and solidify these interactions, making the fight against corruption exceptionally difficult. The corruption risk map (see page 7) provides a framework for identifying these stakeholder incentives, potential conflicts of interests and the points along the water value chain that are most vulnerable to capture.

Irrespective of the actors involved, corruption flourishes whenever the short-term benefits outweigh the expected losses. The calculation of costs and benefits will depend on the risk of getting caught and being held accountable. A key element of any sustainable anti-corruption strategy is to change these trade-offs so that stakeholders are no longer motivated towards corrupt behaviour – whether for national policy-makers allocating sector funding or the actors (politicians, managers and community leaders) involved in a community irrigation project. Shifting incentives involves minimising the frequency of transactions, reducing the potential gain from each one, raising the probability of detection and increasing the magnitude of penalties.[35]

Incentives need careful diagnosis in each setting. The corruption map can be used to identify the incentives of all actors along the value chain but these are highly context-specific. The incentive structures for officials managing utilities in Russia, for instance, are very different from those affecting the operation of irrigation channels in remote areas of Pakistan, or from the logic that determines how international contractors, financiers or policy-makers in industrialised countries respond to corruption risks. This demands knowledge of local settings, particularly of social and institutional norms, and engaging local actors is key.

The chapters that follow provide illustrations of how these incentives make water and corruption such a destructive partnership. Each chapter examines one dimension of the sector and profiles the specific corruption risks, their impacts and the possible policies and instruments to tackle them. Although interlinked, the sub-sectors come with their own particular characteristics, stakeholders, governance challenges and corruption risks. Analysing them individually permits a better comprehension of the challenges each confronts and a broader vision of the obstacles the sector faces.

Chapter 2 focuses on water resources management and outlines the fundamental concerns for the sector. It examines how corruption affects the basic parameters of water availability, sustainability and allocation between different uses and users. It addresses the role of corruption in water shortages, water pollution and inequitable distribution.

Chapter 3 considers the problem of corruption in water supply, the water that people need to live. It describes how corruption affects the way people, particularly the poor, access and pay

35 J. Huther and A. Shah, 'Anti-corruption Policies and Programs: A Framework for Evaluation', Working Paper no. 2501 (Washington, DC: World Bank, 2000).

for adequate and safe water services. It also analyses how corruption risks differ between industrialised and developing countries, and between public and private providers.

Chapter 4 provides key insights into the impact of corruption on food security and agriculture. Agricultural production accounts for one the largest uses of water around the world. Irrigation processes – both sophisticated and simple – feed water to the fields of large-scale and small farmers alike. When corruption is present, food security, poverty reduction and equity are compromised, allocations are distorted and limited water resources are often captured by commercial agriculture producers at the expense of small farmers.

Chapter 5 covers another dimension of the sector: water for energy use. It describes how corruption in hydroelectric power comes with a unique set of characteristics that reflect the size of projects and funding. To turn water into power, dams must be built, and, inevitably, individuals, communities and the environment are subject to involuntary change.

Chapter 6 provides a summary of the policy lessons highlighted in the report. It illustrates how accountability can be created and anti-corruption reforms established. Recommendations draw on the experiences profiled in the report and selected best practices from the sector. By looking at how each actor can make a difference, the chapter sets forth approaches for discussion and future action.

This *Global Corruption Report*, focused on water, aims to provide information on the practicalities of corruption and anti-corruption activity in a sector that is critical for people, food, energy and the environment. The first step in the process of tackling the many and varied forms of corruption in water, however, is to improve our understanding of it. Much more effort is needed to develop knowledge about the nature and scope of corruption in the water sector, and to improve knowledge and awareness of its impact. Change will not come about without first establishing the demand for action. This report is an important step forward in building the commitment that is so urgently needed to fight against 'corrupt water'.

Corruption in water – a matter of life and death

Charles Kenny[1]

Everyone needs water to live. Yet many households in the developing world are without access to piped water – either because they are outside the reach of networks, or the systems have fallen into collapse. Maintaining and building water supply systems are the clear responses. But, even when hard-to-find funding is made available, corruption exerts a tax that distorts allocation decisions, wastes resources and, ultimately, takes lives.

A survey of corruption in water provision in South Asia suggests that contractors have frequently paid bribes to win contracts, in addition to the petty corruption that occurs at the point of service delivery. The study, which was done between 2001 and 2002, shows that the cost to companies and the sector represents a sizable burden and loss of resources when the bill is finally tabulated. Bribes on average ranged from 1 to 6 per cent of the contract values. Kickbacks paid during construction escalated the costs to companies by up to another 11 per cent of the contract value. The formation of 'sanctioned' cartels added to the problem of inflated costs, since they helped to push prices 15 to 20 per cent higher than what the market would have demanded. What is worse, these payments actually facilitated companies' failure to meet contract obligations. Kickbacks tended to cover low-quality work and the non-delivery of goods. Materials worth between 3 and 5 per cent of the contract value were never supplied.[2] The economic cost of each dollar of missing materials can be calculated at US$3 to 4 as a result of the water network's shorter life and limited capacity. These costs add up to another 20 per cent on top of already inflated contract prices. This double impact of corruption in the construction of water networks may raise the price of access by 25 to 45 per cent.

What is the economic and social cost of this corruption? An analysis of household survey data for forty-three developing countries suggests a strong correlation between access to water and child mortality. For each additional percentage point of household access, there was a reduction in the under-five mortality rate: a decline of one death for every 2,000 children born.[3]

Comparative country work suggests that the cost for a household water connection is around US$400.[4] Taking the high-end estimate for the cost of corruption in water provision,

1 Charles Kenny is a senior economist at the World Bank, Washington, DC. The opinions expressed are those of the author and do not necessarily reflect those of the World Bank, its executive directors or the countries that they represent.

2 J. Davis, 'Corruption in Public Service Delivery: Experience from South Asia's Water and Sanitation Sector', *World Development*, vol. 32, no. 1 (2004).

3 D. Leipziger *et al.*, 'Achieving the Millennium Development Goals: The Role of Infrastructure', Policy Research Working Paper no. 3163 (Washington, DC: World Bank, 2003). It is worth noting that this estimate is open to dispute: see M. Ravallion, 'Achieving Child-Health-Related Millennium Development Goals: The Role of Infrastructure – A Comment', *World Development*, vol. 35, no. 5 (2007).

4 M. Fay and T. Yepes, 'Investing in Infrastructure: What is Needed from 2000 to 2010', Policy Research Working Paper no. 3102 (Washington, DC: World Bank, 2003).

the price for households would increase by 45 per cent to US$580. As this case demonstrates, the failure to combat corruption results in fewer households being connected, tempered progress on lowering child mortality and increased challenges for achieving the Millennium Development Goals related to water, health and poverty.

Taking the estimate of connection costs being US$400 per household, an investment of US$1 million in piped water projects in countries with under-serviced water needs would benefit 2,500 families and might save nineteen children per year.[5] Having access to water would have other positive impacts, such as on household health, education, women's empowerment and poverty. Yet the costs imposed by corruption over twenty years would mean that from the same investment nearly 30 per cent fewer households would gain access, perhaps 113 fewer children would survive and the related development affects would be undermined.

One recent estimate to assess investment costs based on past trends indicates that low-income countries would have to invest US$29 billion in water projects to meet user demand over the decade ending in 2010.[6] The impacts of corruption would inevitably create leakages and lost resources, undermining the effectiveness of such investment. Assuming a context of low corruption, each year the global toll of child deaths could be 540,000 lower thanks to a decade's investment in water access. A high-corruption environment would save 30 per cent fewer lives.

This is only a partial estimate. As signalled, the impacts of corruption on household access to water go beyond increased childhood mortality. Access affects illness and death among older children and adults as well. Less water and more illness means missed days at school and work. The pass-through effects of reduced water access leave lasting marks on household educational outcomes and income generation. Other household members have to take time away from economically productive activities to care for sick family members. When there is no household access, considerably more time is spent collecting water from elsewhere. Women and children often bear these responsibilities and are forced to make trade-offs between education and other activities.[7] Weak governance and high levels of corruption combine in different forms that affect households and undermine their livelihoods through multiple channels. Yet the most startling impact remains the cost they exert in matters of life and death.

5 Based on an average household size of five people and a crude birth rate of thirty per 1,000 people (the average for low-income countries). The exact estimates are 18.75 and 12.93 deaths averted, respectively. The calculation for the low-cost case is as follows: each US$1 million invested connects 2,500 (US$1,000,000/US$400) households containing 12,500 people (2,500 × 5). These households give birth to 375 children each year (0.03 × 12,500). For these households, coverage increased from 0 to 100 per cent, resulting in 100 fewer child deaths per 2,000 children born. This suggests each US$1 million can save an average of 18.75 children per year (375 × 100/2000).

6 M. Fay and T. Yepes, 2003. The cost estimates are for the period from 2000 to 2010 in order to increase and adequately maintain water infrastructure networks. It is not based on the infrastructure needed for MDG achievement.

7 See article starting on page 40.

2 Water resources management

Kristen Lewis and Roberto Lenton introduce the major challenges for water resources management in their lead piece, sketching out the different forms of corruption in the sector and presenting their consequences with a set of case studies that cover water pollution and environmental sustainability, watershed management and water allocation. They conclude with a set of recommendations on how to tackle corruption in the sector. Transparency International explores how corruption in the water sector affects the mitigation and adaptation efforts with regard to climate change. John Butterworth discusses under what circumstances integrated water resources management (IWRM) offers a promising framework for making water resources management more accountable. Drewery Dyke presents a case study from Afghanistan that shows how local power plays and corruption seize water resources. Enriqueta Abad's contribution on Spain underscores that corruption in industrialised countries can also have serious consequences on local water availability. A final contribution by TI to this section explores the important transboundary dimension of water resources management and examines how corruption in this area runs the risk of undermining regional cooperation and security.

Corruption and water resources management: threats to quality, equitable access and environmental sustainability

Kristen Lewis and Roberto Lenton[1]

Few things are more fundamental to sustainable development than ensuring that the management of the world's water resources is sustainable, equitable, efficient and free from significant governance failures, including corruption. Unfortunately, this ideal has yet to be realised. Water resources management (WRM) means *all actions required to manage and control freshwater to meet human and environmental needs*. These actions include not only an array of governance and management measures but also investment in physical infrastructure for storing, extracting, conveying, controlling and treating water. WRM also includes efforts to protect groundwater, control salinity and promote water conservation.

In short, water resources management is about the fundamental rules of the game. How should water resources be shared among agricultural, industrial, environmental and recreational uses?

1 Kristen Lewis is the co-director of the *American Human Development Report* and an independent consultant specialising in international development and environment issues. Roberto Lenton is currently chair of the Technical Committee of the Global Water Partnership and chair of the Water Supply and Sanitation Collaborative Council; he co-authored this chapter in his individual capacity.

How should water sustainability, quality and aesthetic appeal be valued, and to what extent should they be traded off against competing uses? Who is entitled to use how much? Given the defining role of water for health, livelihoods, economic development, settlement patterns, food production, competitive industrial advantage and, increasingly, tourism, these questions are intimately linked to fundamental decisions about national development strategies and urban planning, as well as political alignments, social equity and cohesion.

The challenges for WRM are formidable: in many places in the world, a large gap between water supply and demand has opened, and it is expected to grow dramatically in the near future.

Competition for water is heating up everywhere. Continuing population growth and urbanisation, shifting dietary habits towards more water-intensive foods, spiralling demand for new fuel crops and expanding water-intensive industries all contribute to ever-growing demand. At the same time, water pollution, degraded ecosystems and global warming[2] endanger local water recharge, quality and sustainable supply around the world.
The numbers speak for themselves.

- Over the past 100 years the world's population has quadrupled while water consumption has risen sevenfold. Water scarcity already affects local regions on every continent, in particular South Asia, China, sub-Saharan Africa, the Middle East, Australia, the western United States and South America's Andean region. By 2025 more than 3 billion people could be living in water-stressed countries. Most distressingly, some of the most affected countries already exhibit a high incidence of poverty and population growth.[3]
- One-fourth of the African population faces chronic water stress,[4] and by 2025 the population in water-stressed regions in sub-Saharan Africa is expected to rise from 30 to 85 per cent.[5]
- Due to overuse and pollution, water-based ecosystems are considered the world's most degraded natural resource. In northern China, 25 per cent of the Yellow River's flow is needed to maintain the ecosystem around it, but human overuse only leaves 10 per cent for one of the greatest arteries of life in East Asia.[6] In Africa, the ecosystem of Lake Victoria, the second largest lake in the world, is in serious decline partly due to pollution.[7]
- Overuse and deterioration of surface water resources has led to a pumping race for groundwater, rapidly depleting aquifers and often leading to saltwater intrusion that makes them unusable. In Yemen, parts of India and northern China, water tables are falling at more than one metre a year, and in Mexico extraction from a quarter of all aquifers exceeds sustainable levels.[8]

Competition for water, already intense, will worsen still with climate change. This competition revolves around water systems that are increasingly vulnerable to pollution, overexploitation and desiccation. Tackling corruption in such a context is as difficult as it is imperative.

---

2 See article starting on page 28.
3 United Nations Development Programme (UNDP), *Human Development Report 2006. Beyond Scarcity: Power, Poverty and the Global Water Crisis* (New York: Palgrave Macmillan, 2006).
4 World Water Assessment Programme, United Nations Educational, Scientific, and Cultural Organization (UNESCO), 'Water, a Shared Responsibility', World Water Development Report no. 2 (New York: UNESCO, 2006).
5 UNDP, 2006.
6 Ibid.
7 World Water Assessment Programme, 2006.
8 UNDP, 2006.

An overview of corruption in WRM

It is important to begin with a caveat: the nature, extent and effects of corruption in irrigation and drinking water supply are well documented, but there have been few systematic inquiries into corruption in water resources management. Nonetheless, it is clear that factors that allow for corruption to take hold in water service sectors also exist in WRM, and, indeed, many cases of corruption in WRM have come to light in recent years.

Corruption in water resources management appears to be closely interlinked with a range of other unethical practices, as well as with governance failures. It is difficult and of limited practical value to draw a strict line between corruption on the one hand and the lack of laws, frameworks, resources, awareness and capacity on the other. Indeed, corruption can be a cause for, consequence of and contributing factor to wider policy failures. Corruption in WRM can therefore be broadly grouped into three areas.[9]

- Corruption related to *water allocation and sharing*, including bribes to obtain water permits and cover up overuse of water resources; patronage or policy capture to skew decisions on water transfers; and allocations favouring specific interests in exchange for money or political support.
- Corruption related to *water pollution*, including kickbacks to regulatory officials to cover up pollution or to distort environmental assessments; and policy capture or bribes to enable deforestation in watersheds.
- Corruption related to *public works and management*, including bid-rigging and collusion among contractors, embezzling WRM funds, buying appointments and promotion in WRM bureaucracies, and favouring construction of large infrastructure projects over other options because of policy-makers' corruption opportunities.

Importantly, corruption and policy failures indirectly related to water resources management often have a strong impact on water quality, availability and distribution. Allowing illegal logging in watersheds, for example, can affect watershed management, modifying streamflows, hurting downstream water users, harming wildlife and causing soil erosion. Unauthorised urban development can adversely affect local water regimes. And allowing overdevelopment of coastal resorts can impact on local water sustainability, for example by exacerbating salinity intrusion. Corruption-fuelled overdevelopment along Spain's coast has aggravated concerns about water shortages while landing dozens of politicians and officials in jail.[10]

The effects of corruption in WRM also have three components.

- *Impacts on economic efficiency.* Water is an important input factor in many economic sectors, including agriculture, fisheries, industry, transport and, in its recreational function, tourism. Corruption can distort the most productive allocation of water among these competing uses while generally inflating the overall cost of supplying and treating water.

9 Examples drawn from P. Stålgren, 'Corruption in the Water Sector: Causes, Consequences and Potential Reform', Policy Brief no. 4 (Stockholm: Swedish Water House, 2006).

10 See article starting on page 35.

- *Impacts on social equity, cohesion and poverty reduction.* Water allocation equals power, and policy capture can instrumentalise WRM to favour specific ethnic groups or business interests – with adverse consequences for poverty reduction, social equality and political stability.
- *Impacts on environmental sustainability and health.* Corruption that leads to water pollution and overexploitation not only has serious consequences for human and animal health and sustainable water supply, it also contributes to degradation of wetlands and other valuable ecosystems, with long-term consequences for livelihoods, development prospects, and wildlife preservation and restoration.

What makes WRM vulnerable to corruption?

Corruption can find fertile ground in water resources management for a number of reasons. First, some stakeholders cannot raise their voices to demand accountability. The fight against corruption in irrigation, drinking water supply and hydropower finds natural allies in those corruption affects most: farmers, households, and communities to be resettled. But in WRM, some important stakeholders are not directly represented in the domestic political arena and thus go unheard: the environment, future generations and, in the case of transboundary waters, water users in foreign countries.

Second, water resources management is extremely complicated, both conceptually and practically. WRM is interlinked in complex ways with environmental systems that themselves are highly complex and often poorly understood by decision-makers and the general public. Similarly, WRM is tasked to deal with a resource that sometimes stretches across vast areas and crosses borders, literally often underground in the form of aquifers, generating multifaceted hydrological interconnections between uses and users that are far from being fully mapped.[11] The resulting veil of obscurity breaks the direct link between a corrupt act and its impact, making it difficult to apportion blame and helping corruption go undetected and unpunished. And, to a much greater extent than in water service sub-sectors, systematic research on corruption in WRM is in short supply, raising doubts about its nature and extent and further contributing to its low profile on the policy agenda.

The large scale and technical complexity of many WRM infrastructure projects can make oversight difficult, rendering the sector vulnerable to corruption. Many large water management projects, such as water storage or inter-basin transfers, are customised engineering endeavours that require expert input for environmental, hydrological and geological questions, as well as for socio-economic, legal and financial issues. Private sector experts – consulting firms, financiers and specialised building contractors – are called upon to help implement such projects. But public authorities in many countries may find it difficult to muster the breadth and depth of expertise to oversee such multifaceted projects effectively.

11 World Water Assessment Programme, 2006.

Big-ticket, fast-paced public construction works offer many opportunities for personal enrichment, and WRM includes many such projects. Such projects require numerous layers of official approval, use large amounts of tax money and face various risks of delay and overruns. These factors offer multiple opportunities and incentives for hold-ups, extortion and collusion in awarding contracts, granting permits and concealing poor-quality work.

In addition, a weak framework for environmental protection and flimsy enforcement mechanisms often let corruption in WRM off the hook. Legal and regulatory weakness is pronounced in the environmental area in many countries, and corruption contributes to environmental degradation. Limited monitoring capacities and toothless punishments for environmental pollution offer little deterrence to water polluters. Developing countries in particular face serious resource and capacity issues with regard to their legal and regulatory framework for addressing environmental issues, including water and watershed management. Even those with strong laws on the books can find themselves hamstrung by a lack of resources when it comes to enforcement.

Mobilising against corruption in WRM is also not easy. The diversity of stakeholders and interests that are involved in WRM makes it difficult to find common ground. In water resources management, many different and often competing actors and sectors vie for the same resources. But they are not pursuing common ends, they operate on very different value frameworks and they often have very few connections and shared organisational structures. These factors make establishing a cross-cutting anti-corruption platform very difficult. Common professional standards, values and organisational structures to discuss and negotiate frameworks for resource sharing can help instil anti-corruption norms and community pressure for responsible behaviour and prevent a corrupt free-for-all.

Finally, WRM has many public masters and often insufficient coordination. Domestic water supply often resides in the health ministry, and irrigation in the agricultural ministry. But water resources management often falls between the stools in terms of institutional responsibility and accountability. Responsibility for water resources is sometimes housed in environment ministries or paired with forestry – but this arrangement leaves out water for household use, water for agriculture, water for energy, water for industry and water for transport, all important aspects of WRM. This lack of clear accountability can create opportunities for corruption to take hold.

Sustainability, water-sharing and corruption: where things have gone awry

Enrichment in watershed management: India

In India, watershed management programmes were launched by the government at a significant scale in the early 1970s. Research[12] shows that, in the early stages of the programmes' development when the main implementing agencies were government departments, financial

12 C. Lobo, 'Reducing Rent Seeking and Dissipative Payments: Introducing Accountability Mechanisms in Watershed Development Programs in India', presentation at World Water Week, Stockholm, August 2005.

'leakages' were of the order of 30–45 per cent of approved amounts. Approved plans included costs that were overestimated by at least 15–25 per cent through the overdesign of structures and misrepresentation of labour requirements, deceptions that then set the stage for the diversion of funds during implementation. This was achieved in several ways, such as forcing labourers to pay a fee in order to gain entry into the workforce, or not adhering to design specifications – using less cement than required, digging trenches to less than the specified depth, planting fewer saplings than the design called for, etc. The net result was not only an increase in implementation costs but also a reduction in capacity to mitigate droughts, augment usable water resources and improve productivity. Later on, when the government actively involved people in implementation, devolved funds to a village body and issued new guidelines, financial leakages were reduced to 20–35 per cent of approved amounts – largely because villagers became more aware of how much money was received and for what purpose.

Water pollution and corruption: China

China's water pollution problems have reached shocking levels. Estimates suggest that aquifers in 90 per cent of Chinese cities are polluted, more than 75 per cent of river water flowing through urban areas is considered unsuitable for drinking or fishing and 30 per cent of river water throughout the country is regarded as unfit for agricultural or industrial use.

The consequences are equally devastating. Two-thirds of China's approximately 660 cities have less water than they need and 110 of them suffer severe shortages. About 700 million people drink water contaminated with animal and human waste. Water pollution has sickened 190 million Chinese and it causes an estimated 60,000 premature deaths every year. Environmental degradation and pollution is believed to cut into China's GDP by 8–12 per cent annually.

The situation is not surprising, given that 13,000 petrochemical factories out of the national total of 21,000 were built along the Yangtze and Yellow rivers, and an estimated 41 per cent of China's wastewater is dumped in the Yangtze alone.

Corruption is a significant factor in the problem. Although China has more than 1,200 anti-corruption laws, bribery, kickbacks and theft account for an estimated 10 per cent of government spending and transactions, with infrastructure projects and procurement among the hot spots. Only a half of the money earmarked for environmental protection between 2001 and 2005 was judged to have been spent on legitimate projects.[13]

Laws and regulations against environmental pollution do exist,[14] but they are weak, poorly monitored and rarely enforced. Only a fourth of factories in 509 cities properly treat sewage before disposing of it, according to a 2005 survey. A company owner admitted in an interview

---

13 M. Pei, 'Corruption Threatens China's Future', Policy Brief no. 55 (Washington, DC: Carnegie Endowment for International Peace, 2007); E. C. Economy, 'The Great Leap Backward?', *Foreign Affairs*, vol. 86, no. 2 (2007).

14 L. Buckley, 'Valuing Ecosystem Services: An Answer for China's Watersheds?', Worldwatch Institute, 11 September 2007.

he would ignore guidelines to install cleaner technologies since they would cost as much as fifteen years' worth of fines. The national environmental protection agency (SEPA) tries to enforce regulations with fewer than 1,000 full-time employees, less than one-tenth the staff at the disposal of its US counterpart. This makes environmental protection an uphill battle. SEPA director Zhou Shengxian, as reported by Xinhua News Agency, put it the following way: 'The failure to abide by the law, lax law enforcement, and allowing lawbreakers to go free are still serious problems in many places.'[15] He further complained that some local government leaders directly interfere in environmental enforcement by threatening to remove, demote and retaliate against environmental officials. Local enforcement agencies usually report to local officials, who often have personal or financial relations with polluting factories. And these officials have been found in many cases to put pressure on courts, the media or even hospitals to cover up pollution.[16]

Bribery and bid-rigging in water transfer projects: Lesotho

Managing water resources includes massive investments in infrastructure for storage, extraction, conveyance and control. 'Grand corruption' in WRM can arise in the design and construction of such big-ticket projects.

Perhaps the best-known case is the Lesotho Highlands Water Project, a US$8 billion project involving the construction of dams and canals for water transfer and supply, hydroelectric power generation and rural development. The chief executive of the Lesotho Highlands Development Agency was found guilty of accepting more than US$6 million in bribes from multinational companies to secure tenders, and in 2002 he was sentenced to eighteen years in prison. Multinationals from the United Kingdom, France, Germany, Italy, Canada and other countries were also prosecuted for seeking to influence the tendering procedure.[17]

The Lesotho case raises two issues of particular relevance to WRM. The lure of milking big-ticket projects for private gain may keep officials from exploring a wider range of alternatives, such as water conservation. In particular, corrupt decision-makers may favour projects where corruption rents are concentrated, and can be easily appropriated by them or their chosen cronies, over smaller projects, which disperse corruption rents more widely.

Second, because the Lesotho case occurred in the context of a large international water-sharing arrangement, the question is whether these agreements may offer incentives or disincentives for corrupt behaviour. Admittedly, these arrangements can be highly complex – technically, financially and administratively – and thereby provide potential entry points for corruption.[18] But this means comparing them to a situation without any joint governance

---

15 Statement made by Zhou Shengxian on 26 December 2006, reported in many sources including www.chinadaily.com.cn/china/2006-12/27/content_768328.htm.

16 *Financial Times* (UK), 5 July 2007; E. C. Economy, 2007; *Financial Times* (UK), 24 July 2007.

17 The Lesotho case has been extensively documented. For more, see *Global Corruption Report 2007*. Examples of media reports include *Business Day* (South Africa), 23 August 2004, and *Pambazuka News* (Africa), 8 August 2004.

18 See article starting on page 37.

frameworks and thus without the mutual gains from joint projects and without any regulation of excessive water abstraction or pollution across borders.

In addition, water-sharing arrangements can also open new opportunities for keeping corruption more effectively in check. In essence they are power-sharing agreements that give each party a strong incentive to watch the others to ensure they do not take more than their fair share. As such, 'competitive oversight' among riparian nations, coupled with assistance in capacity building provided by supporting governments and international institutions, can create an environment less conducive to corruption. Indeed, one could argue the Lesotho scandal came to light because of the involvement of other interested and engaged countries.

Practical measures to prevent and limit corruption in WRM

The fight against corruption in water resources management can be advanced through a mix of initiatives.

Institutional reform

Governments can undertake institutional reforms that clarify the WRM responsibilities of different agencies and establish formal mechanisms for public participation, as well as transparency for the entire decision-making process. They can lay down clear criteria for decision-making that also recognise social and environmental factors, such as the need to maintain *environmental flow,* the minimum volume of water throughput required to safeguard the basic functioning of a hydrological system. Water resource agencies should adopt policies and procedures that require the systematic analysis of project alternatives prior to decision-making.[19] Such policies would help ensure that major investment decisions are made based on clear economic, social and environmental criteria, and reduce the opportunities for decisions to be made because of their potential for private gain. Such policies would need to be complemented by clear policies on such issues as procurement of both goods and services.

Such reforms need not reinvent the wheel but can be guided by established principles and models for water resources management spelled out by the 1992 Dublin Principles (see Box 1), and by transparency and participation standards included in the 1998 UN Economic Commission for Europe's Aarhus Convention.[20] With regard to water-sharing across states, the 1997 UN Convention on the Law of the Non-navigational Uses of International Watercourses provides an important template for cooperation and equitable transboundary water-sharing.[21]

19 One example of such a policy is the World Bank's operational directive 4.01, which states that the analysis of alternatives should include 'a systematic comparison of the proposed investment design, site, technology and operational alternatives in terms of their potential environment impacts, capital and recurrent costs'.

20 UN Economic Commission for Europe, 'Convention on Access to Information, Public Participation in Decision-making and Access to Justice in Environmental Matters', 25 June 1998. See www.unece.org/env/pp/documents/cep43e.pdf.

21 Convention on the Law of the Non-navigational Uses of International Watercourses, adopted by the General Assembly of the United Nations on 21 May 1997. See untreaty.un.org/ilc/texts/instruments/english/conventions/8_3_1997.pdf.

Donors and international financial institutions can also do their share by adhering to proactive information disclosure and consultation for WRM projects they finance and commission, and by putting in place effective sanctions against corrupt employees and contractors. Development projects can be designed so they do not reinforce local power structures that underpin corrupt water-sharing arrangements.[22]

Box 1 Integrated water resources management and the Dublin Principles

IWRM is a process that promotes the coordinated development and management of water, land and related resources with a view to maximising economic and social welfare in an equitable manner without compromising the sustainability of vital ecosystems.[23] IWRM has three goals: environmental and ecological sustainability, economic efficiency in water use, and equity and participation.[24]

At the heart of IWRM lie the Dublin Principles,[25] established at the 1992 International Conference on Water and the Environment in Dublin, which was held in preparation for the 1992 Rio Earth Summit.

- Principle no. 1: fresh water is a finite and vulnerable resource, essential to sustain life, development and the environment.
- Principle no. 2: water development and management should be based on a participatory, public approach, involving users, planners and policy-makers.
- Principle no. 3: women play a central part in providing, managing and safeguarding water.
- Principle no. 4: water has an economic value in all its competing uses and should be recognised as an economic good.

A second set of approaches recognises that a larger constellation of stakeholders are essential for tackling corruption in WRM.

Shining the spotlight on irresponsible WRM

A better understanding of water flows, interdependencies and environmental dynamics such as recharge rates and critical thresholds is required. This will make the implications of WRM

22 See article starting on page 33.
23 Global Water Partnership, 'Integrated Water Resources Management', Technical Advisory Committee (TAC) Background Paper no. 4 (Stockholm: Global Water Partnership, 2000).
24 See article starting on page 31.
25 M. Solanes and F. Gonzalez-Villarreal, 'The Dublin Principles for Water as Reflected in a Comparative Assessment of Institutional and Legal Arrangements for Integrated Water Resource Management', TAC Background Paper no. 3 (Stockholm: Global Water Partnership, 1999).

choices more visible and encourage decision-making that considers all stakeholders in a shared river basin context.[26] The research community can make an important contribution by developing and implementing more refined indicators for equitable and sustainable water sharing and modelling the implications of specific decisions on all involved stakeholders. These steps would provide important information tools for consultation and inclusive WRM decision-making.

An instructive example is the eco-regional assessment of the upper Yangtze River, which combines detailed hydrological, environmental and socio-economic datasets. The resulting simulation model not only informs WRM decisions but also provides a planning platform to bring together different stakeholders and forge a consensus around specific WRM strategies. All these measures make policy capture more difficult.[27]

Shaming water polluters into cleaning up their act

Civil society initiatives and the media can help put the spotlight on environmental polluters. This can be particularly effective where powerful local corruption networks thwart attempts by weak regulators to enforce environmental regulations. In 2006 the Institute of Public and Environmental Affairs in Beijing launched the China Water Pollution Map, a public, searchable, online database that meticulously records water pollution by more than 2,500 polluting enterprises, including some foreign-owned ones. Similar disclosure and shaming initiatives, such as the Toxic Release Inventory established in 1986 in the United States, have successfully contributed to a sharp reduction in environmental pollution.[28]

Strengthening communities for more accountable watershed management

The public at large is critical in the fight against corruption in a number of ways, from voting corrupt politicians out of office, to demanding greater accountability, to becoming involved in environmental monitoring and protection. In response to the corruption in Maharashtra, India, in watershed management, the Watershed Organisation Trust in Maharashtra has developed an approach based on participation, transparency and accountability that has shown promising results. The NGO's initiatives include support for establishing self-help groups for local groups and villagers, and participatory impact monitoring and peer group reviews, in which villagers visit watershed projects in other villages to compare experience and performance. In addition to strengthening accountability of watershed management, the participating villagers have developed greater confidence and ability to deal with officialdom – which has translated into a

26 World Water Assessment Programme, 2006.

27 S. Zhang, 'China Blueprint: Eco-Regional Assessment of the Upper Yangtze River', presentation at World Water Week, Stockholm, August 2007.

28 P. H. Sand, 'The Right to Know: Environmental Information Disclosure by Government and Industry', in F. Biermann, R. Brohm and K. Dingwerth (eds.), *Proceedings of the 2001 Conference on the Human Dimensions of Global Environmental Change*, Report no. 80 (Potsdam: Potsdam Institute for Climate Impact Research, 2002).

lower tolerance to being short-changed. In addition, several of these tools have by now been adopted by government and donor-funded watershed programmes in the country.[29]

Filling the research and awareness void

Finally, developing practical ways forward is clearly hampered by the paucity of research on corruption in the context of water resources management. There is a virtual absence of rigorous studies documenting the scope and impacts of corruption across the spectrum of water resources management, despite the clear evidence that some types of water management actions are prone to corruption. This situation undoubtedly reflects the relative lack of detailed field-based research on how water resources management actually works and the practicalities of administering and financing it. It needs to be remedied, however, if we are to understand more fully the role of corruption in the management of water resources and put in place measures to prevent and limit corrupt practices.

29 C. Lobo, 2005.

Climate change: raising the stakes for cleaning up corruption in water governance

Transparency International

Few informed people doubt climate change poses the single most important policy challenge to global human development, world peace and prosperity – even the sheer survival of societies in their current form. It is little wonder, then, that this far-reaching problem would affect the issue of water and corruption.

For starters, if global warming continues on its current trajectory, it is expected to change our hydrological systems fundamentally – altering rainfall patterns and river flows, diminishing water storage in the polar ice caps and driving up sea levels, leading to saltwater intrusion into the precious supplies of big cities. The world will see more and larger storms, floods and droughts. Climate change will thus alter the basic properties of water systems around the world and therefore the basic properties on which water governance is built.

More droughts and local water scarcity will increase competition for water – raising risks of corruption

By 2020 between 75 and 250 million people in Africa alone are projected to be exposed to increased water stress due to climate change. This comes on top of already severe local water shortages throughout the world and ever-intensifying competition for water due to population growth and rising industrial and agricultural demand.[1] When water flows more sparsely, powerful farmers, rich urban dwellers and water-dependent industries will have strong incentives to secure a larger share and continuous supply through bribes at the service level and political lobbying at the policy level.

Less water means more corruption, both grand and petty. And water shortage in conjunction with corrupted water governance increases the risk of social and political conflict. The abysmal conflict in Darfur has been convincingly linked to corrupted governance and local water shortages intensified by climate change.[2] Many more such conflicts can be expected in the future, if global warming continues to unfold.

More extreme weather requires building new water infrastructure – raising the scale of construction and exposing corruption hot spots

Climate change creates additional urgent demands for upgrading existing water infrastructure and building new facilities. Rising sea levels are estimated to create tens or even hundreds of millions more flood victims each year. This will increase demands for coastal protection systems in many parts of the world.[3] Climate change is also expected to require the modification of many existing dams and therefore additional investment in this sector.[4] Global warming could also shrink yields of rain-fed crops in many regions by up to 50 per cent by 2020, raising demand for more irrigation systems.[5] And, in urban areas, more frequent and intense flooding means overflowing sewers and the risk of contamination of shallow groundwater resources. These effects will make investments in floodproof water networks and adequate sanitation infrastructures more urgent.[6]

Given all these predicted implications, global warming is likely to trigger additional demand for new water infrastructures from flood controls and urban water systems to irrigation and hydropower projects. The United Nations Development Programme estimates that at least US$86 billion need to be allocated annually for climate-proofing infrastructure and building the resilience of the poor to the effects of climate change.[7] This makes it even more urgent to tackle corruption in the water sector, so that valuable resources are not squandered.

1 Intergovernmental Panel on Climate Change (IPCC), Working Group II, 'Climate Change 2007: Climate Change Impacts, Adaptation and Vulnerability', Summary for Policymakers, April 2007.

2 United Nations Environment Programme (UNEP), 'Sudan: Post-Conflict Environmental Assessment' (Nairobi: UNEP, 2007).

3 N. Stern, *The Economics of Climate Change: The Stern Review* (Cambridge: Cambridge University Press, 2007).

4 World Conservation Union (IUCN), 'Adaptation Framework for Action for the Mediterranean Region: Views from the Athens Roundtable' (Gland, Switzerland: IUCN, 2002).

5 Intergovernmental Panel on Climate Change, 2007.

6 ActionAid International, 'Unjust Waters – Climate Change, Flooding and the Protection of Poor Urban Communities: Experiences from Six African Cities' (2006); see www.actionaid.org/assets/pdf/Unjust Waters5HI%20(2).pdf.

7 United Nations Development Programme (UNDP), *Human Development Report 2007/2008. Fighting Climate Change: Human Solidarity in a Divided World* (New York: Palgrave Macmillan, 2007).

Changing water flows and more floods require resettlement at a massive scale and more frequent emergency relief – both particularly vulnerable to corruption

Even cautious climate change estimates suggest 200 million people may become permanently displaced due to rising sea levels, heavier floods and more intense droughts.[8] As chapter 5 shows, resettlement is a hot spot of corruption, inviting fraud, bribery and embezzlement in reimbursement schemes and land transfers on a massive scale.[9] Emergency relief efforts for floods and storms are equally prone to corruption, as mobilising short-notice help often results in suspending sound procurement rules.

In Bihar, India, eleven government and bank officials and a private contractor were charged with embezzling some US$2.5 million in state funds designated for flood relief efforts in 2005.[10] Similarly, Hurricane Katrina, whose devastation of New Orleans may have been intensified by global warming, spawned scandalously corrupt relief and clean-up efforts. Up to US$2 billion in assistance may have been lost to fraud and waste, more than 250 people have been convicted of fraud and some 22,000 reports of fraud, abuse and waste have flooded into the US Hurricane Fraud Hotline.[11]

Climate change aggravates the global water crisis, and corruption slows down mitigation efforts

Not only does climate change increase corruption risks in the water sector, the relationship also works the other way round: corruption makes it more difficult to tackle climate change and thus further exacerbates the global water crisis.

Attempts of science and policy capture

Arriving at a robust scientific and policy agreement on the existence, effects and urgency of climate change was exceedingly difficult because of the complexity of the subject matter. But the scientific pursuit was also bogged down and inexcusably delayed by the rather dubious activities of some industry players and their government allies. They sponsored and promoted pseudo-scientific claims casting doubt on the reality of global warming in the face of overwhelming evidence to the contrary. And they ruthlessly pushed a special interest policy agenda at a time when the disastrous implications for low-level island countries and future generations were already plain to see. These activities have delayed the timely development of an international policy response to global warming, thereby aggravating the global water crisis.[12]

8 N. Stern, 2007.

9 See article starting on page 85.

10 *Wall Street Journal* (US), 16 August 2007.

11 M. Worth, 'New Orleans-Style Corruption Taints Katrina Recovery', Water Integrity Network, 15 March 2007. Available at www.waterintegritynetwork.net/page/375/#_edn4#_edn4.

12 G. Monbiot, *Heat: How to Stop the Planet Burning* (London: Allen Lane, 2006).

Emissions tradings: the corruption risks of a new currency

Curbing greenhouse gas emissions is an integral part of tackling climate change. Emissions trading – trade in 'permits' for generating carbon dioxide and other greenhouse gases – is becoming an important incentive to reduce emissions. But, as with any new currency and market mechanism, this system can be corrupted at several levels. Creating and certifying emission credits must be transparent and follow independently verifiable criteria. The infant market for emissions must be carefully established and regulated to avoid price manipulations. And the consumption of permits requires credible monitoring and sanctions in case of violations. All these considerable governance challenges have already been subject to fraud and corruption.[13]

The many linkages between climate change, corruption and the water sector have potentially grave implications that demand our prompt attention. Global warming is already exacerbating the global water crisis and amplifying related corruption risks, pushing water governance at many places to the brink of collapse. Climate change makes tackling corruption in the water sector even more urgent and will continue to raise the stakes even further in the coming decades.

13 *Times* (UK), 25 April 2007; *Financial Times* (UK), 28 June 2007.

Can integrated water resources management prevent corruption?

John Butterworth[1]

Reforms based upon a strategy known as integrated water resources management (WRM) are well under way in much of the developed and developing world. They aim to address water scarcity crises, especially in the developing world, and water quality problems, particularly in post-industrial societies such as Europe. IWRM's key feature is promoting decentralisation and user participation while enhancing the regulatory role of states.

Measures typically include appropriate basin or catchment institutions; integrated planning to meet agreed-upon water quantity to quality targets; a system of formal administrative water rights, such as licences to extract or pollute water; cost recovery and water pricing (the 'user pays' principle); market-based mechanisms for reallocating water; and better environmental protection, such as reserving water for ecological purposes and the 'polluter pays' principle.

Can IWRM open the door to corruption risks? What happens when informal water providers, which still probably supply most of the world's water users,[2] transition to more formalised, and supposedly more transparent and accountable, public administration systems?

1 John Butterworth is a programme officer at IRC International Water and Sanitation Centre, Delft, Netherlands.

2 J. Butterworth *et al.*, *Community-based Water Law and Water Resource Management Reform in Developing Countries* (Wallingford, UK: CABI Publishing, 2007).

IWRM calls for intensive coordination and cooperation among previously independent government agencies.[3] Along the way, IWRM also introduces complexity. And, by adding another administrative layer that prolongs the decision-making chain, it may open up new opportunities for rent-seeking. Research suggests corruption risks increase at the interface between actors without a previous history of interaction. This is because the level of social control and administrative monitoring decreases as interactions occur outside or on the margins of established organisational systems. Catchment agencies, for example, tend to be new, frequently understaffed in the developing world, and lacking established checks and balances that help to prevent corruption.

Tanzania is an instructive, if worst-case, example. Water resources management reforms have been introduced to address problems related to a large number of rural water users and a relatively weak government infrastructure. With World Bank assistance, the Tanzanian government has introduced a new water permit system over the past decade that aims to improve basin-level management, reduce conflict and improve cost recovery of water resources management services. It sits alongside, but is eroding, a wide variety of customary or traditional systems for locally controlling access to water by farmers. These reforms amount to 'corruption by design'.[4]

A lack of objectivity and transparency creates conditions in which corruption can occur within the Tanzanian system in several ways. Permits based upon agreed extraction volumes may seem objective and fair, but in practice they can be highly subjective. Irrigation systems do not allow for volumetric measurements and delivery; enforcement of fee payments is difficult and costly because of limited staff and large distances; and handling permit funds by water officers is not subject to the same checks as government investments. Some argue that water taxes should focus instead on large-scale users, because the current system costs more to run than it raises in revenue.[5]

A key lesson from Tanzania is that 'modern' governance cannot be easily imposed in rural settings dominated by small-scale water use. In such a setting it may be more effective to amend customary systems carefully and strengthen the position of marginalised smallholders, such as women or the poor. Better water laws and regulations along IWRM principles for larger users are needed in many countries, including Tanzania and other African countries, as well as in Latin American countries such as Guatemala and Bolivia. In these countries, traditional systems without effective alternatives struggle to control some large water users.

Along with new laws and agencies, IWRM can be prevented from opening the door to corruption with the help of strong capacity building among traditional institutions and

3 P. Stålgren, 'Corruption in the Water Sector: Causes, Consequences and Potential Reform', Policy Brief no.4 (Stockholm: Swedish Water House, 2006).

4 B. van Koppen *et al.*, 'Formal Water Rights in Rural Tanzania: Deepening the Dichotomy?', Working Paper no. 71 (Colombo: International Water Management Institute [IWMI], 2004).

5 Ibid.

regulatory bodies, well-resourced and transparent administrative systems, and checks and balances, including mechanisms for citizen complaints.

Afghanistan's upstream powers, downstream woes

Drewery Dyke[1]

For downstream villages in much of rural Afghanistan, access to water is hampered by more than just sub-par infrastructure and other resource limitations. They are also disadvantaged by upstream villages' better access, as well as by local power brokers who either dictate the terms of water usage or induce officials to ignore complaints of people living downstream.

A traditional system under stress and vulnerable to corruption

In much of Afghanistan, managing water from the point it enters an irrigation system is generally supervised by a water master, or *mirab*.[2] Pivotal figures to say the least, *mirabs* are responsible for nothing less than safeguarding the equitable distribution of water. The process of choosing a *mirab*, whether by election or appointment by local councils, or *shoura*, has been described as 'opaque'.[3] How a *mirab* goes about distributing water can also be questionable. He can come under the influence – possibly corrupting – of large landowners (*arbab*), community elders or other powerful figures. A *mirab* may even hold land benefiting from the very irrigation system he controls.

Studies conducted in northern Afghanistan after Hamed Karzai established his first government in December 2001 draw attention to the severe strain facing *mirabs* and traditional water management techniques.[4] Customary rules for distributing common resources among villages have, in various instances, 'completely broken down'.[5] Additionally, canal-head communities are in a stronger bargaining position when it comes to allocation, as they can block canals and illegally divert water.[6]

1 Drewery Dyke is a researcher at Amnesty International's International Secretariat in London. This article contains the views of the author and does not represent those of Amnesty International.

2 The term *mirab* is Persian; there are cognates in other languages, such as *kök basi*, or head of source, in Turkmen. In Herat, the controller of a primary canal is called a *wakil*, or deputy.

3 A. Pain, 'Understanding Village Institutions: Case Studies on Water Management from Faryab and Saripul' (Kabul: Afghanistan Research and Evaluation Unit [AREU], 2004).

4 Principally these include studies published by the AREU, including A. Pain, 2004; J. Lee, 'Water Management, Livestock and the Opium Economy: The Performance of Community Water Management Systems' (Kabul: AREU, 2007).

5 A. Pain, 2004.

6 J. Lee, 2007. There are, however, other reasons in other places why traditional water distribution mechanisms are failing. These may include an absence or failures of governance, change of technologies, such as in regions where wells with handpumps or subsurface dams have been built, and changing economic relations and water use.

Other sources of strain to traditional distribution mechanisms include encroachment by migrant communities, theft, diversion communities and the absence or failure of governance. In Daulatabad district, Faryab province, downstream water consumers endured a continuous-flow irrigation system that supplied higher flow rates at the top than at the lower end.[7] District officials acknowledged the inequities, but their response was, 'These are armed people. We can do nothing.'[8] In Kunduz, an upstream community illegally dammed a canal and diverted irrigation water onto its fields, then bribed a *mirab* with cash to ensure additional water for a rice paddy.[9] The *mirab* was later replaced.

Downstreamers have developed several coping mechanisms in response to these inequities: attempting to negotiate with upstreamers; requesting provincial authorities to intervene; bribing their *mirab*, possibly for additional irrigation; stealing water; fighting neighbours who steal water; or persuading a *mirab* or *shoura* to reduce a neighbour's allocation. In 2007 a study found that *mirabs* abused their position by accepting bribes to deliver additional water to landowners or communities. Greed, threats from power brokers, community pressure or personal financial distress motivated these corrupt acts.[10]

Instances of unequal participation also occur when armed militia leaders, well-connected figures and large landowners force the election of their own nominee as water master and skew water distribution in their favour. In one settlement near the Atishan canal, a single absentee landowner had the right to 95 per cent of the water in a secondary canal, and all decisions regarding allocation lay solely with him or his representatives.[11]

Despite international pledges to combat such corruption,[12] the Afghan government and its leading donor countries have been slow to develop mechanisms to prevent these practices in large swathes of both rural and urban Afghanistan. Yet, policy planners on the ground are increasingly able to differentiate between traditional practices harmful to sharecroppers, women and the landless peasantry and practices that provide social cohesion and development.

Through information exchange, targeted financial support, water user groups or, on a higher level, district development assemblies, it remains possible to limit the scope of corruption or compulsion that upstream communities can impose on downstream water users in the country. Such interventions promise not only to make water governance less corrupt but also to restore some trust to an embattled government.

---

7 A. Pain, 2004.

8 Ibid. The author notes that another official stated that there were no armed power holders in the district.

9 J. Lee, 2007.

10 Ibid.

11 Ibid.

12 The ninth 'Principle of Cooperation' set out in the Afghanistan Compact, a multilateral accord concluded in London on 1 February 2006, states that the Afghan government and international community will '[c]ombat corruption and ensure public transparency and accountability'. The full text of the Afghanistan Compact is available at www.unama-afg.org/news/_londonConf/_docs/06jan30-AfghanistanCompact-Final.pdf.

Corruption fuels housing boom and water stress along Spain's coast

Enriqueta Abad[1]

In Spain, where housing construction accounts for up to 10 per cent of the national economy,[2] plans for new residential development along the coast have doubled in just one year. By mid-2006 communities along the prized Mediterranean coast had approved 1.5 million new homes – along with more than 300 golf courses and 100 leisure craft harbours.[3] An estimated 40 per cent of all new construction in Europe is now taking place in Spain, even though its population makes up less than 10 per cent of the European Union (EU) total.[4]

None of this would be possible without spiralling demand and speculation. But it would not be happening in such wild proportions without a sizable dose of corruption. The authorities have launched dozens of criminal investigations against elected officials and developers. According to Greenpeace, thirty cases have been opened in the eastern province of Valencia and twenty-one are under way in the southern region of Andalucia, where 70,000 illegal houses have sprouted up along the coast.[5]

Most shocking is the story of Marbella, a lavish Andalucian seaside resort near Gibraltar. In 2006 'Operación Malaya' led to the arrest of the mayor, two previous mayors and dozens of city officials after the authorities learnt that 30,000 homes had been built illegally – including 1,600 on parkland. Police froze 1,000 bank accounts and seized more than US$3 billion in villas, thoroughbred horses, fighting bulls and works of art from politicians, attorneys and planning officials accused of taking bribes to approve building permits and re-zonings.[6]

In many parts of Spain, development and corruption go hand in hand. Once a 'greased' construction project is approved, elected officials can use money reaped from licences, land sales and property taxes to fund popular, vote-winning projects. Construction-related income provides upwards of 70 per cent of municipal budgets for towns in the Marbella area.[7] This underground economy thrives where democracy and transparency do not. Town councils have grown immensely rich in the process.[8]

It is a win-win scenario, except for the cause of water resources management. This corruption-fuelled free-for-all in one of Europe's driest regions has severely challenged the

1 Enriqueta Abad is an MSc student at the School of Oriental and African Studies, London.
2 *The Economist* (UK), 3 May 2007.
3 *El Mundo* (Spain), 6 July 2006.
4 *Washington* Post (US), 25 October 2006.
5 *El Mundo* (Spain), 6 July 2006.
6 *Washington Post* (US), 25 October 2006; *El Mundo* (Spain), April 2006.
7 *The Economist* (UK), 16 September 2006.
8 *Financial Times* (UK), 25 May 2007.

ability of planners to provide water services. As it is, 4.3 million people living in 273 coastal towns have no wastewater treatment, according to Greenpeace.[9]

Scandal has also struck the Andalucian city of Ronda, famed for its picturesque cliffs and canyons. With the blessing of city officials, developers want to build a resort called Los Merinos that includes 800 homes, two luxury hotels and two golf courses. The dispute over the project's legality, the area's ability to provide water and the risk of pollution has created a tangled governance crisis. According to the Ministry of Environment, Los Merinos is one of 200 planned urban developments in Spain with no certain water supply.[10] 'I only want to warn people intending to buy whatever type of home at Los Merinos there is no guarantee of water,' said regional environment chief Ignacio Trillo.[11]

According to Cuenca Mediterranea Andaluza, a regional organisation created by the Andalucían government to tackle water corruption, the project is illegal because it does not abide by regulations related to water protection.[12] Developers plan to extract water from an aquifer under the Sierra de las Nieves, a mountainous woodland designated a 'Reserva de la Biosfera' (Biosphere Reserve) by UNESCO. Builders want to supply each Los Merinos resident with more water per day than the maximum level established by local planners. Because the sierra and its fauna, as well as surrounding villages, already rely on the aquifer, overtapping could put citizens and the environment at risk.

The Andalucian government filed an appeal with the Malaga regional court in hopes of blocking Ronda's approval of Los Merinos, claiming 69 per cent of the 800-hectare area is being developed illegally. A judge rejected the appeal in July 2007, declaring the project would not cause 'serious, irreversible destruction of the environment' and that developers have a sufficient water supply.[13]

Like elsewhere in Spain, Ronda's government stands to benefit from licences, land sales and property taxes. Los Merinos represents a vote-winning project, as it would stimulate 'long-term and qualified employment', according to a local golf advocacy group.[14] Civil society groups in Ronda have organised several demonstrations against Los Merinos. In hopes of resolving the controversy, the European Commission has begun a review of the development's approval process.[15]

As of mid-2007 the Spanish parliament had not discussed the issue of corruption in water management for Ronda or similar projects elsewhere. Whether the parliament is unable or unwilling, the link between lucrative development projects and the pressure on scarce water resources may be either too inconvenient or too complex to address.

9 *Washington Post* (US), 25 October 2006.
10 *El País* (Spain), 16 April 2007.
11 *The Olive Press* (Spain), 2 August 2007.
12 *El Mundo* (Spain), 26 January 2006.
13 *The Olive Press* (Spain), 2 August 2007.
14 *El País* (Spain), 25 February 2007.
15 *El País* (Spain), 19 February 2007; *El País* (Spain), 11 February 2007.

Corruption without borders: the challenges of transboundary water management

Transparency International

Water not only crosses different regulatory regimes and legal classifications, it also crosses borders. The extent of global water interdependence is stunning.

Two in every five people in the world today live in international water basins – catchments or watersheds – which account for 60 per cent of global river flows. In Africa, 90 per cent of surface water and more than 75 per cent of the population are located in transboundary river basins. Around the world, water sources for 800 million people living in thirty-nine countries originate beyond their national borders.[1]

Transboundary water issues affect almost everyone. And this hydrological interdependence adds another layer of complexity to the fight against corruption in water resources management.

But is it really possible to speak of corruption – the abuse of entrusted power for personal gain – when water conflicts transcend the domestic legal sphere and occur in the context of power politics between sovereign states? It is. The 'entrusted power' need not be tied to a domestic political system. In transnational water management, it can derive from commitments states enter into through multilateral water treaties, 200 of which have been signed in the last fifty years.[2] Or it can be tied to fiduciary duties to govern water responsibly and sustainably, in accordance with established international norms and agreements such as the Dublin Principles or Agenda 21.

For two reasons, tackling corruption in transboundary water-sharing is more difficult and even more urgent than national water resources management. It is harder to prevent and punish, and it has very grave consequences.

Corruption in transboundary water can cause international conflict, destabilise entire regions and lead to ecological disaster[3]

Over the last fifty years countries have engaged in more than 500 conflictive events over water. Almost 90 per cent were disagreements over infrastructure and quantity allocation.[4] The main

1 United Nations Development Programme (UNDP), Human Development Report 2006. *Beyond Scarcity: Power Poverty and the Global Water Crisis* (New York: Palagrave Macmillan, 2006). World Water Assessment Programme, United Nations Educational, Scientific, and Cultural Organization (UNESCO), 'Water, a Shared Responsibility', World Water Development Report no. 2 (New York: UNESCO, 2006).

2 A. Wolf, 'Conflict and Cooperation over Transboundary Waters', Human Development Report Office occasional paper (New York: UNDP, 2006).

3 Ibid.; S. Postel and A. Wolf, 'Dehydrating Conflict', *Foreign Policy*, no. 126 (2001); World Water Assessment Programme, 2006.

4 A. Wolf, 2006. It is important to note, however, that no outright wars have been fought over water during this period.

trigger for conflict is usually not water scarcity per se, but unilateral construction of a dam or diversion of a river. Both such projects can be heavily influenced by corruption from powerful vested interests.

What is more, many important transboundary water-sharing arrangements coincide with long-standing flashpoints for regional conflict, such as in the Middle East. This makes corrupt water grabs particularly damaging to regional stability. Even when corruption does not lead competition for water to escalate into conflict, it can precipitate the collapse or block the establishment of water-sharing arrangements.

Preserving and sharing the benefits of a common good such as a river basin is vulnerable to a serious free-rider problem: everyone has a strong incentive to take more than their fair share if there is suspicion that others also do so. Trust in the effective enforcement of commitments on all sides is essential to sustaining such agreements. But water corruption fatally undermines this trust by thwarting enforcement and opening the door to irresponsible water grabs or water pollution. The result is not only that countries forfeit opportunities to realise gains from joint water management, but also that shared water ecosystems are vulnerable to overuse and ecological collapse.

The devastating environmental, social and economic consequences of failing water resources management are plain to see at Lake Chad, the great African river basin that has shrunk to 10 per cent of its former size, and at the Aral Sea, formerly the size of Belgium and now a hypersaline water basin one-fourth its original dimension.

Out of jurisdiction, out of sight: more incentives for corruption in transnational contexts

Even where international water-sharing arrangements are in place, monitoring abuse and enforcing effective sanctions is considerably more difficult than within a national jurisdiction. When the victims of water pollution are outside one's own jurisdiction and excessive water diversion hurts only the farmers in neighbouring countries, such corruption is more likely to go undetected and unpunished and is therefore more difficult to resist. Even when water projects are undertaken jointly by two or more states, the jurisdictional twilight zone in which they are placed fosters corruption. The bi-nationality of the Itaipú Dam, a joint project by Brazil and Paraguay, made it possible for management to operate a parallel account not declared to either authority. The resulting fraud has been estimated at US$2 billion.[5]

Leveraging hydro-diplomacy for the fight against corruption

Though the corruption of transnational water resources is both more tempting and pernicious than the corruption of domestic water resources, sharing waters can also provide opportunities for fighting corruption in water across borders. When domestic laws against excessive water diversion are weak or provisions for wastewater treatment unenforced, international

5 O.-H. Fjeldstad, 'Corruption: Diagnosis and Anti-corruption Strategies', Independent Evaluation Group background paper (Washington, DC: World Bank, 2007).

agreements may provide an additional entry point for public pressure. They can take governments to task to preserve ecosystems and provide consultative mechanisms in water management. And they often come with institutional mechanisms such as river basin committees, which can serve as platforms to shine the spotlight on corruption and mobilise new allies in the fight against domestic polluters or water-guzzling agro-industrialists who capture domestic water policies or bribe local enforcement officials.[6]

The 1997 UN Convention on the Law of the Non-navigational Uses of International Watercourses codifies important principles of *prior notification, equitable and reasonable utilisation* and *no significant harm* for the use of transboundary waters.[7] These principles inform many international water-sharing agreements, although only a few countries have so far signed up to the convention itself.[8]

---

6 World Water Assessment Programme, 2006.

7 Convention on the Law of the Non-navigational Uses of International Watercourses, adopted by the General Assembly of the United Nations on 21 May 1997.

8 Ibid.

3 Water and sanitation

In this chapter Muhammad Sohail and Sue Cavill explore in compelling detail how corruption exacerbates the challenge to provide safe and affordable drinking water and sanitation to the poor. Their section presents a great wealth of case studies that document how corruption makes drinking water inaccessible, unaffordable and unsafe. The authors also explore the underlying dynamics that sustains corruption in this sector and conclude with a comprehensive set of recommendations for action, drawing on inspiring examples of successful initiatives from around the world.

A number of supplementary contributions further deepen the analysis of different aspects of corruption in drinking water and sanitation, which is the water sub-sector most closely linked to health and human development. Bernard Collignon adds case evidence on corruption in water as it affects the urban poor. Jack Moss explains from the industry perspective how corruption affects the day-to-day operations of private water operators. Per Ljung examines the significant corruption risks for drinking water and sewage in industrialised countries and Transparency International discusses the corruption risks for private and public operators. Virginia Lencina, Lucila Polzinetti and Alma Rocío Balcázar report on a successful initiative to strengthen anti-corruption provisions in the public procurement of water infrastructure. Venkatesh Nayak describes how freedom of information legislation is used in India to make water governance more accountable to the poor.

Water for the poor: corruption in water supply and sanitation

Muhammad Sohail and Sue Cavill[1]

The slum is overcrowded, noisy and polluted. Most of its residents live in shacks that hardly resemble decent homes. Ajay has lived in the slum with his wife and children for five years. Getting enough water every day is a constant problem. The Slum Department was supposed to have implemented a water project for the slum-dwellers, but the project exists on paper only; in reality the area is still without water and sanitation. No one knows where all the money went.

1 M. Sohail is Professor of Sustainable Infrastructure and the leader of Research and Consultancy at the Water, Engineering and Development Centre (WEDC) at Loughborough University. S. Cavill is a researcher at WEDC. This section is based on a research project conducted by the authors entitled 'Accountability Arrangements to Combat Corruption' (initially funded by the Department for International Development, UK). For more, see wedc.lboro.ac.uk/projects/new_projects3.php?id=191.

A few years ago an NGO set up a water and sanitation project in the slum: Ajay and the other residents formed a committee to look after the water point and sanitation block and collected money to pay the water bills. The mastaan *(muscle man) – who also happens to be the local ward councillor – saw the project as competition to his water-vending business, however. The pump was vandalised one night and hasn't worked since.*

Nowadays Ajay's wife gets up early every morning to collect water for the family: she usually walks to the nearest public water fountain but she also begs for water from the gate staff at the nearby factory or from homes in the wealthy areas of the city. Occasionally she has to buy from water vendors, but she can't afford much because the water is so expensive; the family goes thirsty on those days.

Water, corruption and the poor: a specific challenge

More than any other group, the poor are the main victims of the global water crisis. But water poverty is not just an important cause and characteristic of economic poverty; it is also a consequence of it. There is a causal relationship between poverty and the lack of water that flows both ways. Two-thirds of the roughly 1.2 billion people who do not have access to safe drinking water live on less than US$2 a day. Of the more than 2.6 billion people who lack basic sanitation, a half fall below that same poverty line.[2]

Poor people without water are trapped in a desperate, daily struggle for survival to access water and other basic needs. Without economic resources to improve their situation, poor citizens suffer on multiple levels and become trapped in an inescapable cycle. Corruption is a major force driving these problems and the growing global water crisis. Inadequate access to clean water, combined with the lack of basic sanitation, is a key obstacle to progress and development in the world. Historically water-deprived regions, such as sub-Saharan Africa, are suffering disproportionately under these pressures.

Political voice and patronage dependencies

Income poverty also goes hand in hand with political marginalisation, low social status and unequal power relationships. All these factors limit the tools and space available for poor citizens to take action against corruption. Poor people may feel the need to reduce their own vulnerability and resort to bribery to obtain a modest level of political protection and financial security, making it even more challenging to break the cycle of corruption in the water sector.

Water, poverty, health and gender: close linkages

Access to water and sanitation services is a critical factor in the ability of poor households to generate the income and savings needed to exit poverty. Increased access saves households time. It allows them to do other activities – from entering the labour force to studying more

2 United Nations Development Programme (UNDP), *Human Development Report 2006. Beyond Scarcity: Power, Poverty and the Global Water Crisis* (New York: Palgrave Macmillan, 2006).

in order to get a better-paying job. According to one estimate, some 40 billion hours a year are spent collecting water in sub-Saharan Africa – a figure that is equivalent to the number of hours worked annually by France's entire labour force.[3] Greater access to water and sanitation also means the reduced risk of missing work from waterborne illnesses. Throughout the whole of Africa, an amount equivalent to about 5 per cent of GDP is lost to illness and death caused by dirty water and poor sanitation every year.[4]

In developing countries, about 80 per cent of health problems can be linked back to inadequate water and sanitation.[5] Across the world, water-related ailments such as diarrhoea claim the lives of nearly 1.8 million children every year.[6] These illnesses exact a different toll on the lucky ones who survive. Poor health hampers income-earning potential and cuts down on education. An estimated 443 million school days are lost each year because of water-related ailments.[7] The same diseases are blamed for costing the Indian economy 73 million working days each year.[8] In responding to these health problems people are forced to waste excessive amounts of time and resources, which are already in short supply. Sickness means a loss of work days, output, wages and savings.

In most societies, women have the primary responsibility for collecting and managing water for their households. In the best cases, water may be found at a local standpipe or nearby river. In the worst cases, getting water may be a day-long activity. It is not uncommon for women and girls in Africa to walk more than 10 kilometres to gather water for their families in the dry season.[9] Girls are often tasked to help with the work and are forced to forgo other activities, such as schooling. Improving household access to water services can reduce these burdens placed on women. It also supplies a reliable and safe water source for a family's daily necessities.

Disconnected from the mainstream

The poor often have very limited ability to connect to formal water networks. A legacy of the colonial era in many developing countries, formal water and sewerage networks were often designed to cater to the interests of elites and have outgrown the demand now coming from poor areas. The poor in developing countries typically live in rapidly expanding, poorly planned and illegal settlements that are a manifestation of their political disenfranchisement and corruption's reach.

Getting the poor connected to formal networks is not a simple task. In some countries, water utilities are legally barred from serving informal settlements. Even when water service is available, poor households may be unable to apply for a water connection without proof of a land title. Other communities may find it difficult to connect to water and sewerage networks because

---

3 UNDP, 2006.
4 Ibid.
5 United Nations, *Millennium Report* (New York: UN, 2000).
6 UNDP, 2006.
7 Ibid.
8 P. Swann and A. Cotton, 'Supporting the Achievement of the MDG Sanitation Target', Well Briefing Note for CSD-13 (Loughborough: Loughborough University, 2005).
9 UNDP, 2006.

they are geographically isolated, located on steep hillsides or constructed on marginal land. When expanding networks is possible, the formal private sector may be reluctant to provide service to low-income areas. Their perceptions may be that poor customers fail to pay bills or will vandalise the infrastructure once it is built. And, even if all these hurdles are cleared, the costs for directly connecting households to the water network are often prohibitive for poor families. A utility connection in Manila is equal to about three months of income for the poorest 20 per cent, while the equivalent figure is six months in Kenya and more than a year in Uganda.[10]

Dependence on informal providers

Lack of access to the public water network deprives the poor of what is usually the cheapest source of water. To fill in the gaps, the poor turn to public standpipes or suppliers that include NGOs and informal water vendors. Very often these alternative providers operate in a legal limbo. Their businesses are insufficiently recognised by the authorities, unregulated and dependent on securing access to bulk water resources through informal means.

Being outside the law allows informal providers to charge above public utility rates for water access. A cruel irony results from these circumstances: poor people living in slums unconnected to the water grid frequently pay far more than connected consumers. In Jakarta, Lima, Manila and Nairobi, the poor pay five to ten times more for water than their wealthy counterparts. Residents of Manila without water service rely on kiosks, pushcart vendors and tankers to meet their needs. At a cost of US$10–20 per month, it is more than what people living in New York, London and Rome pay for water.[11]

The result: the heightened vulnerability of the poor to water corruption

Lack of access to a formal and legal water connection, limited choice and voice, powerlessness, and a heavy dependence on informal and illicit providers make the poor extremely vulnerable to corruption. Locked into dependency and necessity, they are affected by many types of corrupt practices.

Corruption in access, service delivery and maintenance

Country studies provide a graphic overview of how corruption corrupts the provision of water services. A groundbreaking 2004 survey in India found that 40 per cent of water customers had made multiple small payments in the previous six months to falsify meter readings so as to lower their bills. The findings were based on more than 1,400 interviews and meetings with customers, utility staffers, elected officials, development workers, activists and journalists. Customers also said they had paid bribes to speed up repair work (33 per cent of respondents) or expedite new water and sanitation connections (12 per cent of respondents).[12]

10 Ibid.
11 Ibid.
12 J. Davis, 'Corruption in Public Service Delivery: Experience from South Asia's Water and Sanitation Sector', *World Development*, vol. 32, no. 1 (2004).

Other countries have shown a similar extent of corruption occurring at the level of water users. More than 15 per cent of respondents to a national household survey in Guatemala said they paid a bribe when they sought a water connection or reconnection.[13] In Kenya, over 50 per cent of households surveyed in Nairobi felt their bills were unfair, 20 per cent said they paid their bills regardless of the accuracy (in order to avoid disconnection) and 66 per cent said they had had a water-related corruption experience in the past year.[14]

Fee collection is also vulnerable to corruption when additional middlemen are involved. Local water committee members may steal money that has been collected from residential customers to pay the community's water supply and sanitation bill. In the case of Namibia, the result of the theft of fees was that some residents suffered a disconnection in service.[15]

Extortion in the repair and maintenance services is also common. In Zimbabwe, a resident of Harare was told the broken pipe that leaked sewage into his house would not be fixed unless he 'dropped a feather' – paid a bribe. A woman who was wrongly billed sixty times more than her normal monthly rate for water was told that to have her service turned back on she would have to make the full payment. The elderly widow refused and instead began having the renters she took in collect water from a nearby church.[16]

The pressure to extract bribes from customers is further compounded by another form of corruption in the sector: superiors in public services charge 'rents' from their subordinates in exchange for preferential shifts, locations or responsibilities. In Mauritania, standpost (e.g. water point) attendants are known to pay bribes to obtain these important community jobs.[17] The ability of staff to purchase these choice posts in turn depends on their ability to collect bribes from customers. The poor make an easy target.

Collusion to corner the market

In Bangladesh and Ecuador, private vendors, cartels or even water mafias have been known to collude with public water officials to prevent network extension or cause system disruptions. These service breakdowns help to preserve their monopoly over provision and increase the business for private water vendors in specific neighbourhoods.[18]

Collusion limits the choice of the poor and forces them to rely on potentially unsafe and overpriced water from cartels that often are operating illegally. The stark human consequences of this manifestation of corruption are vividly described by one survey respondent in Bangladesh: 'It is really tough for a day labourer to give a high price for . . . water. So, our

---

13 Acción Ciudadana, 'Indicadores de Percepción y Experiencias de Corrupción de Guatemala – IPEC' (Guatemala City: Acción Ciudadana, 2006).

14 TI Kenya, 'Nairobi Water & Sewerage Company Limited: A Survey, April–May 2005' (Nairobi: TI Kenya, 2006).

15 IRC International Water and Sanitation Centre, 'Zimbabwe, Namibia: Examples of Corruption', 21 September 2007.

16 IRIN News, 'Zimbabwe: As Services Collapse, Corruption Flourishes', UN Office for the Coordination of Humanitarian Affairs, 28 May 2007.

17 See article starting on page 52.

18 E. Swyngedouw, *Social Power and the Urbanization of Water: Flows of Power* (Oxford: Oxford University Press, 2004).

budget is strained and we cannot afford to meet our needs. We cannot save anything for our future either.'[19]

Corrupted policy design also hurts the poor

Corruption occurring higher up the water supply chain, where policies are set and infrastructure projects designed and managed, also affects the day-to-day struggle of the poor for water. This grand corruption reinforces inequitable water policies, diverts resources away from pro-poor projects and stymies infrastructure build-outs to meet user demand. The economic and financial costs are difficult to quantify but the sizable amount of funding the water sector receives makes the opportunity for siphoning off resources great.

In 2003 the European Commission, for example, learned that 90 per cent of EU funds intended to help improve water service in fifty communities in Paraguay had been diverted. The funds were eventually traced to a bank account of a foundation that was not involved in the project. As a result of these findings, Paraguay launched a criminal investigation into the affair.[20] Rather than shadow companies, collusion was found to be a problem on a World Bank water project in Albania.[21] In 2005 the multilateral lender debarred six companies and five people after it was found that they had colluded on a project to improve failure-prone pipes, wells and pumping stations across the country.[22]

Fraud in bidding and the award of contracts is another hot spot for grand corruption. Corrupt procurement can take on many forms, including tailoring project specifications to a corrupt bidder, providing insider information, limiting bid advertising, shortening bid periods and breaching confidentiality. Contractors may 'sweeten up' the review committee with lavish entertainment in exchange for certifying their work or turning a blind eye to construction shortcomings.

Political corruption

As in most other public works sectors, political corruption also tarnishes water service. Various forms of corruption may lead to policy capture that sways project selection. Politicians may be bribed to divert resources away from improving rural water supply networks and using them in urban areas where influential constituencies are based. Politicians may back expensive and high-tech infrastructure projects to maximise opportunities for extortion or to steer lucrative business contracts to cronies.

19 Institute for Development Policy Analysis and Advocacy at PROSHIKA, *Accountability Arrangements to Combat Corruption in the Delivery of Infrastructure Services in Bangladesh* (Loughborough: Loughborough University, 2007).
20 European Anti-Fraud Office (OLAF), 'Report of the European Anti-Fraud Office, Fifth Activity Report for the Year Ending June 2004' (Brussels: European Commission, 2004).
21 IRC International Water and Sanitation Centre, 'Albania: World Bank Debars Fraudulent Firms Involved in Water Project', 8 April 2005.
22 World Bank, 'Albania: Water Supply Urgent Rehabilitation Project', (Washington, DC: World Bank, 2004).

Bribes can also be used as a means to shore up the political power of individuals and groups. Contracts with private sector companies for building and managing water networks can be padded to provide slush funds for political campaigns and parties. Contracts may also be awarded in order to favour a specific constituency or friend in return for votes.[23]

When projects are built, there is an all too common mismatch between their design and sector needs that leads to poor management and infrastructural maintenance. The resulting infrastructures are likely to fall quickly into disrepair, neglect and irrelevance. A study in one rural district in Malawi showed that three-quarters of new village water points relied on expensive drilling technologies even though two-thirds of the population lived in high water table areas where hand digging and other simple technologies could have been used.[24]

How corruption in water and sanitation can be tackled

Fighting water corruption while focusing on the needs of the poor presents a tremendous challenge. It means changing a system that favours powerful vested interests and making it more – if not primarily – accountable to the needs of society's weakest citizens (economically, politically and socially). It also requires designing anti-corruption strategies carefully to ensure that they do not harm the intended beneficiaries in the process.

Approaches also must be targeted to break the cycle of corruption. Grand corruption at the sectoral level nurtures petty corruption at the street level. Manipulated policies and botched infrastructure create and perpetuate the very shortages and lack of choice, voice and accountability the poor face in dealing with water suppliers. To ensure anti-corruption reforms work for the poor, action is needed both upstream and downstream and at different levels along the supply chain.

Strategies must build and match the capabilities of all water stakeholders

The effective linking of capabilities to anti-corruption activities is essential at all levels and among different players.

At the national level, anti-corruption work needs to match governance capabilities. For certain countries, general government reforms may be a more useful starting point than establishing anti-corruption commissions. If overall governance is weak and the incidence of policy capture high, setting up regulatory or oversight agencies could leave them vulnerable to the corruption they were created to combat.[25]

At the sector level, the sequencing of private sector engagement must be assessed. Private sector involvement has been found to be less effective and accountable when it is brought in

23 See article starting on page 55.
24 S. Sugden, 'Indicators for the Water Sector: Examples from Malawi' (London: WaterAid, 2003).
25 A. Shah and M. Schacter, 'Look before You Leap', *Finance and Development,* vol. 41, no. 4 (2004); J. Plummer and P. Cross, 'Tackling Corruption in the Water and Sanitation Sector in Africa: Starting the Dialogue', in E. Campos and S. Pradhan (eds.), *The Many Faces of Corruption* (Washington, DC: World Bank, 2007).

too early or if strong regulatory capacities are not yet in place.[26] The negative experiences that many developing countries have had with privatising their water services signal what can happen when proper government oversight powers are not established.

At the local level, creating transparency and consultative mechanisms will work only if poor communities have the resources, information and mobilisation structures to take advantage of them. This highlights the need for complementary capacity-building efforts. Watchdog functions in South Africa, for example, were found to be neither premised on partnerships with the poor nor geared to reporting at this level.[27]

Anti-corruption efforts for the water sector need to be intentionally pro-poor

Most successful anti-corruption measures in the water sector directly or indirectly benefit the poor. But some initiatives need to be designed more carefully to ensure that the intended beneficiaries are not hurt in the process of combating corruption.[28] Cost recovery, for example, can strengthen budgetary discipline and the financial independence of water providers – important building blocks for more accountability which have been successfully deployed in many reform projects. Nevertheless, this strategy can work only if pro-poor targets for expanding networks and keeping tariffs affordable are clearly recognised and incorporated into financing plans and tariff-setting schedules.

The OECD estimates that, in the absence of targeted subsidies, increased cost recovery through tariffs would force more than a half of households in many Eastern European and Central Asian countries to spend more than 4 per cent of their income on water. This is considered the maximum sustainable level of household spending on water.[29] In Bolivia, Honduras and Nicaragua, the UNDP anticipates affordability problems for more than a half of the population, and for a staggering 70 per cent of households in sub-Saharan Africa, if cost recovery were introduced without accommodating measures.[30]

Some corruption in water is best fought through legalisation[31]

Informal providers offer important bridging services – as well as capital and expertise – that make water and sanitation available where official networks fail the poor. In many

26 C. Kenny, 'Infrastructure Governance and Corruption: Where Next?', Policy Research Working Paper no. 4331 (Washington, DC: World Bank, 2007).

27 G. Hollands and Mbumba Development Services, 'Corruption in Infrastructure Delivery: South Africa', case study (Loughborough: Loughborough University, 2007).

28 J. Plummer, 'Making Anti-corruption Approaches Work for the Poor: Issues for Consideration in the Development of Pro-poor Anti-corruption Strategies in Water Services and Irrigation', Report no. 22 (Stockholm: Swedish Water House, 2007).

29 OECD, 'Keeping Water Safe to Drink', Policy Brief (Paris: OECD, 2006).

30 UNDP, 2006.

31 T. M. Solo, 'Independent Water Entrepreneurs in Latin America: The Other Private Sector in Water Services' (Washington, DC: World Bank, 2003); M. Kjellén and G. McGranahan, *Informal Water Vendors and the Urban Poor* (London: International Institute for Environment and Development, 2006); S. Trémolet and C. Hunt, 'Taking Account of the Poor in Water Sector Regulation', Water Supply and Sanitation Working Note no. 11 (Washington, DC: World Bank, 2006).

developing countries, 20–30 per cent of urban households depend on independent vendors as their main water providers. Attempting to stamp out these indispensable yet informal services would drive them deeper into illegality and hurt their main clients: poor communities.

Bringing informal providers into the legal fold – through licences, 'light touch' regulations and their formal recognition as alternative suppliers – is a more viable strategy. This could protect both vendors and customers from corruption and exploitation.[32] Authorities in countries as diverse as Senegal, Vietnam, Mozambique and Ghana have already licensed informal vendors (or are considering doing so) and established guidelines for tanker operators and independent entrepreneurs.[33]

Box 2 System reform: routes to accountable water utilities

The Phnom Penh Water Supply Authority has achieved significant progress in combating a culture of corruption and improving service delivery to the poor. Among the poorest families in the city, the number of household connections rose from 100 in 1999 to 15,000 in 2006.[34] Key components of Phnom Penh's success include the following.

- Replacing often corrupt bill collectors with public offices where customers can pay their bills directly.
- Offering training and performance-related bonuses for staff, fast-track promotion for young dynamic staffers and profit-sharing.
- Subsidising connection fees and bills for the poorest people.
- Installing meters for all connections.
- Establishing inspection teams and stiff penalties for illegal connections.

Serious challenges remained in the area of procurement, however. Due to corruption, the World Bank suspended a contract and withheld US$1.8 million (€1.4 million) in June 2006 from a water project in Phnom Penh intended to expand water service to targeted towns and peri-urban communities. The suspension was lifted only after the authorities agreed to delegate procurement of World-Bank-financed projects to an international firm.[35]

32 Competition is found more important than ownership for performance in many sectors. See D. Parker and C. Kirkpatrick, 'Privatisation in Developing Countries: A Review of the Evidence and Policy Lessons', *Journal of Development Studies*, vol. 41, no. 4 (2005).

33 UNDP, 2006; S. Trémolet and C. Hunt, 2006.

34 M. C. Dueñas, 'Phnom Penh's War-torn Water System Now Leads by Example', *Asian Development Bank Review*, vol. 38, no. 4 (2006); World Bank, 'Rehabilitating the Urban Water Sector in Cambodia', 2006; see go.worldbank.org/DRCGF75J80.

35 World Bank, 'World Bank Lifts Suspension of Projects', 7 February 2007; World Bank, 'Rehabilitating the Urban Water Sector in Cambodia'; World Bank, 'Cambodia: World Bank Releases New Statement and Update', 6 June 2006.

Towards integrity and professionalism for water services

Service providers can promote codes of conduct and citizen charters as a means of improving the professionalism and integrity of their operations. Once finalised, these commitments should be publicly displayed in local languages and in a way that respects community norms. In the Indian state of Tamil Nadu, efforts to promote the sector's integrity have involved engaging the community in the decision-making process. Internal reforms of the water utility are being led using a *koodam,* a traditional body that treats everyone equally, including women and Dalits (or 'untouchables'). As a result of involving local citizens, water access has increased by 10 per cent each year and efficiency measures have driven down investment costs by more than 40 per cent. Tamil Nadu's experience is now helping other public utilities in India replicate their success.[36]

Making the right to water an enforceable entitlement

Rights are the ultimate guarantor of equality. When enforced, a legal right to water can be an important mechanism for poorer communities. It can help them outflank local power relations and hold authorities to account for corrupt water policies and dysfunctional delivery systems.

Existing international mechanisms are already in place that outline the obligation of countries to provide water for their citizens. Access to sufficient, safe and affordable water for personal and domestic use is recognised as a human right by the United Nations. The UN Committee on Economic, Social and Cultural Rights gave access to water this status in 2002 and outlined the duties of governments to respect, protect and fulfil their commitments. To date, however, no international treaty exists to enforce or monitor compliance.

At the country level, states can create their own legal commitments by incorporating the right to water into specific sectoral policies and government laws. Once passed, the court system can be used as the channel for enforcement. In Argentina, for example, community members, with the help of a human rights NGO, took the municipality and state of Cordoba to court over failing to stop daily spillage from a sewage treatment plant that contaminated their drinking water. In 2004 a court ruled in the citizens' favour and both the state and municipality were forced to take action.[37]

Shedding light on corruption in the water sector through access to information

As in many other sectors, making corruption – or at least its impact – visible can provide a strong impetus for change.

In Malawi, geographic information systems (GIS) have been used to show how much water spending actually reaches the poor. The results are startling and graphically simple to understand. The mapping of new water points constructed between 1998 and 2002 found that

36 See World Development Movement, www.wdm.org.uk/campaigns/water/public/india.htm.
37 M. Gorsboth, 'Identifying and Addressing Violations of the Human Right to Water', (Stuttgart: Brot für die Welt, 2005).

a half of them were in areas that had already reached the recommended coverage density and that more equitable siting could have lifted almost all districts above this threshold. In some communities, this disparity in coverage was linked to political affiliations determining whether and where water points would be built.[38]

Sectoral budget analyses also provide a quick overview to show who actually benefits from water subsidies intended for the poor. After examining Tanzania's water budget, the World Bank found that a poor rural citizen received only one-fifth of the water subsidy that a rich urban resident garnered. Moreover, up to 41 per cent of all subsidies went to the country's wealthiest 20 per cent of households.[39] Likewise, in Bangalore, India, and Kathmandu, Nepal, the richest 10 per cent of households were found to receive more than twice as much in water subsidies as the poorest 10 per cent.[40]

Techniques and tools that shine the spotlight on corrupt policies are straightforward, but the resources to apply them at regular intervals or greater scale are difficult to mobilise. And excessive secrecy on the part of governments hinders their application. A survey of fifty-nine countries found that more than a half do not release to the public budgetary information produced for their own internal use or for donors.[41]

Strengthening the voice and participation of the poor in water governance

A variety of innovative initiatives show how empowerment can translate into greater participation and a more powerful voice for the poor. At the same time, special efforts are needed to overcome the traditional exclusion of women and other vulnerable citizens from participatory processes. Their inclusion in activities needs to be targeted and a common respect created for their contributions.

Setting water policy and budget priorities is one area for a more inclusive approach. Greater public participation and transparency in budget-setting activities can contribute to a more equitable distribution of resources for the poor. In Porto Alegre, Brazil, citizens are directly involved in participatory budgeting and spending reviews on water and sanitation. Within seven years of adopting these measures access to water increased from 80 per cent in 1989 to near-universal coverage by 1996, and access to the city's sanitation system expanded from less than a half to 85 per cent of all citizens over the same period. To ensure a pro-poor focus, the votes of the poorest people were weighted to give them greater voting power in budget-setting and spending reviews.[42]

---

38 S. Sugden, 2003.

39 'World Bank, 'Tanzania: The Challenge of Reforms: Growth, Incomes and Welfare' Report no. 14982-TA, cited in F. Naschold and A. Fozzard, *How, When and Why does Poverty Get Budget Priority: Poverty Reduction Strategy and Public Expenditure in Tanzania* (London: Overseas Development Institute, 2002).

40 C. Brocklehurst, 'Reaching out to Consumers: Making Sure We Know what People Really Think and Want, and Acting upon It' (Washington, DC: World Bank, 2003).

41 The International Budget Project, Open Budget Initiative, Open Budget Index, Survey Questionnaire 2005/6.

42 O. M. Viero, 'Water Supply and Sanitation in Porto Alegre, Brazil', presentation at WaterTime workshop, Cordoba, Spain, October 2003.

Participatory approaches have also been used and found successful among rural communities. Ghana has experienced a dramatic improvement in rural water service by decentralising responsibilities and funding from the central government down to the village level. Communities have established village water committees to decide how best to manage their water systems to meet local needs.[43]

Tracking and auditing expenditures for water can also be carried out with community input. To ensure that budget priorities are implemented fairly and transparently, public expenditure tracking and service delivery surveys have become the favoured tools for diagnosing corruption and other problems in developing countries. They were pioneered in 1996 to assess Uganda's primary education system and resulted in exposing the theft of funds and inspiring a wave of effective anti-corruption reforms in the country. Community involvement in audits can also be useful when corruption is suspected in public works. An analysis of corruption in village-level infrastructure projects in Indonesia has confirmed that audits can be highly effective in curbing corruption, but that auditors also need auditing.[44] In the Philippines, public auditing has been taken a step further. Civil society organisations, such as the Concerned Citizens of Abra for Good Governance, have partnered with government agencies to monitor public works projects. In one instance, monitoring discovered a river control structure was being built on an unstable foundation and helped to avert a potential disaster.[45]

Monitoring the performance and impact of water provision is another important area for civil society engagement. First used in Bangalore in 1993 and since replicated in more than twenty countries, citizen report cards capture feedback from the poor and other marginalised groups about the quality of public service delivery. This focus allows personal stories about corruption to be scaled up into a powerful collective body of evidence that an endemic problem exists. Report cards have helped to benchmark the performance of Bangalore's water board and other public utilities and produce significant improvements in service provision since the first round of surveys.[46]

Towards a new future: the least should come first

For water and sanitation services to be effective and accountable to all, poor citizens must be placed at the centre of service provision. Poor citizens must be enabled to monitor and discipline service providers. There must be space for them to raise and have their concerns heard. Poor people's greatest strength lies in their numbers. Combining their limited time and resources (skills, labour and money) has been shown to have a positive impact on combating corruption. At the same time, incentives must be strengthened for service providers to engage

43 UNDP, 2006.

44 S. Guggenheim, 'The Kecamatan Development Project: Fighting Corruption at the Grassroots', presentation at World Water Week, Stockholm, August 2007.

45 M. Sohail and S. Cavill, *Accountability Arrangements to Combat Corruption: Synthesis Report and Case Study Survey Reports*, WEDC (Loughborough: Loughborough University, 2007).

46 See article starting on page 106.

with the poor. Both private and public utilities should be encouraged to take steps that increase transparency and the role of independent oversight by auditors and regulators.

While the solutions seem simple, these have not been easy tasks in the past nor will they be in the future. Water corruption that harms the interests of the poor is based on a complex system of unequal power relationships and interlocking incentives that is difficult to tackle. It took many years for this system to be built, and it will likely take many years to tear it down.

A wide range of promising initiatives and instruments are at hand. None of them can single-handedly stamp out water corruption and make the system more accountable to the poor. But together they can provide the mix of incentives and sanctions, choice and voice, and checks and balances that will help to break corrupt power relationships and make water more accessible and affordable for the poor.

Corruption in urban water use by the poor

Bernard Collignon[1]

In addition to a host of day-to-day insecurities, the informal status of most slum dwellers makes them especially vulnerable to corruption. Though they have the right to vote and the responsibility to pay taxes, they are often denied the official documents and legal standing they need to compete with other customers for access to water. A simple way to overcome these handicaps is to pay an overhead.

In most large cities in developing countries, water is normally provided either by standpipes or household connections – both of which present many corruption challenges for the poor.

Securing an individual in-house connection can be an almost insurmountable challenge for the poor, as described in chapter 3 of this report.[2] Poor households, especially in slum areas, lack not only legal entitlements and political clout, but also the money to pay for or bribe their way into obtaining a household connection. This leaves public standpipes and informal providers as the main water source for millions of poor households in the developing world. The incentives for corruption are as diverse as they are powerful.

Corruption to capture the market and ways to counter it

Securing a local water monopoly can boost profits at the expense of the poor, and operators often resort to corrupt practices to stave off competition. Such 'water mafias' have been reported in South and South-east Asia, but rarely documented in detail.[3]

1 Bernard Collignon is the chairman of Hydroconseil, a consulting firm in the water sector (Avignon, France).
2 See article starting on page 40.
3 Regional Institute for Research on Human Settlements Technology, 'Small Scale Water Providers in Metropolitan Jakarta', PPIAF-funded study for WASPOLA Working Group, 2005; *BBC* (UK), 19 August 2004.

Nonetheless, effective competition can grow from the informal sector. In Maputo, Mozambique, inefficiencies on the part of the main utility have given rise to a flourishing informal water market. More than 200 small-scale alternative suppliers channel water from private, unregulated boreholes through self-built networks to thousands of clients, covering 40 per cent of all city districts. Most providers are competing for additional customers, and networks commonly overlap.[4] These competitive, alternative markets can play an important role in extending network coverage and curbing predatory water pricing.

Competition between customers when resources are under stress

When water becomes scarce, customers compete to obtain as much of it as possible. This creates more incentives to resort to corruption to grab more than one's fair share. This problem is common in Kathmandu, Delhi, Algiers, Nairobi, Port-au-Prince (Haiti) and many other large cities in the developing world with water shortages.

When water companies are unable to provide sufficient water pressure throughout the entire city at the same time, they resort to rationing – making water available only for portions of the day or week in each district. Utility staffers charged with opening valves and distributing water are in a very sensitive position, and find themselves with very good opportunities to pad their income illicitly. High-income households and water resellers that serve slums are prepared to pay bribes for access, driving up prices and skewing water allocation further towards the rich and influential.

Water shortages are normal in Port-au-Prince. A group of valve attendants traverses the city every day, opening and closing valves to distribute water – district by district and even street by street. Along the way, rich people bribe them in order to get more water. But they also compete with slum water associations (*comités de l'eau*), which also bribe valve attendants to fill their storage tanks for resale. The final payers of the bribes are the slum dwellers – those who, obviously, have the least money to spare.[5]

Local jobs for loyal voters

Filling local water jobs provides yet another opportunity for corruption. Standpipe attendants, sometimes known as fontainiers, who resell water to local communities have low turnover (US$3–10 per day) and very low net revenue (US$1–4 per day). Nevertheless, as job opportunities in the slums are limited, competition for the position is intense.

Because a late bill payment can result in a water company swiftly cancelling a fontainier's contract, they have been known to offer bribes to keep their jobs.[6] In addition, in Mauritania,

4 Seureca and Hydroconseil, 'Projecto de Reabilitação das Redes de Água Potável da Aglomeração de Maputo', Final Feasibility Report to FIPAG, Government of Mozambique, 2005.

5 B. Collignon and B. Valfrey, 'La Restructuration du Service de l'Eau dans les Bidonvilles de Port-au-Prince', presentation at the second Rencontre Dynamiques Sociales et Environnement, Bordeaux, 9–11 September 1998.

6 B. Collignon and M. Vézina, 'Independent Water and Sanitation Providers in African Cities: Full Report of a Ten-country Study', Water and Sanitation Program (Washington, DC: World Bank, 2000).

fontainiers have been known to get their jobs in return for a bribe.[7] One way or another, these bribes are ultimately paid by standpipe customers. Finally, many water companies allow local governments to select standpipe attendants, opening the door for these officials to abuse their power by providing friends and 'good voters' with jobs. This practice has been reported in Indonesia, Mali and Senegal.

7 Hydroconseil, 'La Gestion des Bornes-fontaines Publiques dans la Commune d'El Mina', l'Atelier de l'Agence de Développement Urbain de Nouakchott, Mauritania, 2003.

Building water integrity: private water operators' perspective

Jack Moss[1]

From a business perspective, corruption increases costs, reduces efficiency and threatens the ability to deliver required results. A private operator's *raison d'être* is to deliver high-quality water services in a businesslike and committed manner. This means understanding and satisfying the needs of its customers and meeting the obligations set by clients and regulators, while ensuring adequate returns to investors and owners. Keen to escape the scourges of coercion and corruption that limit their performance, operators have taken action to combat these practices.

What corruption risks do private water operators face in their day-to-day operations?

Legacy practices of corruption in dealing with customers, subcontractors and suppliers can be a challenge. Tracking, monitoring and quality control systems, as well as training for subcontractors and a separation of functions such as decision-making, operations and cash management, are designed to eliminate opportunities for petty corruption. But implementing adequate processes often calls for strong management at the start of contracts, in order to change the staff culture inherited from former management and eradicate corrupt internal practices.

Companies also have adopted codes that usually start with a clear and simple set of ethical principles. These principles are supported by operational procedures that generate audit trails, and also may contain web-based checks and whistleblower protection.

Another difficult challenge is to resist extortion by low-level officials responsible for issuing local permits and licences or approving completed work. This involves issuing documents such as 'digging permits' or 'works completion certificates'. Combating this kind of corruption often

1 Jack Moss is senior water adviser for AquaFed, the International Federation of Private Water Operators.

puts operators at risk of non-compliance with contractual or regulatory targets. This challenge can be even more difficult when petty officials are seeking bribes in collusion with senior officials who may be involved with the operator's client. Preventing an operator acting alone from engaging in this kind of coercion and extortion can be very difficult. Support is needed from the community, the industry and organisations such as the Water Integrity Network.

In all these ways, private operators are engaged in the fight against corruption for the benefit of the communities they serve. This is especially the case for low-income customers, who suffer the most from corrupt practices. Aquafed's Code of Ethics encourages member companies to take care of vulnerable groups,[2] and the organisation supports the Right to Water for all.[3]

2 See www.aquafed.org/ethics.html.
3 AquaFed, 'Water and Sanitation for Women', 8 March 2007. Available online at www.aquafed.org/documents.html.

Water corruption in industrialised countries: not so petty?

Per Ljung[1]

Western Europeans and US citizens, who generally enjoy high-quality water service, might only rarely have to consider paying a bribe for a falsified water meter reading, an expedited repair or an illegal connection.[2] But the virtual absence of petty corruption does not mean that the water and sanitation sector in industrialised countries is free from governance problems and corruption. It takes place at another level.

Rigging competition in building water infrastructure

Water and sanitation networks require more than double the capital investment relative to revenue than other utilities such as electricity, gas or telecommunications. In 2007 total worldwide capital expenditures for municipal water and sanitation were estimated at US$140 billion.[3] These investments primarily involve public works construction, a sector in which corruption risks are high.[4]

1 Per Ljung is chief executive of PM Global Infrastructure.
2 A notable exception occurred in New Jersey in 2007, when a water agency employee pleaded guilty to colluding with a landlord to extract money from poor households that sought to avoid water disconnection due to outstanding bills. See A. MacInnes, 'A 6th Official in Passaic Corruption Sweep is Guilty', *Bergen County Record*, 27 July 2007.
3 Global Water Intelligence, *Global Water Market 2008: Opportunities in Scarcity and Environmental Regulation* (Oxford: Global Water Intelligence, 2007).
4 In the United Kingdom, for example, a two-year investigation by the Office of Fair Trading (OFT) had by 2007 uncovered evidence of bid-rigging in thousands of tenders in the construction industry. See OFT, Press Release, 22 March 2007, www.oft.gov.uk/news/press/2007/49-07.

Collusion among bidders appears to be the most prevalent corrupt practice in industrialised countries. Perhaps the best known is Japan's 'dango' system, in which bidders for public works projects politely decide amongst themselves who will win contracts. The 'winning' firm as well as its 'rivals' submit choreographed bids to public agencies to maintain the illusion of competition.[5]

In Australia, three suppliers of valves and fittings used for water, irrigation and sewage systems were fined a total of A$2.85 million (US$2.5 million) in 2000 for engaging in price-fixing, tender-rigging and market-sharing.[6] Two years later three Swedish suppliers of water and sewage pipes were convicted of price-fixing and market-sharing.[7]

Corruption in awarding water contracts

Water agencies often award high-budget contracts to private companies to operate and maintain public water and wastewater systems. The larger of these contracts have long durations and involve complex provisions, making the tailoring of contracts to preferred suppliers hard to detect. Moreover, such contracts are often awarded in the context of soft budget constraints. The possibility of drawing on public subsidies or adjusting user fees emancipates water managers from strictly commercial cost pressures and provides additional discretion in designing and awarding contracts.

As many well-documented cases show, the temptation to engage in corrupt practices in such a context is very strong. Not only are industrialised countries not immune from these problems, many of the more notorious corruption cases have occurred in Europe and the United States.

In cities as diverse as Grenoble, Milan, New Orleans and Atlanta,[8] officials were allegedly wined and dined, treated to lavish holiday trips and even apartments and given large cash amounts, all for the purpose of awarding or influencing the design of water and sanitation contracts.

In Milan, for example, an executive of a private water company was imprisoned in 2001 for planning to bribe local politicians with L4 billion (US$2.9 million) to win a L200 billion (US$145 million) wastewater treatment contract. The city council president was also convicted and jailed.[9]

---

5 J. McMillan, 'Dango: Japan's Price-fixing Conspiracies', *Economics and Politics*, vol. 3, no. 3 (1991).

6 Australian Competition and Consumer Commission, 'Penalty of $100,000 against Watergear Brings Penalty Total to $2.85 Million for Collusion in Fittings, Valves for DICL Pipes', 21 July 2000. See www.accc.gov.au/content/index.phtml/itemId/87433.

7 Swedish Competition Authority, *Konkurrens Nytt* (newsletter), no. 1 (2002).

8 M. Sohail and S. Cavill, *Accountability Arrangements to Combat Corruption: Synthesis Report and Case Study Survey Reports*, WEDC (Loughborough: Loughborough University, 2007); J. Godoy, 'Water and Power: The French Connection', (Washington, DC: Center for Public Integrity, 2003); 'Ex-New Orleans Political Figure Pleads Guilty', Associated Press, 5 January 2006; US Department of Justice (Northern District of Georgia), 'Former Atlanta Mayor Sentenced to Prison on Federal Felony Tax Charges', 13 June 2006.

9 M. Sohail and S. Cavill, 2007; Public Citizen, 'Veolia Environment: A Corporate Profile', Washington, DC, February 2005.

Corruption for political power

Bribes can also be used to shore up political power through supporting political campaigns and parties, steering contracts to political cronies or making sure water policies favour a specific constituency. In San Diego, for example, an audit in 2006 found that households were improperly overcharged on their monthly sewage bills, with the excess being unlawfully used to subsidise the sewage costs of large industrial users.[10] In Chicago, the head of the water department was found guilty in a scheme to extort campaign contributions from subcontractors and use employees from his department to do campaign work.[11]

What's at stake?

In developing countries, the main effects of corruption are reduced access for the poor and low-quality service for those who have it. Though less related to death and disease, corruption in industrialised countries is no less real. Cost escalation due to corruption is borne primarily by consumers and, to some extent, by local and/or national taxpayers. These direct costs are difficult to quantify, but the stakes are huge. Western Europe, North America and Japan spent an estimated US$210 billion on municipal water provision and wastewater treatment in 2007, and this will climb to more than US$280 billion by 2016.[12] Even a small corruption factor can translate into formidable losses for the public.

But the real social costs of corruption cannot simply be boiled down to money. When corruption raises the price of water provision and utilities face severe budget constraints, 'less urgent' environmental investments, primarily in sewage treatment, may be cancelled or postponed. This shifts the burden to future generations. Perhaps more gravely, corruption to secure political power fuels widespread public cynicism about local institutions and undermines the trust in political legitimacy.

10 A. Levitt Jr. *et al.*, 'Report of the Audit Committee of the City of San Diego: Investigation into the San Diego City Employees' Retirement System and the City of San Diego Sewer Rate Structure' (New York: Kroll Inc., 2006).

11 US Internal Revenue Service, 'Former Chicago Department of Water Management Official Sentenced in Federal Corruption', FY2007 Examples of Public Corruption Crimes Investigations (see www.irs.gov/compliance/enforcement/article/0,,id=163040,00.htm); *Chicago Sun-Times* (US), 30 July 2005.

12 Global Water Intelligence, 2007.

The public and private faces of corruption in water

Transparency International

Does business or government do a better job supplying water to the people and keeping corruption in the sector low? In almost no other policy area has the public versus private controversy been waged with as much fervour and ideological zeal. This is not surprising. No other resource is so fundamental to our notion of life and living on what is aptly called the 'blue planet'. For some, this makes water the ultimate social good, a moral no-go area for private profit. For others, the very urgency of the global water crises calls for efficient management and a mobilisation of capital that, in their view, the private sector can provide best.

But there is more agreement in this debate than initially meets the eye. First, affordable, effective access to a sufficient amount of safe drinking water is an uncontested human right that establishes a clear responsibility for governments, and, if they fail, the wider international community must ensure that the social minimum is incorporated in any kind of water provision system, be it public or private.[1] Second, basic decisions about water supply, allocation, cost, quality and use directly or indirectly affect everyone in society in fundamental ways. This establishes a clear right for every citizen to have a say in these decision-making processes and a duty for the state, donors and private players to put such mechanisms in place.

In the 1990s the failure of large-scale, state-led infrastructure development to deliver accountable water systems and resolve water crises led to an upsurge in water privatisation. But many of the more exuberant hopes have been frustrated. Several large privatisation initiatives collapsed amidst high-profile political acrimony. They failed in the daunting task of aligning their own commercial interests with the public sensibilities, social objectives or changing economic contexts of water policies. By 2006 the investment volume of cancelled or 'distressed' private water contracts had risen to almost a third of all private sector participation in low- and middle-income countries between 1990 and 2006.[2]

Growing pragmatism in the debate

Two lessons have been learned. First, effective water provision depends more on the quality of governance, both for the provider and the sector, than on the ownership structure. Second, no one can go it alone. Even if water infrastructure is financed and managed by the public sector, the system will still depend on products and services delivered by private entrepreneurs. The task is to harness the private sector's expertise and capital for a specific local context. As table 2 shows, there are many different ways to do this.

The public and private faces of corruption

Public and private operators share many common corruption challenges. Any large-scale organisation that interacts with multiple suppliers and customers must ensure that employees do not take advantage of their entrusted powers and solicit bribes. Codes of conduct and promoting integrity – alongside effective customer complaint, whistleblowing and financial tracking systems – have been applied successfully in both settings.[3] Incentives for reform may

1 United Nations, Economic and Social Council, Committee on Economic, Social and Cultural Rights, General Comment no. 15, E/C.12/2002/11, November 2002.
2 World Bank, 'Private Activity in Water Sector Shows Mixed Results in 2006', PPI data update note no. 4, July 2007.
3 M. Sohail and S. Cavill, *Accountability Arrangement to Combat Corruption: Synthesis Report and Case Study Survey Reports*, WEDC (Loughborough: Loughborough University, 2007); and see articles starting on pages 40 and 54.

Table 2 Public–private sharing of water provision

Option	Ownership	Management	Investment	Risk	Duration (years)	Examples
Service contract	Public	Shared	Public	Public	1–2	Finland, Maharashtra (India)
Management contract	Public	Private	Public	Public	3–5	Johannesburg (South Africa) Monagas (Venezuela), Atlanta (United States)
Lease (affermage)	Public	Private	Public	Shared	8–15	Abidjan (Côte d'Ivoire), Dakar (Senegal)
Concession	Public	Private	Private	Private	20–30	Manila (Philippines), Buenos Aires (Argentina), Durban (South Africa), La Paz-El Alto (Bolivia), Jakarta (Indonesia)
Privatisation (state divestiture)	Private	Private	Private	Private	Unlimited	Chile, United Kingdom

Source: Adapted from United Nations Development Programme (UNDP), *Human Development Report 2006. Beyond Scarcity: Power, Poverty and the Global Water Crisis* (New York: Palgrave Macmillan, 2006).

be stronger when internal corruption directly hurts the profits of private owners, rather than when losses are dispersed across a larger community of public taxpayers.

Public utilities are very vulnerable to political interference by corrupt policy-makers intent on awarding lucrative public sector jobs to cronies, tweaking water provision and pricing in favour of influential supporters or diverting money from public budgets into their own pockets.[4]

With private sector involvement, corruption hot spots include bid-rigging, collusion and bribery. These practices occur when private contractors vie for large water contracts and infrastructure assets are privatised in complex deals.

Be it public or private, strategic collusion can game the system and exploit corruption opportunities if additional checks and balances are weak.

Achieving transparent and accountable water provision

Developing contracts for private sector involvement faces the challenge of *double delegation* – shifting the responsibility for water provision from public provider one step further away from citizens to a private operator. But such contracts also provide an opportunity to lay down

4 See article starting on page 55.

transparency objectives and clear lines of responsibility, information that may be buried in a patchwork of administrative rules in a public agency. Unfortunately, such agreements often remain under lock, making collusive behaviour and manipulation difficult to detect.[5]

What is more, private operators' penchant for commercial confidentiality limits the public's access to key operational information. Clarifying disclosure obligations is therefore essential.[6] Investment plans, management contracts, rate-setting data and financial and operational performance indicators must be open to public inspection and monitoring.

To make public management more transparent and autonomous, and prevent political interference, water utilities should be incorporated as separate entities. Their budgets and operational management should be clearly separated from the wider administration, overseen by a multi-stakeholder board and audited independently.

Water utilities in Porto Alegre, Brazil,[7] Phnom Penh, Cambodia,[8] and Dakar, Senegal,[9] have improved performance and network coverage significantly with this strategy. Likewise, a study of more than twenty water utilities in Africa, Asia and the Middle East found that more autonomy typically comes with better performance.[10]

Strong regulatory oversight and performance-based monitoring: a must for both public and private

Both private and public utilities must abide by clear pro-poor objectives, and be subject to independent oversight by auditors and regulators with investigative authority and enforcement power. Straightforward as these requirements sound, much remains to be done. By 2004 not even a fourth of developing countries had introduced independent regulatory agencies in the water and sanitation sector, lagging far behind electricity and telecommunications.[11] And, where regulators are in place, their dealings are often not very transparent. In 2005 fewer than a third of water regulators assessed in a survey published contracts and licences, and only a half published results of consultations.[12]

5 In Malaysia, for example, the government even classified a water concession as an official secret, in order to keep it from public scrutiny; see *Malaysiakini*, 14 June 2007.

6 P. Nelson, 'Multilateral Development Banks, Transparency and Corporate Clients: "Public–Private Partnerships" and Public Access to Information', *Public Administration and Development*, vol. 23, no. 3 (2003).

7 United Nations Development Programme, *Human Development Report 2006. Beyond Scarcity: Power, Poverty and the Global Water Crisis* (New York: Palgrave Macmillan, 2006).

8 See article starting on page 40.

9 In Dakar, the newly incorporated public water utility has gone one step further and engaged a private operator, also with great success in expanding coverage and efficiency. C. Brockelhurst and J. Janssens, 'Innovative Contracts, Sound Relationships: Urban Water Sector Reform in Senegal', Water Supply and Sanitation Sector Discussion Paper no. 1 (Washington, DC: World Bank, 2004).

10 O. Braadbaart *et al.*, 'Managerial Autonomy: Does It Matter for the Performance of Water Utilities?', *Public Administration and Development*, vol. 27, no. 2 (2007).

11 A. Estache and A. Goicoechea, 'A "Research" Database on Infrastructure Economic Performance', Policy Research Working Paper no. 3643 (Washington, DC: World Bank, 2005).

12 L. Bertolini, 'How to Improve Regulatory Transparency: Emerging Lessons from an International Assessment', Public–Private Infrastructure Advisory Facility Gridlines Note no. 11 (Washington, DC: World Bank, 2006).

Excessive secrecy also limits the benefits of audits. Nearly a half of fifty-nine surveyed countries delayed publishing their public sector audit findings by more than two years, if they published them at all. For ten countries, audit findings were not even made available to legislators.[13]

Performance indicators are a prerequisite for output-oriented accountability. In the water sector, indicators for operating efficiency, equity and service effectiveness are well established, widely recognised and rather easy to benchmark.[14] They include coverage rates, portion of system leakages and uncollected fees, employee per connection ratio, service uptime and water quality indicators.

But operations are not always governed by clear *performance* targets. A study for Australia found, for example, that the contractual arrangements for public water utilities on average include fewer performance criteria than contracts with outsourced private providers.[15] Even worse, performance is often difficult to inspect by the public, even in industrialised countries. For example, both in the privatised water sector in the United Kingdom and in the publicly organised sector in Germany, information on water quality is collected and published online. But in both cases the information is very difficult to find, understand or compare, limiting its usefulness for public oversight.[16]

Mechanisms for citizen participation and monitoring

Citizens can provide essential input to water policies and check the performance of both private and public water utilities. Local initiatives range from social contracts between providers and citizens to social scorecards, citizen surveys and social audits.[17] More grassroots water democracy, through formal institutional mechanisms for public hearings and participation in water regulation, would appear to be easier to establish where utilities are publicly owned and operated. But reality points to formidable challenges in either setting.

Despite some shining examples,[18] formal mechanisms for consultation and participation are still an exception in both spheres (see table 3). Even in Colombia and Peru, where such measures are in place, they are rarely implemented.[19]

The conditions for corruption in water have both a public and a private face. Official secrecy and commercial confidentiality can both make it difficult to create the transparency that is

13 V. Ramkumar, 'Expanding Collaboration Between Public Audit Institutions and Civil Society', International Budget Project (2007).
14 See, for example, the International Benchmarking Network of Water Utilities: www.ib-net.org.
15 J. Davis and G. Cashin, 'Public or Private "Ownership": What's in a Name?', *Water Science and Technology: Water Supply*, vol. 3, no. 1/2 (2003).
16 D. Zinnbauer, 'Vital Environmental Information at your Fingertips?' (Berlin: Anglo-German Foundation, 2005).
17 See article starting on page 40.
18 Ibid.
19 V. Foster, 'Ten Years of Water Service Reform in Latin America: Toward an Anglo-French Model', Water Supply and Sanitation Sector Board, Discussion Paper no. 3 (Washington, DC: World Bank, 2005).

Table 3 Mechanisms for participation and consultation

Country	Complaints office	Public hearings	Consultative committees
Argentina	Yes	None	None
Bolivia	Yes	Optional	None
Chile	Yes	None	None
Colombia	Yes	None	Comités de Desarrollo y Control Social
Panama	Yes	Optional	None
Peru	Yes	None	Comités Consultivos Regionales

Source: V. Foster, 2005.

needed for accountable water provision. Regulatory oversight often lags behind other sectors, and limited means for broader public consultation further hamper accountability. This often fuels public suspicion that, no matter who calls the shots, corruption will continue to influence the supply of water.

Pipe manufacturers in Colombia and Argentina take the anti-corruption pledge

Virginia Lencina, Lucila Polzinetti and Alma Rocío Balcázar[1]

Lacking transparency and plagued by mistrust, Colombia's pipe manufacturing industry faced a crisis of confidence in the 1990s. Several factors were conspiring to intensify corruption pressures. Because of unethical overpricing and substandard work quality, pipe companies were losing public projects. This, combined with a recession, pushed companies to boost revenues by any means – 'to the extent that the limits between commercial and corrupt practices blurred'. In the government sphere, job instability and low salaries made public employees more inclined to solicit bribes. By 2000 the situation had become unmanageable.[2]

1 Virginia Lencina is the co-ordinator of the Business Sector Programme at Poder Ciudadano Foundation; Lucila Polzinetti is a programme assistant at Poder Ciudadano Foundation; Alma Rocío Balcázar is director of the Private Sector Programme at Corporación Transparencia por Colombia.

2 A. R. Balcázar. 'The Establishment of an Anti-Corruption Agreement with Pipe Manufacturing Companies: A Colombian Experience', presentation at World Water Week, Stockholm, August 2005; P. Stålgren, 'Corruption in the Water Sector: Causes, Consequences and Potential Reform', Policy Brief no. 4 (Stockholm: Swedish Water House, 2006).

Generally, corruption in Colombia is no small problem. More than two-thirds of entrepreneurs surveyed recently said public procurement processes have little or almost no transparency. On average, a competitor must pay an additional 12 per cent of a contract's value in order to win the deal.[3]

In 2003 the Colombian Sanitary and Environmental Engineering Association approached Transparency International's local chapter, Transparencia por Colombia, to try to find a remedy. The organisation, known as ACODAL, represents pipe manufacturing companies that account for 95 per cent of the national pipe market and 100 per cent of the public bids for water supply and sewer projects.

Negotiations ensued between Transparencia por Colombia and eleven of ACODAL's seventeen affiliated companies, which have combined annual revenues of more than P540 trillion (US$266 million). Problems on the table included the lack of a corporate anti-corruption culture, an absence of internal ethical standards, the permitting of bribery and a lack of transparency in public procurement. In April 2005, after a year of talks, the parties signed an Anti-corruption Sectoral Agreement.

By signing the pact, the companies agreed to define clear rules of the game among competitors, set minimum ethical standards, prevent corrupt practices, promote a culture of transparency and contribute to society by consolidating the country's economic and social development. Based on TI's Business Principles for Countering Bribery (BPCB), the agreement contains specific measures to deal with bribery, facilitation payments, political contributions, pricing and purchasing, and internal controls and audits. Protection for whistleblowers was also instituted.

To help ensure compliance, an Ethics Committee was established to act as an arbitrator in the event of a conflict. Its decisions are binding on all parties, and those who fail to abide by the committee's rulings can be reprimanded or suspended from bidding on contracts.[4]

Improvement was swift. By 2006 bid award prices had dropped significantly, reducing the scope for paying bribes. 'We never before have had a code to guide us. Now we have parameters for action,' said one of the signatories. 'With this agreement, we . . . will act differently amongst ourselves, since the same rules and regulations apply to all.'[5]

Seven months after the Colombian pact took effect, pipe manufacturers in Argentina signed a similar agreement with the help of TI's local chapter there, Poder Ciudadano. In December 2005 nine companies, representing 80 per cent of the nation's water and drainage infrastructure market, signed the first Business Sector Transparency Agreement in the country.

As in Colombia, the agreement is based on TI's Business Principles for Countering Bribery. The companies agreed to implement an internal transparency policy to guide business

3 A. R Balcázar, 2005.
4 Ibid.
5 P. Stålgren, 2006.

transactions and their dealings with the government. Specifically, the companies have pledged to:

- promote transparency in bidding;
- refrain from all forms of corruption and bribery;
- make no political contributions;
- deal with sales intermediaries in a clear, transparent manner; and
- fight tax evasion.

In addition to the pipe companies, the agreement was also signed by the Argentinean Association for Sanitation Engineering and Environmental Sciences (AIDIS). And it was supported by the Avina Foundation, an alliance of social and business leaders working to promote sustainable development in Latin America.[6] As in Colombia, an Ethics Committee will be formed to monitor compliance and sanction companies that breach the agreement.

The parties have also agreed to present a consensus regarding transparent biddings to state and public organisations and multilateral organisations that participate in this kind of public bidding process and as financiers to achieve its adhesion and present proposals of modifications in the procedure. As a result, the local government of Rosario, in Santa Fe Province, has signed a Framework Agreement recognising the agreement for future activities in public bids and purchasing.

Hoping to build on their success, the companies that signed the agreements in Colombia and Argentina may submit similar proposals elsewhere in Latin America.

6 TI, 'Leading Argentinian Water-sector Companies Say No to Bribery', 15 December 2005.

Clearing muddied waters: groups in India fight corruption with information

Venkatesh Nayak[1]

Throughout India, citizens are using the power of public information not only to fight corruption, but to enhance their stake in the political system.

In the small village of Keolari in the central state of Madhya Pradesh, citizens used India's new transparency law, the Right to Information Act (RTI Act) of 2005, to prevent a local politician from claiming a public water well for his own personal use. The man, an elected *Pancha* (member) of the local government, was building a home in December 2006 when he erected a wall around a well that his father had donated to the community nine years earlier. The well is one of only two sources of potable water available to the village's 2,500 residents.

1 Venkatesh Nayak works for the Commonwealth Human Rights Initiative.

Local citizens asked the *Pancha* not to cut off their access to the well, but he refused. They then filed complaints with the village chief and higher levels of government, to no avail. Not even getting local newspapers to write about the problem was enough to move officials to action.

A few weeks later, while attending an awareness camp organised by a regional transparency group, one of the citizens learned about India's new Right to Information Act. The group, Madhya Pradesh Suchana Adhikar Abhiyan (MPSAA), along with the Commonwealth Human Rights Initiative, provides free help to citizens trying to obtain public information under the law. With the group's help, citizens requested copies of the gift deed for the well signed by the *Pancha*'s father, as well as information on any public money spent to maintain the well. Within two days citizens obtained documents confirming the gift and showing that the local government had spent Rs11,608 (US$293) to strengthen its platform and walls.

Residents then wanted to use the RTI Act to find out what had happened to their original complaint. But, when they went to the local government office, they were told the information was exempt from the law, so there was no point filing the request. When an MPSAA representative returned and asked for the refusal in writing, he was told the matter would be investigated.

As for the well, when residents went there in February 2007 they saw revenue officials inspecting the disputed property and measuring the *Pancha*'s encroachment. They confirmed that the *Pancha*'s wall was illegal and ordered him to demolish it within a week. Today the wall is gone, and villagers once again are able to draw water from the well.[2]

This is not an isolated case. Freedom of information legislation is also being used as a way to fight for greater transparency by many other groups in India. In Delhi, a transparency group called Parivartan is using the power of information and employing Gandhian tactics to fight corruption in local public works projects.

Parivartan uses the RTI Act to obtain documents on water, sanitation, electricity, road, waste management and other projects – from work orders to sketches to completion certificates. Then they hold street-corner meetings to tell residents how much money has been spent on local projects and they inspect the projects to see if the money went toward its intended purpose.

Finally, Parivartan holds public hearings (*jan sunwai*), at which government officials have the opportunity to explain where the money went. In several cases, they had trouble coming up with an explanation. When residents of Patparganj fell ill from drinking sewage-fouled water, Parivartan asked for the status of residents' complaints and the names of responsible officials. Repairs were made two days later and water testing was conducted throughout the area. Parivartan obtained similar results in the case of a leaking water pipe, which was fixed three days after the group filed an information request.

---

2 A more detailed version of this story may be accessed on the website of the Central Information Commission of India; see cic.gov.in/Best%20Practices/rti_restores_peoples_right.htm.

When the government refuses to release information, Parivartan members engage in *satya-graha* – a form of passive resistance developed by Gandhi. Citizens wait at government offices as long as necessary, until officials give them the information they want.[3]

A similar organisation that pioneered this strategy has long been active in the state of Rajasthan. There, Mazdoor Kisan Shakti Sangathan – or Workers and Farmers Grass Roots Power Organisation – exposes fraud by obtaining balance sheets, tenders, bills, employment records and other government records. The group discovered, for example, that local officials were overbilling the central government for work on a water project in a drought-prone area. They also found out that people listed as labourers on public works projects never got paid, and that large payments were made for construction projects that were never built.[4]

3 M. Sohail and S. Cavill, *Accountability Arrangements to Combat Corruption: Synthesis Report and Case Study Survey Reports*, WEDC (Loughborough: Loughborough University, 2007).

4 J. Plummer, 'Making Anti-corruption Approaches Work for the Poor: Issues for Consideration in the Development of Pro-poor Anti-corruption Strategies in Water Services and Irrigation', Report no. 22 (Stockholm: Swedish Water House, 2007).

4 Water for food

Frank Rijsberman introduces the different forms of corruption that are prevalent in agriculture and irrigation, where water consumption is high and where food security is at stake. He documents how sophisticated systems of 'trickle-up' bribery divert resources from the sector and how large-scale users benefit from biased policies, offering a number of practical suggestions on how to make irrigation systems less vulnerable to corruption. Jean-Daniel Rinaudo further illustrates the interlocking incentive systems that underpin corruption in irrigation services in Pakistan. Sonny Africa shows how failing irrigation projects squander public money and deprive farmers of much needed water resources in the Philippines. In the final contribution to this section, Grit Martinez and Kathleen Shordt elaborate the role and responsibilities of donors in the fight against corruption in the water sector.

Water for food: corruption in irrigation systems

Frank R. Rijsberman[1]

Food for the world: why irrigation matters

The vast majority of the world's farmers still rely on rainfall to grow their crops. In some parts of the world almost all rain falls within such a short period of time that it is either impossible or very risky to try to farm on rainfall only. In large parts of South Asia's monsoon region, more than 90 per cent of the annual rainfall comes in less than 100 hours. The answer for millions of farmers over the millennia has been irrigation. Since pre-Roman times, communities in dry places from Iran to Morocco have built underground canal systems to channel water from the mountains to fertile, but dry, valley floors. Kings in Sri Lanka built ancient hydro-civilisations on cascades of small reservoirs or tanks.

Of all the water that humans take out of nature, some 70 per cent goes to irrigation – even more in countries with large irrigation sectors such as Australia, China, Egypt, India, Iran, Mexico, Turkey and Uzbekistan. Though only one-sixth of the world's farmed area is irrigated, these farms produce 40 per cent of the world's food. Food security fears have spawned massive investments in dams and irrigation canal systems in Asia, North America and Australia. While

1 Frank R. Rijsberman is the former director general of the International Water Management Institute, Colombo, Sri Lanka, and now works at Google.org, the philanthropic arm of Google Inc.

the world population more than tripled in the twentieth century, water use for human purposes grew sixfold, with the bulk of that water going to irrigation.

Irrigation, done well, is a critical factor in lifting poor farmers out of poverty.[2] Combined with high-yielding grain varieties and fertiliser, irrigation has also been key to preventing the famines predicted for Asia and pushing down world food prices to the lowest levels ever. Some of the world's most important cash crops, particularly cotton and sugar cane, also depend heavily on irrigation. But irrigation is not always done well.

- Farmers at the tail end of canals sometimes do not get their fair share of water because upstream farmers take out too much.
- Irrigation systems have been greatly delayed or built at grossly inflated costs.
- Often no more than 30–40 per cent of the water is actually used by the crops it was intended to help grow, the remainder leaking from canals, seeping into groundwater or running into drains.
- Silted-up canals, broken measuring devices and other problems require costly repairs.
- When farmers do not pay irrigation charges, systems do not have enough money for operation and maintenance.

The poor performance of irrigation systems has some major consequences. For the 70 per cent of all dollar-poor people who live in rural areas, agriculture is in most cases still the only way out of poverty. Not surprisingly, it is poor farmers, particularly those at the tail end of irrigation canals, who bear the brunt of irrigation failures. In addition, where irrigation systems have dominated government infrastructure investments in irrigation-dependent countries, poorly performing systems have an immediate impact on overall investment performance. And, as water scarcity is becoming a global crisis, the inefficient performance of the dominant water user – irrigation – is the gorilla in the room.

Assessing the risk of corruption in irrigation

In countries where agriculture matters most, overall control of corruption is judged to be particularly weak, presenting a challenging backdrop for tackling corruption in the sector.[3] Specific corruption risks in irrigation are driven by many factors.

- The availability of irrigation water depends directly on rainfall, and even in well-established irrigation systems this is uncertain by its very nature. Particularly in multi-reservoir systems with hydroelectric, irrigation and flood control functions, it is almost impossible for irrigators to assess water availability independently. Irrigation management agencies are not accustomed to sharing information that might make their systems more transparent to the user. With irrigation officials in firm control of information not accessible to irrigators, opportunities open up for rent-seeking and corruption.

2 I. Hussain, 'Pro-Poor Intervention Strategies in Irrigated Agriculture in Asia – Poverty in Irrigated Agriculture: Issues, Lessons, Options and Guidelines', Final Synthesis Report (Colombo: IWMI, 2005).

3 World Bank, *World Development Report 2008: Agriculture for Development* (Washington, DC: World Bank, 2007).

- Large irrigation systems require specialised maintenance, management and equipment negotiated through individual (one-off) contracts that are difficult to monitor and thereby susceptible to corruption.
- Large public subsidies for both construction and operation are generally provided to bureaucracies as budgets without a direct link to performance or output. This lack of accountability can foster corruption. As a solution, drought-prone Australia has begun 'benchmarking' irrigation system performance.[4]
- Irrigation as a profession is almost exclusively the domain of engineers, whether in system construction, management or research. Engineers tend to respond to low-performing systems with technical solutions. But addressing technical problems with purely technical solutions is unlikely to be successful if the corruption incentives of all stakeholders are not reduced. An irrigation engineer in South Asia once said that, because 'water management is 25 per cent water and 75 per cent people, you have to soothe people and you have not to displease politicians'.[5]

Forms of corruption in irrigation

A recent and promising approach to understanding corruption in irrigation is to look at it as the provision of a service that requires effective institutions and the alignment of stakeholder interests to function properly.[6] Addressing rent-seeking and corruption then becomes a matter of redesigning institutions in order to remove deficiencies and uncertainties in agreements among stakeholders while increasing transparency and incentives for compliance.

From such a perspective, the major entry points for corruption in surface or canal irrigation include the following.

(1) *Subsidy capture*. Public irrigation subsidies are usually justified on the grounds that irrigation supports national food security and farmers who are unable to pay market prices for water. For individual farmers or landowners, irrigation is attractive as long as their personal financial benefits outweigh the much lower subsidised costs they face. This leads to the temptation for farmers and their representatives and cronies to overestimate projected benefits, underestimate construction costs and lobby governments to pay for projects that do not necessarily deliver net benefits to society, but that deliver a major subsidy to landowners. Businesses that design, build and operate systems can also be tempted to

4 H. M. Malano and P. J. M. van Hofwegen, *Management of Irrigation and Drainage Systems: A Service Approach* (Rotterdam: Balkema Publishers, 1999).

5 R. Wade, 'The System of Administrative and Political Corruption: Canal Irrigation in South India', *Journal of Development Studies*, vol. 18, no. 3 (1982).

6 H. M. Malano and P. J. M. Hofwegen, 1999; J. Renger and B. Wolff, 'Rent Seeking in Irrigated Agriculture: Institutional Problem Areas in Operation and Maintenance', MAINTAIN Thematic Paper no. 9 (Eschborn: Deutsche Gesellschaft für Technische Zusammenarbeit [GTZ] 2000); W. Huppert *et al.*, 'Governing Maintenance Provision in Irrigation: A Guide to Institutionally Viable Maintenance Strategies', (Eschborn: GTZ, 2001); W. Huppert and B. Wolff, 'Principal–Agent Problems in Irrigation: Inviting Rentseeking and Corruption', *Quarterly Journal of International Agriculture*, vol. 41, no. 1/2 (2002). This last describes rent-seeking and corruption in irrigation as typical 'principal–agent' problems – as deficiencies in the contracts and agreements between the partners in an exchange relationship – that may well be in the interest of the most influential stakeholders in the system.

bribe key government officials. Policy capture is difficult to prove, but the existence of powerful, politically well-connected large-scale farmers who manage to secure the bulk of irrigation subsidies in many countries makes policy capture a plausible premise.[7]

(2) *Corruption in construction.* Procurement and tendering are particularly prone to corruption when products cannot be standardised, as is the case with constructing large-scale irrigation projects. Because every large dam is essentially a one-off product, cost estimates among competing contractors can vary greatly, offering the opportunity to include bribes in quotations with little risk of detection. As with all construction projects, corruption in irrigation can result in favoured contractors winning contracts, contractors not being held accountable for poor performance and inferior work, and contractors colluding to overcharge.[8]

(3) *Corruption in maintenance.* Though the amounts may be smaller and more standardised than new construction projects, irrigation maintenance tends to be much less stringently monitored. Some forms of maintenance, such as de-silting a canal, are extremely difficult to monitor, since the results can be literally 'under water'. So the corruption risks are in fact greater.[9] In addition, since maintenance funds are usually provided as part of an agency's annual budget cycle and are subject to the discretion of maintenance engineers, spending can be based on corruption opportunities rather than actual maintenance needs.

(4) *Corruption in operation.* Opportunities for corruption depend on how irrigation systems are organised. Irrigation researchers tend to recommend systems that have more opportunities for manipulations, in order to allocate water more precisely to where it is needed. At the same time, manipulation translates into corruption opportunities. Officials or ditch riders who operate gates can be bribed to open gates further or keep them open longer than intended. Systems with fixed structures can also be manipulated by widening ostensibly permanent outlets, though the 'evidence' of tampering remains visible to inspectors passing by. Some farmers may bribe officials in order to increase their water allocation. But they are also vulnerable to hold-up and extortion by the same officials, since they have a major stake in seeing the crop through. Water shortages caused by drought and other factors can motivate irrigation officials to extract side payments from farmers.

Fee collection is another entry point for corruption. When charges are based on the surface area irrigated, field-level officials can be tempted to charge for the full area but only record part of it in the official records. Because government records of irrigated areas tend not to be public, and the government does not have the capacity to audit collection officials on a large scale, such fraud can easily go undetected. And, when the government decides which areas can be irrigated through zoning processes, officials can be bribed to turn a blind eye to the illegal irrigation of land outside proper zones.[10]

Corruption is not confined to the field level. Enrichment from corruption can significantly boost incomes for local irrigation officials. Appointments to these lucrative jobs then become

7 See page 72.
8 H. Elshorst and D. O'Leary, 'Corruption in the Water Sector: Opportunities for Addressing a Pervasive Problem', presentation at World Water Week, Stockholm, August 2005.
9 R. Wade, 1982; W. Huppert *et al.*, 2001.
10 R. Wade, 1982.

coveted and themselves vulnerable to corruption. Higher-level officials sell jobs to the highest bidders, and appointees have little choice but to extract side payments from farmers in order to recoup their 'investments'. Patronage for irrigation jobs thereby perpetuates corruption and trickles up the administrative hierarchy.

Hidden harm: corruption in groundwater irrigation

In addition to the corruption risks associated with surface water and canal-based systems are those arising from groundwater irrigation. The private provision of well and groundwater irrigation has been fostered by the introduction of small, inexpensive diesel and electric water pumps, combined with subsidised electricity and diesel.

Since groundwater irrigation is financed largely by farmers and other private sector players, rather than the government, it tends to be underreported in government irrigation statistics. Regulation is also a great challenge, particularly in the case of India's estimated 20 million irrigation wells. And, while research on corruption in canal irrigation is scarce, even less has been published about corruption and rent-seeking in groundwater irrigation.[11]

To some extent, fuel and electricity subsidies to groundwater irrigators are comparable to construction and operation subsidies to canal irrigators. Strong farm lobbies react against any proposed changes in energy prices in irrigation-dependent countries, such as India.

Some argue that groundwater irrigation subsidies are more effective because water is delivered on demand and is fully under farmers' control.[12] The implications for equitable access and sustainability are grave, however. The groundwater irrigation boom is leading to rapidly falling groundwater levels and dwindling supplies for smaller farmers, who cannot compete in the pumping race. In Gujarat, India, groundwater levels in key aquifers have dropped from 10 metres to 150 metres below the surface within one generation. In many parts of India, China and Mexico, groundwater levels have dropped 20–40 metres.

As sustainability is put at risk, governments are attempting to regulate groundwater use by requiring a permit to drill a well. This opens up the risk that applicants can bribe officials. Sri Lanka and other countries have attempted to stimulate groundwater by subsidising 'agro-wells', large-diameter, brick- and concrete-lined wells that serve as both short-term storage reservoirs and groundwater extraction points. Even these practices are subject to corruption, however, depending on the design.

---

11 Exceptions are, for example, V. Narain, 'Towards a New Groundwater Institution for India', *Water Policy*, vol. 1, no. 3 (1998) and A. Prakash and V. Ballabh, 'A Win-some Lose-all Game! Social Differentiation and Politics of Groundwater Markets in North Gujarat', Institute of Rural Management, Anand, Working Paper no. 183, 2004.

12 T. Shah, 'Sustainable Groundwater Management', in M. Giordano *et al.* (eds.), *More Crop per Drop: Revisiting a Research Paradigm – Results and Synthesis of IWMI's Research 1996–2005* (London: International Water Association Publishing, 2006); T. Shah *et al.*, 'Sustaining Asia's Groundwater Boom: An Overview of Issues and Evidence', *Natural Resources Forum*, vol. 27, no. 2 (2003); T. Shah, *Groundwater Markets and Irrigation Development: Political Economy and Practical Policy* (Bombay: Oxford University Press, 1993).

In sum, less government involvement in the mainly privately organised and distributed system of groundwater irrigation means, on the surface at least, fewer opportunities for corruption than canal irrigation. The consequences of unchecked groundwater exploitation are grave, however, and regulation is largely absent because overtapping is almost impossible to police. Excessive groundwater use without consideration for sustainability and equity may not be corruption by the letter, but it is a failure of accountable water governance, with serious consequences for secure livelihoods and the environment.

What is the scale of corruption in irrigation?

Whether in the form of bribes, kickbacks, fraud, patronage or undue political influence, corruption in irrigation is a significant problem that disproportionately harms those without enough money or power to compete in this underground economy.

Irrigation subsidies: systematic policy capture

US 'pork barrel politics' for irrigation has been described as 'probably the best-known example of rent-seeking in the public expenditure domain'. Coalitions of farmers, their political representatives and the key irrigation agency, the US Bureau of Reclamation, have combined to expand the federal irrigation subsidy to cover 83 per cent of project costs. Moreover, while the subsidies were intended to support small, economically disadvantaged farmers, a study of eighteen projects showed that the largest 5 per cent of farmers (with 1,280 or more acres) collected a half of the subsidies, while the smallest 60 per cent (with 160 acres or less) received only 11 per cent.[13]

While much less has been written about this phenomenon in other countries, World Bank assessments of China, India, Bangladesh, Pakistan and Mexico show similar trends at work elsewhere. In Mexico the largest 20 per cent of farmers reap more than 70 per cent of irrigation subsidies.[14] In general, it is well documented that irrigation projects around the world recover only a fraction of their costs from farmers, frequently not even recovering operation and maintenance expenses, which are generally less than 10 per cent of the total investment,[15] and that a small number of powerful farmers benefit disproportionately.

Operations and maintenance: a common corruption tax

In the most detailed study on irrigation corruption to date, Robert Wade describes a comprehensive, well-entrenched system of corruption in South India's rice paddies, where irrigation officials *not* engaging in corrupt behaviour were the exception rather than the rule. Illicit payments generally assumed three forms. One is a flat rate of cash or grain paid to irrigation

13 R. Repetto, 'Skimming the Water: Rent-seeking and the Performance of Public Irrigation Systems', Research Report no. 4 (Washington, DC: World Resources Institute, 1986).

14 United Nations Development Programme (UNDP), *Human Development Report 2006. Beyond Scarcity: Power, Poverty and the Global Water Crisis* (New York: Palgrave Macmillan, 2006).

15 R. Repetto, 1986.

officials to ensure enough water for an entire growing season. The whole payment is made up front, with no chance for a rebate if crops fail due to lack of water. Payments are also made in return for more water in acute situations, for example when tail-end users run out of water at the end of the season. The third type is a 'gift' of grain after the harvest, which can be equal to three months' salary for an irrigation field staffer.[16]

Kickbacks and other forms of corruption were also documented in connection with obtaining jobs and favourable job transfers, awarding construction and rehabilitation contracts and obtaining out-of-zone irrigation. Informal but well-established rules determine how much is taken and who gets what. For each contract, for example, 8.5 per cent is kicked back to and shared by several officials. Collectively, all these payments funnel upwards through the political system to support political parties. In the process, the poor lose out to those who can afford bribes, disparities grow between top- and tail-enders, and production is discouraged by 'creating – often deliberately, to suit the interests of the corrupt – uncertainty about when, where and how much water will show up'.[17]

In Pakistan, similar research found that a quarter of the rural population is engaged in a hidden though well-known system of side payments to obtain irrigation water.[18] Here the corruption tax on farmers for obtaining more water than their entitlements was estimated at 2.5 per cent of their income per hectare.[19]

Construction: negotiating low quality

In addition to the minimum corruption tax on contracts, the system described in South India also includes 'savings on the ground' from contractors delivering fewer or lower-quality products and services than mandated by their contracts, and when engineers sign off on poor performance. Such haggling can bring the total rake-off to 25–50 per cent.[20] In India, the Comptroller and Auditor General estimated that, over a seven-year period, as much as 32 per cent of total payments in the state of Orissa under a programme to accelerate the completion

16 R. Wade, 1982.

17 M. Lipton. 'Approaches to Rural Poverty Alleviation in Developing Asia: Role of Water Resources', plenary address at the IWMI Regional Workshop and Policy Roundtable 'Pro-poor Intervention Strategies in Irrigated Agriculture in Asia', Colombo, August 2004; R. Wade, 1982. The system Wade describes for South India is still in effect: see P. P. Mollinga, 'On the Waterfront: Water Distribution, Technology and Agrarian Changes in a South Indian Canal Irrigation System' (Wageningen, Netherlands: Wageningen University, 1998). A similar system in Pakistan is described in M. U. Hassan, 'Maintenance in Pakistani Irrigation and Drainage Systems', MAINTAIN Country Paper no. 2 (Eschborn: GTZ, 1999). A detailed account of corruption in the water supply and sanitation sector in South Asia that confirms Wade's perspective in general terms is in J. Davis, 'Corruption in Public Service Delivery: Experience from South Asia's Water and Sanitation Sector', *World Development*, vol. 32, no. 1 (2004).

18 J.-D. Rinaudo, 'Corruption and Water Allocation: The Case of Public Irrigation in Pakistan', *Water Policy*, vol. 4, no. 5 (2002).

19 J.-P. Azam and J.-D. Rinaudo, 'Encroached Entitlements: Corruption and Appropriation of Irrigation Water in Southern Punjab (Pakistan)', Working Paper no. 252 (Toulouse: Institut d'Économie Industrielle, 2004); and see article starting on page 77.

20 R. Wade, 1982.

of irrigation projects should be characterised as excess or undue payments to contractors, as well as extra, unauthorised and wasteful expenditures. The audit stopped short of pointing the finger directly at corruption, however.[21]

Revenue fraud: massive underreporting

The size of the corruption gap in fee collection due to the underreporting of irrigated areas is difficult to assess, but indications suggest that it is enormous. When responsibility for irrigation management in the Indian state of Andra Pradesh moved from irrigation officials to groups of water users, the officially recorded irrigated area almost quadrupled from 1996 to 1998. Though improved management by users may have fuelled some of this rapid increase, the more likely explanation is that the area was already irrigated but omitted from revenue records by irrigation officials.[22]

Irrigation positions: large-scale enrichment attracts many greedy hands

Corruption gains from irrigation have been found to dwarf officials' above-board incomes. In Pakistan, they were estimated at five to eight times regular salaries, and in India up to ten times.[23] The prospect of such massive enrichments means that corruption did not stop there. In India, these lucrative posts were found to be traded on a well-entrenched market for job transfers. In this de facto trickle-up system, bribes are distributed to other officers and politicians with authority over transfers.

The bottom line is that corruption in irrigation is as rampant as it is elaborate, creating a large-scale shadow economy reaching up from the fields into the higher echelons of irrigation bureaucracies. And this corruption is not limited to South Asia. It has also been documented in Mexico[24] and Central Asia.[25]

The consequences: ineffective, inequitable irrigation

Though they can be seen as victims of corruption, farmers are often willing partners – as long as officials extract usual payments and live up to their (corrupt) promises. From this

21 H. Upadhyaya, 'Accelerated Corruption, a Trickle of Irrigation', *India Together*, 29 January 2005.

22 W. Huppert, 'Water Management in the "Moral Hazard Trap": The Example of Irrigation', presentation at World Water Week, Stockholm, August 2005.

23 R. Wade, 1982; J.-D. Rinaudo *et al.*, 'Distributing Water or Rents? Examples from a Public Irrigation System in Pakistan', *Canadian Journal of Development Studies*, vol. 21, no. 1 (2000).

24 W. H. Kloezen, 'Accounting for Water: Institutional Viability and Impacts of Market-oriented Irrigation Interventions in Central Mexico' (Wageningen, Netherlands: Wageningen University and Research, 2002); E. Rap, 'The Success of a Policy Model: Irrigation Management Transfer in Mexico' (Wageningen, Netherlands: Wageningen University and Research, 2004).

25 K. Wegerich, '"Illicit" Water: Un-accounted, but Paid for. Observations on Rent-seeking as Causes of Drainage Floods in the Lower Amu Darya Basin', Irrigation and Water Engineering Group, (Wageningen, Netherlands: Wageningen University and Research, 2006).

perspective, the system of side payments could even be seen as a form of performance-based remuneration. And the economic impact of corruption on farmers in South India is relatively small, at about 5 per cent of their annual profit.[26]

Irrigation systems do suffer at the hands of corruption, however. Bribes are high when uncertainty is high. And, while irrigation departments are supposed to ensure reliable supply, opportunities to extract revenue increase when supplies are uncertain. Similarly, while maintenance engineers are supposed to ensure that canals are well maintained, the maximum revenue can be extracted from poor maintenance, as this necessitates frequent 'works' to restore performance – each presenting opportunities for side payments. Widespread corruption in construction to cover up low-quality work also contributes to poorly functioning irrigation systems and more uncertain water flows.

When irrigation water becomes scarce, corrupted allocation means that the last in line lose out. A system meant to distribute water equitably morphs into a water funnel for the rich, who can bribe their way to the front of the queue. Two case studies in Pakistan and India showed that small farmers at the tail end of irrigation systems received a fraction of the water flowing to their top-end counterparts. And small tail-end farmers in Pakistan reported that corruption and unaffordable legal costs prevented them from challenging illegal appropriations.[27]

Fixing the flow: what can be done

Fighting corruption in irrigation means strategically restructuring incentive systems rather than piecemeal, out-of-the box reforms.

For policy capture, remedies are tied to broader reforms of political participation and empowering marginalised groups to engage in the political process. The more widespread use of diagnostics that help expose inequities implicit in water subsidies may be a useful sector-specific contribution to this endeavour.

With regard to groundwater overuse, policing is next to impossible. But indirect measures, such as higher prices for electricity and fuel that power pumps, may shift the calculations of large users towards more responsible use while doing little harm to smaller users, who cannot afford large pumps in the first place. Such measures can be expected to be deeply unpopular, however, and hark back to the problem of policy capture, which also besets irrigation subsidies.

Tackling the webs of corruption in canal irrigation requires institutional reform. By far the most common solution to break the hold that irrigation engineers have over operation and maintenance has been transferring irrigation management from the government to groups of farmers, known as water user associations (WUAs). Known as irrigation management transfer (IMT) or participatory irrigation management (PIM), this strategy has gradually become conventional wisdom for World Bank projects that address irrigation system reform. Guidelines

26 R. Wade, 1982.
27 UNDP, 2006.

for the process have been established.[28] All the same, IMT and PIM do not usually address the issue of corruption directly, and few studies exist to demonstrate their impact.[29]

Establishing water user associations is considered a useful tool for addressing corruption.[30] Bundling small, marginalised voices into a collective, formally recognised user group is intended as a step towards empowerment and better protection against extortion and corruption.

Many challenges remain, however. First, corruption may move upstream from the negotiation between farmer and official to the relationship between user association and management agency.[31] Second, technical complexity often requires user associations to hire a skilled manager or engineer. This professional is then in a position to exploit this information advantage. Third, internal WUA governance standards are often low and performance criteria unclear, giving chairpersons discretion to abuse their position for personal gain. Finally, marginalised farmers are in danger of remaining marginalised participants in WUAs. In practice, a group of bundled farmers often contains one or more large farmers who naturally become chairpersons and office-holders, and who use the association to confirm their grip on power.[32]

A number of remedies can help address these problems.

- *Stronger internal governance.* Mandatory rules, including provisions for gender-sensitive participation and auditing procedures for associations, can ensure that farmers have some form of redress and control over association executives to stop corrupt practices.
- *Rotating tasks.* In traditional irrigation systems in the Andes, different management tasks are fulfilled by different age groups within the community. This ensures that, over time, everyone becomes familiar with all tasks in the system and prevents one person from gaining specialised knowledge, thereby preventing the asymmetrical information status that leads to corruption risks.[33]
- *Re-tendering outsourced services at regular intervals.* For irrigation systems that use private service provision, re-tendering every ten years provides some leverage to punish corrupt, low-quality work. Such a system is used in France, but private provision of irrigation services remains relatively rare on a global scale.
- *A transparency offensive.* This can help prevent corrupt practices and reduce various information inequalities that breed corruption. Related measures include strengthening

28 D. L. Vermillion and J. A. Sagardoy, 'Transfer of Irrigation Management Services: Guidelines', Irrigation and Drainage Paper no. 58 (Rome: Food and Agriculture Organization [FAO], 1999).

29 The example cited in W. Huppert, 2005, in Andra Pradesh is extremely interesting but cannot be extrapolated easily to larger scales.

30 K. W. Easter and Y. Liu, 'Cost Recovery and Water Pricing for Irrigation and Drainage Projects', Agriculture and Rural Development Discussion Paper no. 26 (Washington, DC: World Bank, 2004).

31 J.-D. Rinaudo, 2002.

32 B. van Koppen *et al.*, 'Poverty Dimensions of Irrigation Management Transfer in Large-scale Canal Irrigation in Andra Pradesh and Gujarat, India', Research Report no. 61 (Colombo: IWMI, 2002); K. Wegerich, 'Why Blue Prints on Accountability of Water User Associations Do not Work: Illustrations from South Kazakhstan', presentation at fourth Asian Regional Conference and tenth International Seminar on Participatory Irrigation Management, Tehran, May 2007.

33 W. Huppert and K. Urban, 'Analysing Service Provision: Instruments for Development Cooperation Illustrated by Examples of Irrigation', Publication Series no. 263 (Eschborn: GTZ, 1998).

right-to-information provisions and mandatory disclosure of records related to construction, maintenance and management. Performance can be made transparent and comparable by establishing criteria for irrigation performance and publicly benchmarking different irrigation systems.
- *Social audits for collective oversight.* In Andhra Pradesh, the rural employment guarantee scheme of 2006 provides an auditing platform to collectively identify corruption in irrigation works. At a recent district-level meeting attended by more than 1,500 irrigation canal and other public works labourers, village-level social audits unearthed a steady stream of corrupt practices, including payments to deceased villagers, falsified payment lists and side payments to officials. The presiding official took corrective action on the spot and initiated formal inquiries.[34]
- *Standardisation.* Irrigation system design, equipment and services should be standardised to the greatest extent possible, in order to stimulate a market for irrigation products and services and to monitor value for money more easily.

For irrigation, the challenge of curbing corruption rests on the same pillars as in other sectors: increasing transparency, providing publicly available information, establishing stronger accountability for delivering irrigation water services and providing support for marginalised irrigation users to avail themselves of these instruments. A review of more than 300 irrigation projects in fifty countries underscores the fact that better performance requires maximum involvement by farmers in all stages of system development and management, from the beginning.[35]

The key stakeholder to kick-start reform is the government. Donor agencies can play a role by incorporating these recommendations in their projects, but their importance is relatively small, as the sector is dominated by national government investments and budgets.

34 Meeting attended by the author.
35 A. Inocencio *et al.*, 'Costs and Performance of Irrigation Projects: A Comparison of Sub-Saharan Africa and Other Developing Regions', Research Report no. 109 (Colombo: IWMI, 2007).

Power, bribery and fairness in Pakistan's canal irrigation systems

Jean-Daniel Rinaudo[1]

As in much of South Asia, the public canal irrigation systems in Pakistan distribute water to farmers through rationing procedures inherited from the British administration. Despite the efforts of government and international financial agencies, water resources development has not kept pace with the mounting demand caused by population growth and the water-intensive techniques promoted by the Green Revolution.

1 Dr Jean-Daniel Rinaudo is a researcher at the French Geological Survey (BRGM), Water Department, Montpellier, France.

In response to growing scarcity, more farmers are engaging in informal negotiations and extra-legal transactions with irrigation agency officials to obtain water beyond their legal quotas. Usually, a small group of farmers favourably located in the upper reaches of the irrigation system receive extra water at the expense of their downstream counterparts. The system of legal water quotas is generally no longer enforced.

Research conducted in southern and central Punjab between 1995 and 1999 reveals that farmers use political influence to win favour with irrigation officials.[2] Farmers ask local elected politicians to pressure irrigation staffers. In turn, politicians receive political support from these farmers to stay in office. And irrigation officials benefit from promotions and favourable posting.[3] In such a system, everyone wins, apart from the water losers.

Local case studies show about one-fourth of the region's rural population is engaged in this complex system of administrative and political corruption. In one area, a few large farmers were found taking water from nine outlets worth R3,300 (US$55) per hectare annually, while downstreamers spread across forty outlets were losing R600 (US$10) per hectare.[4] The rural elite are not the only beneficiaries of this system. Sharecroppers as well as small- and medium-sized capitalist farmers able to organise collective action also profit. Such arrangements are hardly clandestine. Payments and relationships, which link many types of farmers from different social circles, are common knowledge. Functioning for decades now, this interlocking incentive system is considered by many a well-established 'working rule'.

Equity, though, is often sacrificed. Farmers who take extra water generally use it for water-gulping crops such as rice, sugarcane and high-yield cotton. Meanwhile, downstreamers can hardly produce the minimum amount of staple food and cash crops needed to survive. Downstreamers become fourfold losers. They pay water fees whether or not they get water. They pay bribes to get their rightful quota. Their productivity suffers due to erratic water supplies. And they pay more to support the irrigation system than those who use their influence to avoid paying fees.[5] Corruption also undermines incentives to improve the system – for example, de-silting and reducing flow variability – as this would reduce the power of irrigation officials and influential farmers.

---

2 J.-D. Rinaudo *et al.*, 'Distributing Water or Rents? Examples from a Public Irrigation System in Pakistan', *Canadian Journal of Development Studies*, vol. 21, no. 1 (2000); J.-D. Rinaudo, 'Corruption and Water Allocation: The Case of Public Irrigation in Pakistan', *Water Policy*, vol. 4, no. 5 (2002); D. Mustafa, 'To Each According to His Power? Participation, Access and Vulnerability in Irrigation and Flood Management in Pakistan', *Environment and Planning D: Society and Space*, vol. 20, no. 6 (2002).

3 Similar studies conducted in the Indian subcontinent describe the same dynamics. See R. Wade, 'The System of Administrative and Political Corruption: Canal Irrigation in South India', *Journal of Development Studies*, vol. 18, no. 3 (1982).

4 United Nations Development Programme (UNDP), *Human Development Report 2006. Beyond Scarcity: Power, Poverty and the Global Water Crisis* (New York: Palgrave Macmillan, 2006).

5 M. Ahmad, 'Water Pricing and Markets in the Near East: Policy Issues and Options', *Water Policy*, vol. 2, no. 3 (2000); J.-D. Rinaudo and Z. Tahir, 'The Political Economy of Institutional Reforms in Pakistan's Irrigation Sector', in P. Koundouri *et al.* (eds.), *The Economics of Water Management in Developing Countries* (Cheltenham: Edward Elgar, 2003); World Water Assessment Programme, United Nation's Educational, Scientific and Cultural Organization (UNESCO), *Water: A Shared Responsibility. The United Nations World Water Development Report no. 2* (New York: UN, 2006).

Reform will not come easily. Implementing top-down anti-corruption measures would probably be ineffective at restoring equity in canal irrigation systems. Pakistan set up a system of 'oversight' in the 1960s and 1970s, but this only created a new layer of officials to be bribed.

A better strategy would be to facilitate countervailing actions by those who would lose from perpetuating the corrupt system. For example, the transparency of hydraulic systems could be improved, enabling farmers to detect irregularities in water apportioning among distribution canals. Reliable data on discharge entering the main canal and its distribution canals would be collected and made available to all water users' federations through a 'control panel'.

In 2006 and 2007 the province of Punjab developed a computerised information system that records daily discharges, supplies related information to the public and allows the online registration of complaints.[6] The project was publicised through the mass media with slogans such as 'Computers are guarding water distributions'. Without the concerted involvement of civil society groups, however, this system will probably not lead to significant improvement, as suggested by the numerous complaints for water theft still formulated in 2007 on the Provincial Irrigation Department website.

6 See irrigation.punjab.gov.pk/introduction.aspx.

Questionable irrigation deals ignore plight of Filipino farmers

Sonny Africa[1]

In a country where hand tools, peasant brawn and water buffalo are still the norm, land inequities and traditional farming methods in the Philippines are keeping farm productivity and income low. A third of Filipinos work on farms and more than a half of the population live in rural areas. Yet, despite the economic and social importance of agriculture and rural life, nearly three-fourths of poor families live in rural areas and only 30 per cent of the country's farmland is irrigated.[2]

Hoping to deliver more water and prosperity to the nation's farmers and rural poor, the government's National Irrigation Administration (NIA) has embarked on major irrigation initiatives in recent years. One such effort is the massive Casecnan Multipurpose Irrigation and Power Project in the 'Rice Bowl' area of Nueva Ecija in Central Luzon. The project has two components: a P31 billion (US$675 million) build-operate-transfer hydroelectric dam and a

1 Sonny Africa is head of Research at the IBON Foundation, an independent think tank based in Quezon City, Philippines.

2 National Statistics Office, '2000 Family Income and Expenditure Survey' (Republic of the Philippines: National Statistics Office: 2001); National Statistics Office, '2002 Scenario of the Agriculture Sector in the Philippines', Special Release no. 144, 15 March 2005.

P6.8 billion (US$152 million) irrigation system.[3] Construction of the dam began in 1995 and was completed in 2001, but the irrigation project is another story.

The project was designed to extend irrigation to 53,000 hectares of rice land and rehabilitate systems for an additional 55,100 hectares in the coming decades.[4] Originally scheduled to come online in 2004, the irrigation system is now scheduled for completion in December 2008.[5] As of June 2007 irrigation for only 62,000 hectares has been built or rehabilitated, and the NIA acknowledges that these areas might not necessarily have water yet.[6] Farmers report that canals have been built but remain unused.[7]

Beyond these problems are oddities with the public–private partnership itself. The NIA agreed to pay the contractor, a subsidiary of a US multinational corporation, a guaranteed fee for twenty years whether or not any water is actually delivered or any farmland is actually irrigated. The NIA paid P14.3 billion (US$318.5 million) from 2002 to 2006 for 3.6 billion cubic metres of water,[8] even though most of it never reached farmland because irrigation facilities from the dam had not been built. In order to make these payments, the NIA had to borrow money from the national Treasury.[9]

The project has been rife with anomalies from the outset. An initial government evaluation said the project was not financially viable and would not be able to deliver as much water as promised.[10] And the original agreement was not previously approved by the appropriate government agency.[11] Yet the project was pushed through in the 1990s by then President Fidel Ramos,[12] who reportedly was a close friend of an executive at the contractor's US mother company and a fellow West Point alumnus. Ramos has explicitly denied even knowing the man, however.[13]

---

3 NIA briefing kit on Casecnan Multipurpose Irrigation and Power Project, 31 March 2007.

4 Statement by former Secretary of Agriculture Roberto Sebastian before Senate Committees on Accountability of Public Officers and Investigations (Blue Ribbon) and on Energy, Fifth Joint Public Hearing, 23 May 2003; and from Department of Agriculture, Casecnan Multi-purpose Irrigation and Power Project details.

5 NIA briefing kit.

6 Cited in a letter dated 8 June 2007 from the officer-in-charge of the Casecnan Project in the NIA, in response to a 29 May 2007 request for implementation data; statement in a fact sheet on implementation status as of April 2007 provided by the assistant administrator for PDI of the NIA, in response to a 23 May 2007 request for implementation data.

7 Interview with author, 2 June 2007.

8 CE Casecnan Water and Energy Company, Inc., annual reports for the calendar years ended 31 December 2002 to 2006, FORM 10-K, Securities and Exchange Commission, Washington, DC.

9 Department of Budget and Management, income statements of the NIA in the Budget of Expenditures and Sources of Financing, various years.

10 Inter-agency Committee, Final Report, Annex on Casecnan Project.

11 The Republic Act 7718 or the Amended BOT Law was approved in May 1994 and its implementing rules and regulations took effect in August 1994. Nevertheless, while the Amended Casecnan Project Agreement between the NIA and CE Casecnan was executed on 13 November 1994, the NIA did not submit the project to the ICC until January 1995, and ICC, Cabinet Committee approval of the project in principle and subject to conditions was made only on 5 May 1995. (Chronology as reported to Senate Committees on Accountability of Public Officers and Investigations [Blue Ribbon] and on Energy, First Joint Public Hearing, 8 August 2002).

12 Reported to Senate Committees on Accountability of Public Officers and Investigations (Blue Ribbon) and on Energy, First Joint Public Hearing, 8 August 2002.

13 L. Rimban and S. Samonte-Pesayco, 'Trail of Power Mess Leads to Ramos', Philippine Center for Investigative Journalism, 5–8 August 2002.

If water ever begins flowing through canals and onto their rice fields, many small farmers in Nueva Ecija will be unaware that they are using some of the most expensive water in the country – subsidised by the national Treasury.

Though much smaller than Casecnan, an irrigation project in Talibon in the island province of Bohol is also drenched in controversy. Located 740 kilometres south-east of Nueva Ecija, the Talibon Small Reservoir Irrigation Project is at least delivering some water to farmers. But the 1,000-hectare project remains unfinished, despite an initial completion date of 1999.

Even after the provincial irrigation officer declared the project unviable, construction bids were solicited in 1995. Submitted by a private contractor, the lowest bid was disregarded allegedly because the firm was not qualified and due to lobbying by a local lawmaker.[14] The NIA's own Provincial Irrigation Office then took over the project itself.[15]

An investigative mission by a local anti-corruption group found that, although P165 million (US$2.9 million) had been spent by 2005, there was no sign of a reservoir, dam or an irrigation system.[16] The only progress was some excavations, a row of piping, a bridge-like structure, an office building and abandoned construction equipment. Another inquiry found prima facie evidence that NIA officials had committed construction infractions.[17] A new budget of P280 million (US$5.1 million) was proposed – more than double the private contractor's original low bid.[18]

Small farmers are losing in three ways. They contributed labour towards the construction. They 'voluntarily donated' land and relinquished plants – without compensation – to make way for canals and roads. And they still have not much irrigation to speak of. One farmer commented: 'The dam promised to us to help increase the productivity of our land became just a *damgo* [dream].'

Formal investigations have been launched into both projects. The Senate conducted an investigation about Casecnan in 2002 but its conclusions have not been released. An ombudsman filed a case against local NIA officials in connection with the Talibon project in 2004, but this remains stalled – as does a parliamentary investigation initiated in 2006.[19]

---

14 *Philippine Daily Inquirer*, 20 November 2004.

15 Letter request from Administrator Orlando V. Soriano of the NIA dated 8 January 1998.

16 Panabugkos Kontra K-4 (Panabugkos sa Katawhang Bol-Anon Kontra Kagutom, Kalisod, Korupsyon, Krisis), Investigative Mission Report, 8–9 February 2005.

17 NIA, memorandum dated 6 December 2004 for the NIA administrator from the NIA assistant administrator for project development and implementation on the 'Fact-finding Investigation Report Conducted for Alleged Anomalies in the Construction of Talibon DAM SRIP Project'.

18 Panabugkos Kontra K-4, 2005.

19 House of Representatives, 'Resolution Directing the Appropriate Committee of the House of Representatives to Conduct an Investigation, in Aid of Legislation, on the Reported Irregularities and Anomaly in the Construction of Talibon Small Reservoir Irrigation Project in Talibon, Bohol', House Resolution no. 584, First Regular Session, Thirteenth Congress, Republic of the Philippines.

Sealing water aid against corruption: donor interventions, donor responsibilities

Grit Martinez and Kathleen Shordt[1]

Over the past ten years the recognition of corruption as a major obstacle to development programming has led many donors – bilateral, multilateral and international organisations – to come up with a range of policies, codes and regulations in response to the problem. At the same time, governments and donors have committed themselves to many international agreements and principles, initially focusing on preventing corruption in specific transactions and donor-supported projects. New corruption-fighting strategies related to development assistance are embodied in several international conventions, including the Paris Declaration on Aid Effectiveness (2005), the OECD Principles for Donor Action in Anti-corruption (2006), the Asian Development Bank (ADB)/OECD Anti-corruption Initiative for Asia and the Pacific (2003) and the EU Anti-corruption Policy and Ten Principles for Candidate Countries (2005).

The Paris Declaration and the OECD Principles shift the paradigm away from donor-driven policies towards placing donors in a role that supports developing countries' own efforts to deal with corruption, while fostering a partnership of mutual accountability. These conventions give greater emphasis to the overall enabling environment of development, recognising that donors' practices and internal policies can stimulate or limit corruption within programmes and within countries more generally.

All this matters for corruption in the water sector. Between 2001 and 2005 donor commitments for water and sanitation alone doubled, reaching almost US$6 billion in 2005.[2] But the reach of donor policies and government agreements still does not extend to the lives of people. In part this results from a lack of sector specificity, in that generic corruption-fighting agreements and tools have not yet been tailored to the water sector's specific features or applied at a scale large enough to make a difference.

What are the next steps? Donors can strengthen their own commitment to accountability, build anti-corruption measures more systematically into their water sector programming and harmonise their activities to close loopholes for corruption.

Towards mutual accountability

More transparency is an important step to enhanced donor accountability. Many project-related documents are not made available in a timely and accessible manner to enable

1 Grit Martinez is a fellow with Ecologic, the Institute for International and European Environmental Policy in Berlin. Kathleen Shordt is a senior programme officer at the IRC International Water and Sanitation Centre, Delft, Netherlands.

2 See OECD Development Co-operation Directorate, www.oecd.org/dac/stats.

effective input and oversight by civil society. Stronger sanctions against corrupt staff and contractors can also help. The World Bank has taken a leading role in debarment, levying sanctions against contractors in prestigious water projects, such as the Lesotho Highlands case.[3] Many donors have followed suit, but more coordination of investigation and debarment standards is required, as well as strict sanctions by all donors against their own employees when they are implicated in corrupt activities.

Internal incentive systems still distract from a focus on aid effectiveness, which is essential for accountable water aid. Within donor agencies, performance incentives are often not directly related to project outcomes but, rather, to the number of programmes or volume of funding they process. A commitment to mutual accountability as proclaimed by the Paris Declaration has yet to be put into practice. A progress report on the declaration lamented that by 2006 fewer than a half of the twenty-nine countries surveyed had implemented mechanisms for mutual assessment of progress, and it recommended that donors develop credible monitoring mechanisms.[4]

One promising approach for all donors is output-based aid. Unlike many forms of traditional assistance, output-based aid links payments to the delivery of specified services or outputs. It is being used, for example, to extend water service in Paraguay, where small-scale providers (*aguateros*) are connecting rural and small towns to networks with the help of residents themselves, and in Cambodia, where pilot projects in four towns have identified 3,000 of the poorest households for water service.[5]

Programming against corruption

Donors can use a variety of tools and strategies to tackle corruption in the typical cycle of the development of water services. These tools include transparency in tendering and procurement, audits, independent multi-stakeholder oversight, codes of conduct, anti-corruption agreements and staff training. To address the corruption risk of substandard execution, useful mechanisms include time-bound warranties in implementation and maintenance contracts, sustainability clauses that require partners to submit a monitoring protocol after project implementation, public fault reporting systems and functionality checks on service uptime and water quality.[6]

Coordination of activities to close down opportunities for corruption

In 2007 the European Commission, one of the top donors in the water sector, emphasised the urgent need for a more effective division of labour in development programming. As of 2007

---

3 See article starting on page 18.

4 OECD, '2006 Survey on Monitoring the Paris Declaration: Overview of the Results' (Paris: OECD, 2007).

5 World Bank, 'Output-based Aid: Supporting Infrastructure Delivery through Explicit and Performance-based Subsidies', Global Partnership on Output-Based Aid Working Paper no. 4 (Washington, DC: World Bank, 2005).

6 Grit Martinez, Kathleen Shordt and WIN, 'The Contribution of Netherlands' Development Assistance to Risk Assessment and Mitigation of Corruption in the Water, Sanitation and Hygiene (WASH) Sector', presentation at workshop for the Dutch Foreign Affairs Ministry, The Hague, February 2007.

recipient countries have to deal with an average of 350 donor missions per year.[7] And they often end up with more than 100 donor-installed parallel project implementation units that function outside their bureaucracies,[8] draining scarce management time and talent from the public sector and complicating the budgetary tracking of received funds. This all makes accountable management of aid flows more difficult.

Donor fragmentation also provides opportunities for 'donor arbitrage'. When donor commitments to anti-corruption programming vary, corrupt recipients can pick and choose the funds that provide the best opportunities for personal enrichment. This highlights the need not only to harmonise anti-corruption strategies within the donor community, but also to bring on board more strongly the new crop of increasingly influential donors, such as private foundations and bilateral donors from emerging economies such as China.

7 European Commission, 'EU Code of Conduct on Division of Labour in Development Policy', communication from the Commission to the Council and the European Parliament, COM (2007) 72 final, 2007.

8 OECD, 2007.

5 Water for energy

Lawrence Haas presents a wide array of corruption risks that affect hydropower and outlines practical recommendations for reform by a number of stakeholders. Thayer Scudder reviews the issue of corruption and resettlement, and Gørild Heggelund illustrates related challenges with a case study from China. Kathy Shandling and Reinier Lock examine from an industry perspective the potential of public–private partnerships for tackling corruption in hydropower, while Peter Bosshard and Nicholas Hildyard discuss whether corruption leads to a bias towards large-scale hydropower projects.

Water for energy: corruption in the hydropower sector

Lawrence J. M. Haas[1]

Hydropower and dams: why they matter

One-sixth of the world's electricity comes from hydropower, and it provides at least a half of the supply in more than sixty countries.[2] Electricity will probably occupy an even more prominent place on the global energy scene in the decades to come. As demand for power continues to grow globally, so do pressures to increase the share of electricity generated from non-fossil sources, in order to address the many environmental and socio-political problems associated with oil and coal, cut climate-changing emissions and make electricity more accessible to the more than 1.6 billion people who currently go without.[3] Equitable access to electricity is a central theme in the development debate, and lack of energy services can negatively affect the prospects for realising sustainable development and achieving the Millennium Development Goals.

Any discussion about hydropower invariably leads to the debate about large dams and the role they play in the provision of water, energy and related services. Corruption features

1 Lawrence Haas was team leader in the Secretariat of the World Commission on Dams (WCD) 1998–2000. He currently works in an independent capacity for development organisations including the World Bank, ADB and international non-government organisations including TI, the IUCN and WorldWide Fund for Nature (WWF).

2 WCD, *Dams and Development: A New Framework for Decision-making* (London: Earthscan Publications, 2000).

3 R. T. Watson *et al.*, 'Climate Change 2001: Third Assessment Report of the Intergovernmental Panel on Climate Change' (Geneva: IPCC, 2001); UN-Energy, '*Energy in the United Nations: An Overview of UN-Energy Activities*' (New York: UN-Energy, 2006).

prominently in this debate. There are more than 45,000 large dams in 140 countries, and about two-thirds of them are in the developing world, where new construction is also heavily concentrated. In China's Yangtze River basin alone, 105 large dams are planned or under construction.[4] Of course, dams are not only about electricity, as more than a third have multiple purposes – making the stakes and the corruption risks even higher. Dams help cope with variabilities in rainfall, drought and other hydrological factors, and serve as vital instruments for water supply and flood management. And 30–40 per cent of the 271 million hectares of irrigated land worldwide rely on dams.[5]

Dams are also the infrastructure projects that most fundamentally affect human settlement patterns, livelihoods, health and the environment. They impound about 14 per cent of all global water run-off. And, together with canals and diversions, they fragment 60 per cent of the world's 227 largest rivers, with the remaining free-flowing rivers in the developing world also subject to a high rate of dam construction.[6]

If poorly designed or managed, dams can harm valuable ecosystems and biodiversity as well as provide breeding grounds for waterborne diseases.[7] Irrespective of the benefits, the impacts on human livelihoods are also profound. The World Commission on Dams estimated in 2000 that between 40 and 80 million people had been displaced by dams in the previous fifty years.[8] And governments and project managers have frequently reneged on promises to provide resettlement assistance and other aid to those adversely affected by hydropower projects.[9]

All this makes hydropower and dams central in the debate about the blueprint for a sustainable future. To maximise sustainability and minimise corruption, the building of dams requires that up-front strategic assessments are made that mobilise all the available options to meet today's challenges in water and energy service provision. Hydropower also requires a better integration of governance reforms, to ensure that all stakeholders have a voice to inform decision-making. Improving sustainability in all stages of the infrastructure project cycle will help provide for the security of livelihoods, social and economic well-being, energy, the environment and the climate – while corruption can significantly disrupt this at many levels.

Money and complexity: why hydropower is a high-risk sector for corruption

Huge budgets and opportunities to hide unseemly practices within complex administrative systems are the main drivers of corruption in hydropower projects. Of the US$11.1 trillion the world is predicted to spend on energy infrastructure between 2005 and 2030, US$1.9 trillion

4 United Nations Environment Programme (UNEP), '*Global Environment Outlook no. 4: Environment for Development* (Nairobi: UNEP, 2007).

5 WCD, 2000.

6 UNEP, 2007 (*Global Environment Outlook no. 4*).

7 WCD, 2000.

8 Ibid.

9 See articles starting on pages 96 and 99.

may be expected to go toward hydropower.[10] These large numbers create multiple opportunities for bribery, fraud and other forms of corrupt behaviour.[11] Civil works contracts are typically the largest budget line, accounting on average for 60 per cent or more of total project costs, making dam construction a primary target for corruption. As other contributions to the *Global Corruption Report 2008* indicate,[12] resettlement costs can also be significant and offer entry points for embezzlement and other forms of corruption.

Several ministries are typically involved in hydropower projects, especially in large multipurpose projects with major land acquisition and resettlement components, and related infrastructure such as access roads and tunnels. The result is complexity and opaque oversight mechanisms. Even with a single coordinating body, numerous opportunities exist for miscommunication, institutional disconnect and inadequate cooperation among government departments and agencies. Combined with a lack of transparency, this provides fertile ground for manipulation and abuse.[13]

Complexity on the institutional side is mirrored by complexity in contracting. The many contracts required for equipment, materials, construction, management and consultancies are often joint ventures involving several companies, frequently with a mix of domestic and foreign-based firms. In Laos, for example, the financing consortium for the US$1.45 billion Nam Theun 2 Project involves twenty-six separate financial institutions, including private companies and banks, several public institutions and the Lao government, each with its own accountability requirements.[14]

The risk of policy capture is also very real in hydropower projects, where vested interests unduly influence decisions about the mix of water and energy service options the society chooses. Without adequate compensation measures for affected people benefits and risks stand to be extremely unequally distributed. While urban or industrial consumers and the dam industry gain, often local communities bear a disproportionate share of the cost of hydropower and other large dams. They can be very detrimental to small upstream landowners, displaced communities and other economically and politically disadvantaged people, who often live in remote mountainous rural areas where many potential sites for large dams are located.[15] This requires extra efforts to ensure that all stakeholders are considered in the decision-making process.

---

10 International Energy Agency (IEA), *World Energy Outlook 2006* (Paris: IEA, 2006). If hydropower maintains the current 16.9 per cent share of global energy generation, this translates into a US$76 billion average annual investment in hydropower. This is adjusted downward to use US$50–60 billion due to the cost of hydropower relative to other types of power generation.

11 See also the article starting on page 103 for a bias towards large projects, because they provide better opportunities for high-level officials to extract rents.

12 See articles starting on pages 96 and 99.

13 M. H. Wiehen, 'Transparency and Corruption Prevention on Building Large Dams', paper for WCD, 26 December 1999; see dams.org/docs/kbase/contrib/ins204.pdf.

14 Nam Theun 2 Hydroelectric Project, www.namtheun2.com.

15 For example, with the thirty-four large dams in India, tribal communities – politically marginalised groups that comprise only 8 per cent of India's population – constitute 47 per cent of those displaced. In the Philippines, almost all dams are on the land of indigenous people, who make up less than 10 per cent of the country's population (WCD, 2000).

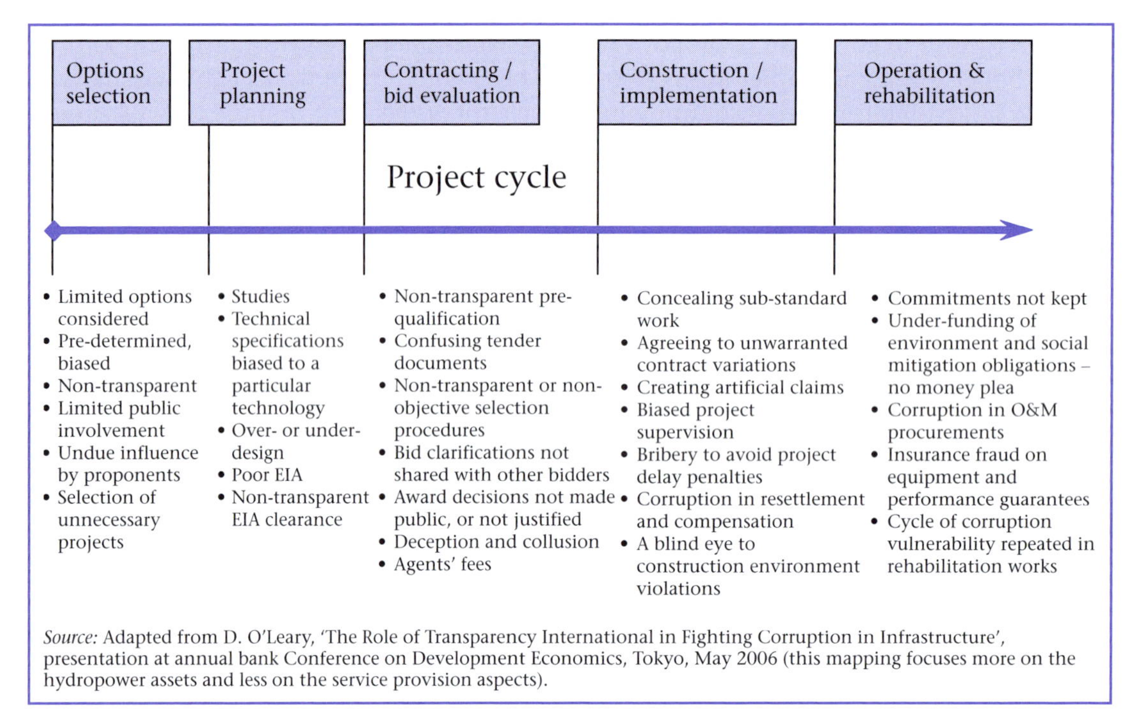

Source: Adapted from D. O'Leary, 'The Role of Transparency International in Fighting Corruption in Infrastructure', presentation at annual bank Conference on Development Economics, Tokyo, May 2006 (this mapping focuses more on the hydropower assets and less on the service provision aspects).

Figure 1 Scope and enabling conditions for corruption in various stages of a project cycle

Forms and effects: what corruption in hydropower looks like

It is widely acknowledged that corruption vulnerabilities in hydropower must be seen through the lens of strategic planning and the project cycle. This means carefully assessing – and tackling – corruption exposures from the early stages of project identification and design, through contractor pre-qualification, tender, construction and operation. Figure 1 illustrates corruption problems that occur along the project cycle.

Corruption risks start with the potential for undue political influence in identifying and selecting hydropower sites, undue outside influence from project developers or inter-departmental collusion in project approval.[16]

Bribes and misappropriation of funds have been reported throughout the world. The cost of the joint Paraguayan–Argentinian Yacyretá Dam, started in 1983 and completed only in 1994, ballooned from US$2.7 billion to US$11.5 billion.[17] It is widely cited as a 'monument to corruption'.[18]

16 Some factors may be considered 'bad practice' rather than direct corruption, but there is a strong overlap with the latter. These also increase the opportunity for corrupt acts, and, equally importantly, they undermine public trust.

17 M. Sohail and S. Cavill, *Accountability Arrangements to Combat Corruption: Synthesis Report and Case Study Survey Reports*, WEDC (Loughborough: Loughborough University, 2007).

18 Ibid.

Grand corruption can occur in the form of bid-rigging and kickbacks in procurement, and kickbacks to accept inflated bills, unit costs and material quantities in contracts. These illicit payments are often disguised by channelling them through agents or subcontractors.

Irregularities with environmental impact assessments can arise during the planning phase. In India, for example, an accounting firm commissioned to conduct an EIA for two dams was caught in 2000 copying 'word for word' large sections of an EIA for a different project 145 kilometres away. After a civil society watch group spotted the plagiarism and posted the information on its website, the contractor said it would rewrite the document.[19]

Vulnerabilities continue during project operation and maintenance. These can include endemic petty corruption related to service access and provision, the misappropriation or misuse of fees, illegal connections, failure to honour social and environmental mitigation commitments, patronage and abuse of funds in resettlement activities, and failure to honour monetary and non-monetary benefit-sharing. The cycle of grand corruption can start all over again with procurement for maintenance, refurbishment and upgrading contracts (see figure 1).

The benefits from tackling corruption that would flow to people and the environment are considerable by any measure. Direct cost savings may start at US$5–6 billion annually, if just the average 10 per cent reduction in contractor bid prices achieved through integrity pacts pioneered by Transparency International were extrapolated to all planned hydropower projects.[20]

If corruption leads to cost overruns that eat into funds originally earmarked for maintenance, proper functioning may be put at risk, reducing the long-term benefits. Corruption can also hamper the expansion of electricity services in developing countries, by driving up costs, delaying projects and lowering service quality and reliability, especially in rural areas considered low priorities. Higher electricity prices disproportionately affect the poor and vulnerable, retarding poverty reduction efforts. In Montenegro, for example, poor households spend more than twice as much of their budget on electricity as higher-income households – 12.9 per cent versus 5.2 per cent.[21] This poverty gap is much greater in Africa and Asia, where the social impacts of tariff increases can spark demonstrations, as in Nepal.[22]

More dramatically, corruption also amplifies the adverse effects that hydropower projects have on ecosystems, which many people at subsistence levels in developing countries rely on for their daily livelihoods and health. In fact, the WCD emphasises negative impacts on ecosystems and affected communities as two of the most serious failings of existing dams.[23]

Finally, chronic corruption ultimately undermines public trust and the political sustainability of hydropower as an option for societies to consider. Many would-be investors melt away

---

19 Public Services International, 'Water Privatisation, Corruption and Exploitation', 20 August 2002; see www.indiaresource.org/issues/water/2003/waterprivatizationpsi.html; Associated Press, 'Ernst & Young rewriting dam report', 3 September 2000.

20 See, for example, the use of integrity pacts in Mexico, from page 95.

21 P. Silva et al., 'Poverty and Environmental Impacts of Electricity Price Reforms in Montenegro', Policy Research Working Paper no. 4127 (Washington, DC: World Bank, 2007).

22 B. Bhadra, 'Hydro-energy for National Development: Small and Medium Hydro Electricity Development Issues', *The Weekly Telegraph* (Nepal), 30 January 2002; *Kathmandu Post*, 'Tariff hike again?', 31 March 2001.

23 WCD, 2000.

Table 4 Why fighting corruption is a long-term interest of all stakeholders

Stakeholder group	Corrosive effects of corruption
Electricity consumers[24]	• Less affordable and reliable electricity • Less access for the poor • Slower pace of service expansion
Impacted communities	• More high-impact or 'bad' projects • Higher adverse livelihood impacts and impoverishment risks • Fewer funds for compensation, mitigation and benefit-sharing • Fewer mitigation commitments for sustainable management
Electricity utilities	• Higher costs of bulk energy or own supply • Higher borrowing and equity costs • Less money for service expansion and improvement • Delayed, overpriced or expensive infrastructure
Governments	• Higher power sector costs • Higher repayments for sovereign loans or guarantees • Setbacks for social policies • Slower economic growth and job creation for projects that depend on improved electricity service
Public hydropower developers/ operators and IPPs[25]	• No level playing field for fair competition • Approvals procured through bribes can be rescinded, terminating the project • Disqualification from office or criminal prosecution
Contractors and equipment suppliers	• Distorted and unfair competition • Higher and wasted tender expenses • Approvals procured through bribes can be rescinded, terminating the project • Criminal prosecution, fines, blacklisting and loss of reputation
Financiers: ECAs, MDBs,[26] commercial banks, credit agencies and insurers	• Higher reputation risks if corruption is proven • Higher than necessary requests for borrowing • Additional costs and fraudulent claims • Financial loss

24 For multi-purpose projects, consumers include irrigators and urban water users in cities, or any groups that would benefit from reducing corruption in water and energy provision from multi-purpose dams.
25 'IPP' stands for 'independent power producer'.
26 'ECA' stands for 'export credit agency' and 'MDB' stands for 'multilateral development bank'.

as concerns about reputation risks and other costs of corruption arise. Table 4 summarises the impact of corruption on hydropower.

Moving towards action

[T]he end of any dam project must be the sustainable improvement of human welfare… If a large dam is the best way to achieve this goal, it deserves support. Where other options offer better solutions, we should favour them over large dams. (World Commission on Dams)

A changing governance landscape

Far-reaching changes in the power and water sector mean that the governance framework for hydropower has also undergone a transition. This provides new risks for corruption but also new entry points for fighting it. In the energy sector, private financiers and operators assume a bigger role. Meanwhile, water resources management has shifted to a more inclusive and participatory approach that recognises more strongly the linkages between hydrology, human geography and the environment.

Taken together, this means more stakeholders around the hydropower table, and more need and opportunities for coordination and participation. It also means more complex risk- and responsibility-sharing arrangements between public and private actors that provide new entry points for corruption, but also new levers to make accountability structures and decision-making more transparent and inclusive.

The case for a common cause

Tackling corruption in such a setting requires forging anti-corruption coalitions between all stakeholders to create momentum for change, as well as establishing a web of checks, balances and trust that makes the fight against corruption effective.

A first step is to demonstrate convincingly that tackling corruption can benefit all stakeholders. Combating corruption is plainly in the interests of electricity consumers, governments, the hydropower industry, public and private financing bodies and, especially, the more than 1.5 billion people who today have no access to affordable electrical services. Although polarised views about hydropower remain part of today's dialogue on sustainable development, a constructive collaboration is building between industry, environment and social interests. For example, the WWF and International Hydropower Association (IHA) intend to work together to improve sustainability guidelines for hydropower projects.[27]

Opening decision-making

The World Commission on Dams remarked in 2000 that 'at the heart of the current debate on dams is the way choices are made, and the different opinions and perspectives that are expressed – or denied expression – in the process'. The WCD proposed a 'rights and risks' approach to identify all legitimate hydropower stakeholders, including involuntary risk absorbers such as displaced communities. Today there is more guidance available on how to undertake inclusive options assessments and move it upstream into strategic planning processes.[28] As the WCD observed, this helps mobilise all possible options not only to meet

27 The IHA and WWF, along with four other partners, are about to announce a two-year initiative to field test and revise the Sustainability Guidelines of Hydropower that the IHA has promoted since 2002.

28 K. Blok *et al.*, 'Stakeholder Involvement in Options Assessment: Promoting Dialogue in Meeting Water and Energy Needs' (Washington, DC: World Bank, 2003).

growing water challenges, but also to address the real and perceived biases in how non-dam options are taken up or rejected.

Fighting corruption from the project finance side

The high capital costs and long payback periods of large-scale hydropower make financing an important factor for success and a powerful lever for fighting corruption. Accountability can be promoted through committed project financiers, adequate financing instruments and sound revenue-sharing governance.

Multilateral development banks and bilateral donors. Though many international donors are stepping up governance and anti-corruption activities, some specific measures have yet to be built into donor-supported hydropower projects, such as governance improvement plans.

Export credit agencies. ECAs provide export credit guarantees and insurance for electrical and mechanical equipment exporters. In 2006 the OECD Council adopted recommendations to deter supply-side bribery in official assistance – including increased disclosure and no-bribery undertakings and sanctions – as a prerequisite for companies to obtain ECA support.[29] This is a good first step, but shared definitions of standards of proof, due diligence and enhanced due diligence, and information disclosure are still needed.[30] Moreover, anti-corruption measures by non-OECD country ECAs must be better harmonised.[31] China's Export-Import Bank, for example, is one of the world's largest ECAs, with primary commercial operations reportedly exceeding those of the United States, Japan and the United Kingdom. It is heavily involved in hydropower projects.[32]

Private commercial banks. Introduced in 2002, the Equator Principles provide a common framework for commercial banks to apply their own corporate responsibility charters, and social and environmental standards in project finance lending to infrastructure, including hydropower.[33] The Equator Principles financial institutions (EPFIs), which represent more than 80 per cent of commercial lending in infrastructure globally, have agreed not to provide

29 OECD, 'OECD Recommendation to Deter Bribery in Officially Supported Export Credits' (Paris: OECD, 2006).

30 Transparency International, 'Export Credit Agencies'; see www.transparency.org/global_priorities/ public_ contracting/instruments/export_credit_agencies.

31 Article 3 of the 2006 OECD Council recommendation encourages non-OECD members that are parties to the OECD Anti-Bribery Convention to adhere to the provisions of the recommendation. The stated aim is to level the playing field among all providers of official export credits. On 21 November 1997 OECD member countries and five non-member countries, Argentina, Brazil, Bulgaria, Chile and the Slovak Republic, adopted the Convention on Combating Bribery of Foreign Public Officials in International Business Transactions.

32 S. Rose, 'China's ExIm Bank Discloses Its Environmental Policy', blog entry, Center for Global Development, 11 May 2007; see blogs.cgdev.org/globaldevelopment/2007/05/chinas_exim_bank_discloses_its.php. See also the website of the Export-Import Bank of China, english.eximbank.gov.cn., and P. Bosshard and M. Chan-Fishel, 'A Case of Environmental Money Laundering', International Rivers Network and Friends of the Earth, 21 July 2005, www.irn.org/programs/finance/index.php?id=050721exim.html.

33 'Equator Principles: A Financial Industry Benchmark for Determining, Assessing and Managing Social and Environmental Risk in Project Financing'; see www.equator-principles.com/principles.shtml.

loans to borrowers that do not comply with the principles.[34] The principles are criticised, however, for their lack of explicit, binding standards that comply with international law in relation to the environment, human rights, indigenous peoples and labour.[35] They also lack transparency in how EPFIs ensure their borrowers actually comply with the principles.[36]

Private equity. Private equity groups are increasingly taking the lead on independent power producer hydropower in Asia and Africa, such as the acceleration of hydropower IPPs in South-east Asia's Mekong region.[37] This is a highly positive trend, because developing countries can attain greater access to financing. But it highlights the growing gap between what a consensus of public international financing bodies require as safeguard policies and what private international equity groups and ECAs of non-OECD countries require – what the media have criticised as a 'no strings' policy for infrastructure lending.[38]

Transparency in contractual arrangements and risk-sharing

New contractual frameworks provide more flexibility for sharing responsibilities and risks in hydropower projects. Transparency on how decisions come about, how risks are calculated and how responsibilities are shared are indispensable for all these new contractual relationships. Clear transparency guidelines are essential not only to prevent and correct corruption, but also to restore the public confidence in responsible hydropower governance that otherwise threatens to make hydropower politically unfeasible. Lessons can be drawn from recent controversies about power purchase agreements (PPAs), such as the 250 MW Bujagali project in Uganda. In 2002 Uganda's High Court had to order the public release of the PPA at the urging of NGOs, because the government had failed to make the information public.[39] The PPA between the new project sponsor and the government of Uganda is now available to the public.[40]

Building transparency and accountability into new financing and revenue-sharing frameworks

Revenue-sharing for hydropower projects and carbon-trading schemes, such as the Kyoto Protocol's Clean Development Mechanism (CDM), are examples of emerging financing mechanisms that are strategically important to advance sustainable forms of hydropower

34 The Equator Principles were revised in 2006 to align with the updated, International Finance Corporation (IFC) Performance Standards on Social and Environmental Sustainability; see www.ifc.org/ifcext/enviro.nsf/Content/EnvSocStandards.
35 Bretton Woods Project, 'From bad to worse: IFC safeguards', 13 June 2005.
36 R. Bailey *et al.*, 'Building Sustainability into Syndication', *Environmental Finance*, July/August 2006.
37 P. King, *et al.*, 'Joint Program on Environment Criteria for Hydropower Development in the Mekong Region', a joint initiative of the Asian Development Bank, Mekong River Commission and World Wildlife Federation, March 2007.
38 *BBC News* (UK), 'China Defends Its Role in Africa', 16 May 2007; S. Rose, 2007.
39 A. T. Balinda and F. C. Oweyegha-Afunaduula, 'Nape's Contribution to Environmental Advocacy in the Nile Basin: Bujagali Power Project, Uganda', presentation at the third World Water Forum, Kyoto, March 2003.
40 See go.worldbank.org/UTHNPOSSD0.

development. The CDM allows industrialised countries with a greenhouse gas reduction commitment to invest in projects that reduce emissions in developing countries, up to certain limits.[41] The CDM has supported hydropower projects that meet eligibility criteria, though some policy and advocacy groups contest the inclusion of large hydropower projects which they consider unsustainable.[42] Rejections of applications to the CDM, such as the Bumbuna Hydropower Project in Sierra Leone, suggest a need to clarify transparency procedures. And the CDM still has no formal appeal mechanism.

Local revenue-sharing

Encouraged by the World Commission on Dams, many countries now allow local communities to receive a monetary share of project revenues when they give up their land or natural resources, but examples are still few and far between.[43] A sustainable financing source to fund environmental and social commitments can go a long way towards addressing many accountability concerns in hydropower, such as governments delivering on promises when they have no real financial capacity to do so. But, at the same time, they can fuel controversy in the absence of adequate provisions for transparency and accountability.

In Sierra Leone, endemic corruption contributed to the eleven-year rebel war that formally ended after national elections in 2002. In post-war reconstruction, proposals to introduce revenue-sharing on the war-delayed Bumbuna hydropower project, mentioned above, were widely endorsed by local people and the newly elected local government. Measures to ensure transparency and social accountability in revenue-sharing arrangements will be evaluated in the set-up phase of the Bumbuna Trust. A multi-stakeholder board will oversee the trust and will help Sierra Leone meet its long-term commitments to affected populations and the environment through a wide range of community projects for poverty reduction, development and environmental protection.[44] Some form of carbon financing and the electricity tariffs will provide sustainable financing for the trust.

Strengthening project and sectoral governance

Governance improvement plans (GIPs) in hydropower projects can help elevate anti-corruption measures to a strategic focus of project management. GIPs can integrate a comprehensive package of anti-corruption tools, including risk-mapping, integrity pacts, formal compliance plans and disclosure standards for all project elements. They have already

41 'Clean Development Mechanism'; see www.cdm.unfccc.int.

42 See SinksWatch, www.sinkswatch.org/pubs/CDM%20Report_English.pdf, and 'Carbon Trading: A Critical Conversation on Climate Change, Privatisation and Power', *Development Dialogue*, no. 48 (Uppsala, Sweden: Dag Hammarskjöld Foundation, 2006).

43 T. Scudder, *The Future of Large Dams: Dealing with Social, Environmental, Institutional and Political Costs* (London: Earthscan Publications, 2005).

44 See www.wds.worldbank.org/external/default/WDSContentServer/WDSP/IB/2005/04/21/000012009_20050421154222/Original/Backup0of0Bumb1praisal0Draft1041505.wbk.doc.

proved effective in other infrastructure sectors, such as road improvement projects in Paraguay and Indonesia.[45]

As far as the donor community is concerned, because only a small number of hydropower projects are donor-supported, action at the national and sectoral levels is crucial. Tools include ethical codes of conduct for key officials, as well as asset declaration and the publication of representation limits for senior staff in public hydropower companies.[46]

Momentum for governmental anti-corruption reforms can also come from international anti-corruption agreements. Most of the top ten hydropower countries are signatories to UN or regional conventions on bribery and corruption. Although often legally binding, implementation remains a big challenge and provides opportunities for more targeted public pressure on governments to live up to their commitments and also recognise hydropower as a high-corruption risk sector.

Private companies working in hydropower can do their share by implementing effective anti-corruption policies, following guidelines such as Transparency International's Business Principles for Countering Bribery.[47] They can also work towards sectoral anti-corruption standards that promote trust in fair play and further reduce corruption risks. The International Hydropower Association (IHA), for example has prepared sustainability guidelines that can serve as a model for developing a voluntary set of anti-corruption guidelines.[48]

Civil society organisations can provide important additional checks and balances through independent monitoring and mobilising community participation in hydropower decision-making.

In order to make public monitoring possible in the first place, the entire hydropower sector must be brought fully under freedom of information regulations to ensure the public disclosure of project documents and budgets.

The power of using the tools at hand

Fortunately, many tools are available to tackle corruption in hydropower – such as corruption risk assessments, integrity pacts, compliance plans and anti-corruption conventions. Too often, however, these tools remain on the shelf.

Integrity pacts for public procurement, for example, have achieved significant savings on several dam projects.[49] In 2002 Mexico's Federal Electricity Commission (Comisión Federal de

---

45 L. Haas *et al.*, 'Setting Standards for Communications and Governance: The Example of Infrastructure Projects', Working Paper no. 121 (Washington, DC: World Bank, 2007).

46 Ibid.

47 Transparency International, 'Business Principles for Countering Bribery', www.transparency.org/global_priorities/private_sector/business_principles.

48 International Hydropower Association, 'IHA Sustainability Guidelines Adopted', www.hydropower.org/sustainable_hydropower/sustainability_ guidelines.html.

49 Integrity pacts are voluntary agreements that identify the steps that all parties in a project will take, individually and collectively, to reduce or eliminate corruption, backed by independent oversight and monitoring. The integrity

Electricidad – CFE) began working with TI Mexico to test an integrity pact for public procurement on the 750 MW El Cajón hydroelectric project. Bidders were required to sign a unilateral declaration of integrity, and similar declarations were made by CFE officials and all the government officials involved in the bidding process. A social witness (*testigo social*) was employed to oversee contracting and report the results to civil society groups and the public. The accepted bid was reduced by 8.5 per cent – P675 million (US$64 million) less than the CFE had expected based on past bidding trends.[50]

Concerted action to roll back corruption in hydropower needs collaboration, a time-bounded strategy and measurable indicators of progress – all of which are quite possible with existing tools and levels of stakeholder commitment. Corruption is not only a serious cost factor. It is a serious blockage to realising the benefits of hydropower for everyone, and it fatally undermines what is already very fragile public confidence in the sector in many countries. Fighting corruption in hydropower is therefore indispensable for a sustainable energy future that maximises the benefits of renewable sources.

Footnote 49 (*cont.*)
pact has shown itself to be adaptable to many legal settings and flexible in its application. See www.transparency.org/global_priorities/public_contracting/integrity_pacts.

50 See L. Haas *et al.*, 2007.

Hydropower corruption and the politics of resettlement

Thayer Scudder[1]

Though the supplier of immense economic resources in the form of water and energy, hydroelectric dams have inflicted a heavy toll on humanity – especially populations with little financial or political power. Up to 80 million people have been displaced by the world's dams, as many as 58 million in China and India between 1950 and 1990 alone.[2]

These resettlers are usually poor ethnic minorities or indigenous people who, rather than benefiting from hydro-projects, become the major risk-takers and are further impoverished economically, institutionally and culturally.[3]

Though it has seldom been documented,[4] corruption is a major cause of impoverishment for resettlers who fail to receive promised compensation and development benefits. These corrupting agents have taken many forms.

1 Thayer Scudder is Professor of Anthropology Emeritus, California Institute of Technology.

2 World Commission on Dams (WCD), *Dams and Development: A New Framework for Decision-making* (London: Earthscan Publications, 2000).

3 On impoverishment, see T. Scudder, *The Future of Large Dams: Dealing with Social, Environmental, Institutional and Political Costs* (London: Earthscan Publications, 2005), and C. McDowell (ed.), *Understanding Impoverishment; The Consequences of Development-induced Displacement* (Oxford: Berghahn Books, 1996).

4 Although the World Bank has published more on development-induced involuntary resettlement than other organisations, sections on corruption do not occur; indeed, the word 'corruption' does not occur in the index of

- Mauritanians living downstream from the Manantali Dam suffered from national land registration laws that ignored their customary tenure, making it easier for their valuable property to be forcibly acquired.[5]
- Governments have failed to observe agreed-upon policies designed to benefit resettler households, such as Sri Lanka's Accelerated Mahaweli Project.[6] They have refused to provide required replacement land, as with India's Sardar Sarovar Project.[7] And they have ignored treaty obligations, as with the Lesotho Highlands Water Project.[8]
- Officials have stolen resettlement funds, as with China's Three Gorges Dam.[9]
- Engineering and other firms have reneged on promises or otherwise cheated resettlers, as with India's Maheshwar Dam.[10]
- Government and private individuals have used corrupt practices to acquire choice reservoir sites reserved for resettlers and/or forest, wildlife and other reserves (Lesotho and hydro-projects in Thailand and Kariba, Zambia) as well as other dam-related opportunities, such as fisheries and aquaculture reserved for resettlers (Indonesia's Cirata reservoir).[11]
- Politically influential resettlers can monopolise community- or kin-based land, as with Sardar Sarovar and Kariba.[12]
- Donors are slow in following research-supported best practices that require their resettler safeguard policies to include both compensation and livelihood development.[13] They have been hesitant to hold staff accountable, through reprimands, salary penalties or demotions. Nor have they cracked down on countries that do not comply with safeguard policies, as with the World Bank's involvement in India's Sardar Sarovar Project, and cases brought before the World Bank's Inspection Panel (Argentina and Paraguay's Yacyretá Dam) and IFC's Compliance Adviser/Ombudsman (Chile's Pangue Dam).[14]

Footnote 4 (*cont.*)
the bank's *Involuntary Resettlement: Comparative Perspectives* (2001) or *Involuntary Resettlement Source Book: Planning and Implementation in Development Projects* (2004).

5 M. M. Horowitz, 'Victims upstream and down', *Journal of Refugee Studies*, vol. 4, no. 2 (1991).

6 T. Scudder, 2005.

7 B. Morse and T. Berger, *Sardar Sarovar: The Report of the Independent Review* (Ottawa: Resource Futures International, 1992).

8 T. Scudder, 'Assessing the Impacts of the LHWP on Resettled Households and Other Affected People 1986–2005', in M. L. Thamae and L. Pottinger (eds), *On the Wrong Side of Development: Lessons Learned from the Lesotho Highlands Water Project* (Maseru, Lesotho: Transformation Resource Centre, 2006); 1989–1991 and 1995 reports prepared by the Panel of Environmental Experts for the Lesotho Highlands Development Authority.

9 WCD, 2000.

10 R. E. Bissell *et al.*, 'Maheshwar Hydroelectric Project: Resettlement and Rehabilitation – An Independent Review Conducted for the Ministry of Economic Cooperation and Development (BMZ), Government of Germany', 15 June 2000.

11 T. Scudder, Field Notes on Lesotho, Thailand and Kariba; for Saguling, see B. A. Costa-Pierce, 'Constraints on the Sustainability of Cage Aquaculture for Resettlement from Hydropower Dams in Asia: An Indonesian Case Study,' *Journal of Environment and Development*, vol. 7, no. 4 (1998).

12 T. Scudder, Field Notes on Sardar Sarovar and Kariba.

13 World Bank, 'Recent Experience with Involuntary Resettlement: Overview', Operations Evaluation Department, Report no. 17538 (Washington, DC: World Bank, 1998).

14 Inspection Panel, World Bank Group, 'Argentina: World Bank Board Discusses Yacyreta Hydroelectric Project', press release, 7 May 2004; IFC, 'Assessment by the Office of the Compliance Adviser/Ombudsman in Relation to a Complaint Filed against IFC's Investment in ENDESA Pangue S.A.' (Washington, DC: IFC, 2003).

No easy fix

Competently resettling displaced people is arguably the most complex and contentious job associated with hydro-projects.[15] As the world's leader, with 22,000 large dams, China has been recognised for its efficient resettlement policies. Nonetheless, the Three Gorges Dam's million-plus-person resettlement project gave rise to the largest such corruption scandal on record, with officials stealing ¥375 million (US$50 million).[16] That said, here are some suggested remedies.

- The World Bank correctly states that the first priority is reducing the number of displaced people. Options assessments must include a risk and distributional analysis to limit the construction of large dams with significant resettlement burdens.
- Resettlement should be financed as a separate project – as with the World Bank's Xiaolangdi Project in China – to increase accountability, improve outcomes and deter corruption.[17]
- Performance bonds and insurance relating specifically to resettlers can deter corruption, as can trust funds created specifically for poverty alleviation.[18]
- When resettlement is necessary, resettlers and their institutions should participate in planning, budgeting, implementing and evaluating compensation and livelihood development programmes. This can improve outcomes significantly[19] and, potentially, reduce corruption.
- Displaced citizens should become major stakeholders in benefit-sharing, such as the co-ownership arrangement with Canada's Minashtuk Dam, China's 'remaining problems fund', which stimulates development with hydropower revenues, Brazil's revenue-sharing and Japan's land-leasing.[20]

15 A. Biswas and C. Tortajada, 'Development and Large Dams: A Global Perspective,' *Water Resources Development*, vol. 17, no. 1 (2001); R. Goodland, 'Ethical Priorities in Environmentally Sustainable Energy Systems: The Case of Tropical Hydropower', in W. R. Shea (ed.), *Energy Needs in the Year 2000: Ethical and Environmental Perspectives* (Canton, MA: Watson Publishing International, 1994).

16 See article starting on page 99.

17 World Bank, 'Implementation Completion Report (IDA-26050) for the Xiaolangdi Resettlement Project', Report no. 29174 (Washington, DC: World Bank, 2004).

18 On performance bonds and trust funds, see WCD, 2000. While performance bonds should address the resettlement process directly, trust funds financed from project revenue, as in the Lesotho Highlands Water Project and Laos's Nam Theun 2 Project, focus more on national poverty alleviation. On insurance modelled on workman's compensation, see T. Downing, 'Avoiding New Poverty: Mining-induced Displacement and Resettlement', *Mining, Minerals and Sustainable Development*, no. 58 (2002). On social insurance resettlement in China, which draws resettlers into the social insurance system by providing medical insurance and old age insurance, see Asian Development Bank (ADB), 'Capacity Building for Resettlement Risk Management: People's Republic of China', PRC Thematic Report no. 3, Improving Resettlement Policies and Practice to Manage Impoverishment Risk' (Manila: ADB, 2006).

19 T. Scudder, 2005.

20 D. Egrè *et al.*, 'Benefit Sharing from Dam Projects – Phase 1: Desk Study' (Montreal: Vincent Roquet & Associates, for the World Bank, 2002); M. M. Cernea, 'Financing for Development: Benefit Sharing Mechanisms in Population Resettlement', *Economic and Political Weekly*, vol. 42, no. 12 (2000).

- Resettlement responsibilities and financing should be delegated to resettler communities and institutions, such as the resettler housing and infrastructure projects associated with Uruguay's Itá Dam. Resettler communities should receive help to develop new institutions, such as cooperatives to invest funds for common property resources (Lesotho) and fisheries co-management (Laos's Nam Theun 2).[21]
- Resettlement policies should require funding for both compensation and development, as with Laos's Nam Theun 2.[22]
- International, national and private financing agencies should levy sanctions against staff and offending countries for failing to comply with best practices. These include independent, publicly reported monitoring and evaluation by experts, NGOs and/or private sector firms. This monitoring must be conducted throughout the project cycle, beginning with pre-project benchmark surveys and continuing into the operational phase.
- An International Arbitration and Compliance Board should be formed, in order for stakeholders to file appeals.

21 C. Bermann, 'Community-managed Resettlement: The Case of Itá Dam', submission abstract for the second WCD regional consultation (São Paulo: WCD, 1999); T. Scudder, Field Notes on Lesotho Highlands Water Project and Laos' Nam Theun 2 Dam Project.

22 Nam Theun 2 Hydroelectric Project, 'Social Development Plan', vol. 2 (2005). Compensation alone lends itself to corruption, since it is usually the responsibility of local officials and difficult to monitor as it involves individual households. The utilisation of development funds for entire communities, social infrastructure, and livelihood is more easily monitored.

The disappearance of homes and money: the case of the Three Gorges Dam

Gørild M. Heggelund[1]

When it is finally completed, perhaps by 2009, the Three Gorges Dam will be the largest river-based hydropower project in the world. Stretching more than 2 kilometres across the Yangtze River, China's longest waterway, the dam also led to the largest resettlement project in dam-building history. Originally estimated at 1.13 million, the number of people displaced by the dam reached 1.4 million in 2007. Resettlement expenditures have been estimated at one-third of the total project cost of ¥200 billion (US$26 billion).

The embezzlement of resettlement funds by Chinese government officials has emerged as one of the main hindrances to resettling displaced people. In 2005 dam officials

1 Gørild M. Heggelund is a Senior Research Fellow at the Fridtjof Nansen Institute, Norway.

announced that 349 people had been convicted for misusing resettlement funds since construction began in 1994. By the end of 2003 ¥58.7 million (US$7.1 million) had been embezzled, misappropriated or illegally used. Of that, ¥43 million (US$5.2 million) had been recovered, and all the embezzlers, including 166 officials, had been 'severely punished'.[2]

This endemic corruption has caused numerous problems. Resettlement compensation has been reduced, the quality of life for displaced people has suffered and migrants have protested at the corruption and a lack of adequate compensation, leading to arrests of demonstrators. In July 2006 residents of Hubei Province protested at a local government office because they had received only ¥5,000 (US$700) of the promised ¥38,000 (US$5,000) in up-front 'settlement fees' for having their land expropriated.[3]

Fighting corruption in resettlement: a steep learning curve

Resettlement regulations approved in 1993 decentralised the Three Gorges resettlement authority, placing responsibility at the provincial, county and local levels.[4] While viewed as a positive step towards improving efficiency, decentralisation has also provided opportunities for local governments to engage in mismanagement and corruption.[5] These challenges prompted the authorities to reform their resettlement policies and take additional measures to strengthen governance.

New resettlement regulations the State Council approved in 2001 banned spending resettlement funds on non-resettlement projects or investments, or on purchasing bonds and stocks.[6] Comprehensive accounting and auditing systems were established, management and expenditure operations were separated, and control of resettlement construction projects was strengthened.

Chongqing Municipality established an auditing network in 2001 consisting of a three-step control system called *shiqian, shizhong, shihou* (meaning before, during and after the event is implemented).

2 *China Daily* (China), 29 March 2005.

3 *China Daily* (China), 29 April 2007; *Chinese Sociology and Anthropology*, 'Popular Petitions Protesting Corruption and Embezzlement by Local Governments in the Regions of the Three Gorges Dam Project, 1997 and 1998', vol. 31, no. 3 (1999); *AsiaNews* (Italy), 12 July 2006; K. Haggart, 'Five Years in Wuhan Women's Prison for Requesting Fair Treatment', Three Gorges Probe, 4 October 2005.

4 World Bank, 'Resettlement and Development: The Bankwide Review of Projects Involving Involuntary Resettlement 1986–1993' (Washington, DC: World Bank, 1996); World Bank, 'Recent Experience with Involuntary Resettlement: China – Shuikou', Report no. 17539 (Washington, DC: World Bank, 1998); G. Heggelund, *Environment and Resettlement Politics in China: The Three Gorges Project* (London: Ashgate Publishing, 2004).

5 See an overview of the disbursement system in L. Heming, 'Population Displacement and Resettlement in the Three Gorges Reservoir Area of the Yangtze River Central China', PhD dissertation, University of Leeds, School of Geography, 2000.

6 Decree of the PRC State Council, 'The Resettlement Regulations of the Three Gorges Project', no. 299, Beijing, 25 February 2001.

New management procedures increased the responsibility of resettlement officials, improved the supervision of funding allotments and established regular meetings with local resettlement directors to increase management control over funds.

The control measures have helped uncover additional instances of corruption and misappropriation, indicating that they are working but that corruption risks persist. In January 2007 the National Audit Office reported the misappropriation of ¥272 million (US$36.4 million) out of ¥9.6 billion (US$1.3 billion) in resettlement funds for Hubei Province and Chongqing Municipality for the years 2004 and 2005.[7] The office ordered local authorities to recover the money or else the officials concerned would be 'held responsible'.[8] The Authorities have also introduced a supervision plan and annual financial reports[9] that require various units to report their spending regularly.

Despite these measures, challenges to successful management remain,[10] including a lack of transparency and participation. Potential solutions include establishing clear communication channels between resettlers and the authorities to solve problems when they arise and to strengthen institutions that provide legal assistance in resettlement. According to a survey of more than 1,000 households in eleven provinces, integration problems persist and displaced people are confronted with lower incomes, a lack of basic social security and poor opportunities to voice their complaints.[11]

Three Gorges has been a continuous learning process for fighting corruption in resettlement. The evolving policy responses, if implemented as intended and found to be effective, are potentially very important for the many future dam projects that China plans to undertake. In the Yangtze River basin alone, 105 large dams were planned or under construction in 2007. First and foremost, fighting corruption in dam-related resettlement means minimising the resettlement disruption of livelihoods.

But, when resettlement is necessary, tackling corruption is essential to ensure that displaced people are not punished twice, turning disruption into long-term despair and poverty. Displaced people must be included in post-resettlement capacity building, have more participation in benefit-sharing schemes and be assisted in re-establishing community networks.

7 The Audit Findings on the Funds for Resident Relocation from the Reservoir Region of the Three Gorges Project, National Audit Office of the PRC, no. 1 of 2007, General Serial no. 19. See also W. Jiao, 'Annual Financial Reports for Dam'; *China Daily* (China), 9 September 2007.

8 *China Daily* (China), 11 February 2007.

9 *China Daily* (China), 9 March 2007.

10 *People's Daily* (China), 1 July 2007.

11 P. Fade *et al.*, 'Study on Social Integration and Impact on Stability of Three Gorges Project Re-settlement', available at www.china-yimin.com/show.asp?id=289.

Industry view: public–private hydropower – minimising the corruption risks

Kathy Shandling and Reinier Lock[1]

Building and financing hydropower projects in developing countries requires massive investments and the mobilisation of private capital. A number of mechanisms – some new, some to be scaled up – promise to help fill this funding gap and attract long-term investments to the sector. These include private equity, local commercial bank financing and local bond funding, as well as increased use of guarantee/credit enhancement instruments provided by international financial institutions (IFIs), bilaterals and, in some cases, private sector financial players.[2] But establishing these mechanisms and attracting financiers for hydropower in developing countries presents unique challenges. And risks related to corruption are a central issue.

Learning from failures: aligning expectations and sharing risks in a transparent manner

Recall the 1990s 'gold rush' of billion-dollar independent power projects in Asia – Dabhol in India, Paiton I & II in Indonesia, and Hub River and Uch in Pakistan. All were structured as quasi-public–private partnerships (PPPs). And they all failed, for a variety of reasons. The key problems they shared were a lack of transparency and well-defined contracts between all relevant parties, lack of proper legal and regulatory frameworks, mismatched expectations between the international developers and host governments, and currency exchange disconnects.

Towards 'PPP plus': transparent roles, transparent sharing of risks and regard for social responsibilities

More is needed than conventional PPPs to overcome these problems. A new 'PPP plus' contract should serve as a template to organise viable business partnerships for hydropower projects, in order to address all those issues that contributed to past power project failures. What should a PPP-plus-style contract include?

1 Kathy Shandling is executive director of the International Private Water Association (IPWA) and Reinier Lock is a programme officer at IPWA.

2 In October 2007 the World Bank, for example, announced the launch of a global emerging markets fund to channel more of the estimated US$200 billion invested in emerging markets assets towards local currency bonds that are more suitable to financing long-term infrastructure projects in developing countries; see www.ifc.org/ ifcext/pressroom/ifcpressroom.nsf/PressRelease?openform&2242E8BB6FF5A5AF8525736A0053CA0B.

- It should enhance the ability of both the private and public sector project participants to meet corporate social responsibility and anti-corruption standards.
- It should provide a well-structured compact between public and private players that defines precisely the respective roles of all stakeholders and their relationships to the business, legal, regulatory and institutional regimes within which the project will operate.
- It should be structured to ensure adequate levels of transparency for identifying and allocating the risks that different stakeholders are expected to shoulder.
- It should strike an effective balance between the public and private interests in a specific infrastructure project, meet established social standards and manage the long time frames and related uncertainties typically associated with developing and implementing large hydropower projects.

As yet, PPP-plus implementations are rare, but the idea is gaining momentum. In 2007 the International Bar Association established a 'PPP Task Force' to bring the relevant disciplines together to develop workable PPP models that include a strong emphasis on transparency and corporate social responsibility.

Conditions for success: sound institutional frameworks and community involvement

Investment partnerships cannot exist in an institutional vacuum. Central to all successful public infrastructure projects, including PPPs, is creating comprehensive and effectively implemented legal, regulatory, financial and institutional frameworks.

Community support is also key to reducing investment and corruption risks and making PPP plus successful. Developing 'greenfield' hydropower projects requires gaining local community support for proposed solutions to the specific environmental, economic and social issues that these projects often present, especially if they involve resettlements of communities.

As the 'rural electrification' model demonstrates so well, local community involvement is also a key element in countering the kinds of corruption and inefficiency that have plagued power industries in many developing countries. Local community control of distribution systems can dramatically reduce theft and technical losses, and remove an important obstacle for sustainable private investments to extend electricity service to previously unserved, often rural, areas. Moreover, failure to garner adequate community support to counter corruption sufficiently early can seriously delay a new project's development, undermine its revenue stream and investment sources and threaten its basic economics and potential for expansion.

Grand projects – grand corruption?

Peter Bosshard and Nicholas Hildyard[1]

In nature, water always flows downstream. In the geography of power relations, clean water tends to flow to the rich and powerful, while wastewater tends to flow to the poor. An important reason for this dynamic is corruption, which has contributed to a political economy that favours large, capital-intensive projects over small-scale approaches.

In recent years, institutions such as the United Nations Development Programme and the UN Millennium Project have advocated reassessing large-scale water infrastructure projects and focusing more on decentralised projects and efficiency improvements to better meet the needs of poor people.

'From India to Bolivia, Kenya to Nepal can be found the ruins of now-defunct water and sanitation programmes that have never yielded more than a fraction of the benefits expected,' the Water Supply and Sanitation Collaborative Council (WSCC) warned in 2004. 'Increasing the funds available for further large-scale, delivery-oriented infrastructure will achieve very little without a re-think of how and for whom such funds are to be spent.'[2]

Even the World Bank has changed its tune about gigantic hydro-projects that displace entire communities and alter landscapes forever. 'The environmental and social consequences of these dams will continue to be contested,' it said in 2006, 'and it is likely that nations will construct relatively few of them.' Instead, the World Bank sees a brighter future for small dams, because they raise fewer social and environmental concerns.[3]

In Pakistan, the World Bank has found that renovating watercourses may be a cheaper way to expand irrigation than new large dams.[4] In spite of this, the country's water bureaucracy has suffered from a 'build-neglect-rebuild' syndrome and prioritised new investments over maintaining existing infrastructure.[5]

Maximising opportunities for corruption is a key factor that creates a bias towards large greenfield investments in the water sector.

- Large new investments award more political prestige and afford more centralised bureau-

1 Peter Bosshard is the policy director of International Rivers, an environmental and human rights organisation in Berkeley, California. Nicholas Hildyard works with the Corner House, a research and advocacy group focusing on human rights, environment and development based in the United Kingdom.

2 '"Listening" Blasts International Community's Failure on Water and Sanitation', WSCC press release, Geneva, 17 March 2004.

3 World Bank, 'Reengaging in Agricultural Water Management: Challenges and Options' (Washington, DC: World Bank, 2006).

4 World Bank, 'Irrigation Investment in Pakistan', Operation Evaluation Department, Précis no. 24 (Washington, DC: World Bank, 1996).

5 World Bank, 'Pakistan's Water Economy: Running Dry', draft, 23 June 2005.

cratic control than decentralised schemes and efficiency improvements, in which control and resource flows are more dispersed.

- Corruption favours large-scale, capital-intensive projects because they are more likely to involve and benefit actors with deep pockets.
- Illicit payments made as part of large international projects can be funnelled into foreign bank accounts, which corrupt officials may consider safer than bribes for local projects because they tend to remain within the local economy.

In sum, corruption is an important factor that influences how vested interests capture government decisions on the type and size of infrastructure projects. The World Commission on Dams arrived at the same conclusion, and noted in its 2000 report: 'Decision-makers may be inclined to favour large infrastructure as they provide opportunities for personal enrichment not afforded by smaller or more diffuse alternatives.'[6]

It is important to note, however, that local investment projects are by no means free of corruption. As Dipak Gyawali, a former Minister for Water Resources in Nepal, points out, 'Corruption affects all projects, small, medium and large,' and government-sponsored projects as well as projects implemented by non-governmental organisations.[7] In order to maintain power, a corrupt government apparatus will tend to offer spoils to bureaucrats and power brokers at the local, regional and central levels.[8] And local patronage systems have been found to divert money successfully from village-level infrastructure projects.[9]

The projects that offer the fewest rents to be captured by higher-level decision-makers are labour-intensive self-help initiatives. And these are precisely the types of approaches that have the largest potential to reduce poverty.

The implications are twofold: safeguards against corruption may differ with project size, but need to be built into water projects of all scales. At the same time, higher-level decision-makers can be expected to favour larger-scale projects that offer them more favourable opportunities to extract corruption rents for their own clientele. This behaviour requires additional safeguards. Transparency and public participation in the planning process for water sector projects, including the assessment of available options at an early stage, are needed to counter this corruption-driven bias towards larger projects.

6 World Commission on Dams, *Dams and Development: A New Framework for Decision-making* (London: Earthscan Publications, 2000).

7 Interview by Nicholas Hildyard, May 2007.

8 Interview with Shekhar Singh, convenor of India's National Campaign for People's Right to Information and a former adviser to the country's Planning Commission, 7 June 2007.

9 Chapter 3 documents a variety of such cases. An analysis of an Indonesian village development programme found that more than a third of almost 2,000 complaints were related to misuse of funds. See S. Wong, 'Indonesia Kecamatan Development Program: Building a Monitoring and Evaluation System for a Large-scale Community-driven Development Program', discussion paper (Washington, DC: World Bank, 2003).

6 Conclusions

Fighting corruption in water: strategies, tools and ways forward

Donal T. O'Leary and Patrik Stålgren[1]

Corruption is draining the water sector. It is distorting the allocation of precious and scarce resources – economic, environmental and social. It is hindering the sector's potential to serve as a catalyst for national development and, instead, has made water the source of its stagnation. Reducing these costs and realising the sector's range of developmental possibilities will require all actors to prioritise actions that can stem corruption. Without changes in the way corruption is prevented and punished, the global promises set out in the Millennium Declaration for improving water and sanitation, for the betterment of people's lives around the world, will be left unfulfilled.

This report has documented different types of corruption in the water sector and the challenges they pose: whether for the operation of a city's water supply network, the construction of irrigation canals for rural farmers or the allocation of land and contracts for big-money dam projects. As signalled in each of the previous chapters, the evidence is conclusive that the costs of corruption are enormous for the sector. They are unequally distributed and disproportionately borne by the poor. Vulnerabilities – due to gender, age or ethnicity, or all of the above – are reinforced and aggravated when the control of water is corrupted. Ecosystems are imperilled and the problems of one country multiply into the challenges for many.

Corruption remains one of the least analysed and recognised problems in the water sector, however. This report provides a first step in filling this gap and understanding why corruption has been able to take root. Each of the previous chapters maps the corruption risks for one specific area of the sector: water resources management, drinking water and sanitation, irrigation and hydropower.

Water resources management is about safeguarding the sustainability of a resource that has no substitute. It involves the most fundamental policy decisions: how to protect water, ensure its positive contribution to the environment and balance the demands for its different uses (e.g.

1 Donal T. O'Leary is a senior adviser to Transparency International (TI). Patrik Stålgren is a researcher at the Department of Political Science, Göteborg University.

human consumption, agriculture, industry and power generation). Around the world, a large gap between water supply and demand has arisen due to population and economic growth, urbanisation, changing dietary habits and the onset of climate change.

Chapter 2 of the *Global Corruption Report 2008* has analysed WRM in detail, showing how local water scarcities and intensified competition for water provide a breeding ground for corruption. In some instances, water subsidies have been hijacked by powerful elites, water pollution has gone unpunished due to bribery and funds for WRM have ended up in the pockets of corrupt officials. In the short run, the losers in this control contest are typically the marginalised, who are denied access to a vital resource for life. In the long run, corruption in WRM paves the way for overexploitation of water resources and unchecked pollution, as well as inefficient distribution and allocation between different uses.

The consequences of corruption are significant for environmental sustainability, the future security of the water supply, social cohesion and even the stability of certain regions. The damage leaves lasting scars on future generations and the environment. Since many of its victims are silent, increased accountability in WRM is difficult to achieve. As yet, government oversight mechanisms are not in place to ensure that it will be provided. The lack of administrative capacity and the division of institutional responsibilities among different agencies within a country and internationally has left the sector in a regulatory lacuna that makes the fight against corruption very difficult.

Nowhere is the crisis of water governance and the challenge for human development more evident than in the areas of *drinking water and sanitation*. Roughly 1.2 billion people do not have access to safe drinking water and more than 2.6 billion people lack adequate sanitation. On any given day, nearly 50 per cent of people living in the developing world suffer from health problems caused by poor water and sanitation.[2] Without water – safe water – the health, livelihoods and development of individuals and countries are undermined.

As chapter 3 of this volume has shown, corruption intensifies these negative impacts and can be found at every point along the water delivery chain: from policy design and budget allocations to operations and billing systems. It drains much-needed investment from the sector and distorts prices and decisions. Corruption affects both private and public water services and hurts developing and developed countries alike. According to some estimates for developing countries, corruption raises the price for connecting a household into a water network by up to 45 per cent. It leads to policies and projects that favour the middle and upper classes and leaves the poor with limited choices and high prices for water access, making them even more vulnerable to corruption.

In chapter 4, the *Global Corruption Report 2008* details how corruption plays a role in the world's *irrigation and agriculture*. Agriculture accounts for 70 per cent of water consumption

---

2 These figures are based on estimates by the United Nations Development Programme: see UNDP, *Human Development Report 2006. Beyond Scarcity: Power, Poverty and the Global Water Crisis* (New York: Palgrave Macmillan, 2006).

and irrigated land helps produce 40 per cent of the world's food. Without the irrigation of fields, many farmers throughout the world would not be able to practise their livelihoods and would be left in poverty. Corruption can put irrigation systems under the capture of large users. And irrigation systems that are difficult to monitor and require experts for their maintenance offer multiple entry points for corruption, leading to wasted funding and more expensive and uncertain irrigation for small farmers. Irrigation with groundwater resources that thousands of private pumps extract from underground aquifers is even more difficult to regulate. As a result, large users in places such as India and Mexico can drain underground aquifers with impunity, depriving smallholders of essential resources for their livelihoods. All this means that corruption in irrigation exacerbates food insecurity and poverty.

Hydropower is another water sector vulnerable to corrupt practices. More than 45,000 large dams in 140 countries supply more than 16 per cent of the world's electricity and provide vital services for flood control, irrigation and navigation. Chapter 5 of the *Global Corruption Report 2008* has demonstrated that dam-building has its own set of challenges – both for corruption and development. Massive investment volumes (US$50–60 billion annually over the coming decades) and highly complex, customised engineering projects attract corruption to the design, tendering and execution of large-scale dam projects. The impact of corruption is not confined just to inflating project costs. Undue influence on energy policies and dam design by those who benefit from large-scale construction and the alteration of water flows can have dramatic consequences for entire communities. Few other public works projects have a comparable impact on the environment and people, making accountable hydropower governance a prerequisite for equitable human development. Large resettlement funds and compensation programmes that accompany dam projects have also been found to be vulnerable to corruption, adding to the challenges faced in the hydropower sector.

Policy lessons for combating corruption in the water sector

The *Global Corruption Report 2008* demonstrates that increased demand for water (whether for drinking, irrigation or energy) can be managed effectively only when dynamics of power and control are adequately addressed. Responses must tackle a wide range of corruption risks and devise ways to ensure that abuses of power do not go undetected and unpunished. The previous chapters in this section of the report review a wealth of case studies and experiences that yield a set of key lessons, as follows.

Prevent corruption in the water sector early whenever possible; cleaning up after it is difficult and expensive. When corruption leads to contaminated drinking water and destroyed ecosystems, the detrimental consequences are often irreversible. When subsidised water gives rise to powerful agricultural industries and lobbies, refocusing subsidies on the poor becomes increasingly more difficult. Once stakeholders engage in illicit activities to access or control water resources, they are further drawn into corruption networks, as is evident in Bangladesh or Ecuador, where water mafias operate corruption rackets.

Understand the local water context; otherwise reforms will fail. One size never fits all in fighting corruption, but this is particularly the case in the water sector, where conditions of supply and demand, existing infrastructures and governance systems vary widely across countries. Before tackling corruption, it is necessary to create an understanding of the specific dynamics that create and sustain the local governance arrangements for the water sector. Every reform measure must be based on a thorough stakeholder assessment that looks at the strengths and interests of incumbent elites, as well as the preferences and specific needs of the poor and other intended beneficiaries.[3] Analytical methods need to be tailored to the local context and can include: surveying the concerns and current status of water users and providers, mapping corruption risks for related institutions and developing baselines and indicators to monitor progress (in access, service and water quality).

Cleaning up corruption should not be at odds with the needs of the poor or the sustainability of the environment. The costs of corruption in the water sector are disproportionately borne by the poor and exacted on the environment. To combat corruption, responses should engage communities in defining solutions and monitoring the outcomes.[4] Inspiring examples in countries such as Brazil (see page 50) show how anti-corruption strategies have been successful when they have worked to involve poor citizens in budgeting and spending reviews.

Other examples point to the risk that some anti-corruption strategies pose when they are badly designed, however. Rather than supporting communities and positive change, they may undercut peoples' basic livelihoods. Chapter 3 highlights how government crackdowns on informal water providers can have negative fallout for the access to water of the poor. Before taking action in an area such as water provision, it is necessary to assess the local context and understand how the poor get their water and how much they are able and willing to pay. This information can be used to focus anti-corruption work on the types of service provision that matter most to them, such as constructing public standpipes or drilling wells in rural areas.

Linking up anti-corruption reform in the water sector – locally, nationally and beyond national borders – is essential to success. Beware of the weakest link: only coordinated and comprehensive reforms will have lasting benefits. Successful measures may stamp out corruption in one place only for it to reappear in others that may be harder to detect and deter. As chapter 4 in this volume shows, for instance, new water user associations – formed to prevent powerful farmers from bribing public officials to capture irrigation resources – can fall prey to the same interests they were set up to control. Similarly, reforms that successfully prevent local contractors from embezzling money may be unsuccessful in ensuring that most of the project funding does not end up in the pockets of national politicians. Corruption in water is dynamic and reforms must be interrelated to reflect its changing

3 P. Stålgren, 'Worlds of Water: Worlds Apart. How Targeted Domestic Actors Transform International Regimes' (Göteborg: Göteborg University, 2006); J. Plummer, 'Making Anti-corruption Approaches Work for the Poor: Issues for Consideration in the Development of Pro-poor Anti-Corruption Strategies in Water Services and Irrigation', Report no. 22 (Stockholm: Swedish Water House, 2007).

4 J. Plummer, 2007.

nature. This calls for coordination of anti-corruption efforts upstream and downstream in the sector and the need to ensure that they complement related initiatives locally, nationally and globally.[5]

Work on reforms that directly and indirectly combat corruption in the water sector. When corruption takes on systemic proportions, tackling it head-on can be difficult.[6] Many examples throughout this report underscore the fact that corruption in the water sector is intertwined with generic governance failures and dysfunctional public institutions. To begin addressing all these different dynamics, one option might be to start with a more indirect approach that involves a general reform of institutions and promotion of broader citizen engagement. Such initiatives can include technical reforms targeting increased water service delivery and citizen empowerment projects that focus on capacity-building and transparency. Other reform areas that are central for anti-corruption efforts include improving financial management, training civil servants and capacity-building for agency administrators.

Build awareness among stakeholders that creates common ground and mobilises coalitions. Ending corruption in the water sector requires overcoming overlapping interests and altering 'the rules of the game'. There needs to be 'buy-in' by the different groups involved to break the pattern and relationships that are perpetuating the problem. This is particularly difficult in the water sector, however, where the number and diversity of stakeholders is exceptionally high. The *Global Corruption Report 2008* has profiled how fighting corruption in water is in the interests of many different stakeholders – but this common purpose may not always be clear at the outset to everyone involved.

Based on experiences from water resources management in Southern Africa, differences in incentives and perceptions can be overcome through effective communication and mutual learning between stakeholders.[7] Farmers, for example, may see water simply as an input to producing their harvests. They may not make the link that the environment and climate affect the availability of water and may be uninterested in partnering with stakeholders working on these issues. Encouraging collaboration between the groups will rely on building an understanding of how protecting water for farming means protecting the environment. Haas (chapter 5) points out that effective anti-corruption approaches typically follow this formula and build on mutually reinforcing efforts by the public, private and civil society sectors. The Water Integrity Network, a group of international water experts and practitioners dedicated to fighting corruption in the sector, has been involved in striking up such partnerships and provides a good resource base for countries to share good practices.

Build pressure for water reform from above and below. It is also necessary to reconcile top-down and bottom-up approaches. Political leadership from the top is necessary to create momentum and legitimacy to drive institutional reforms. A good example is the case study on how

5 J. Plummer and P. Cross, 'Tackling Corruption in the water and Sanitation Sector in Africa: Starting the Dialogue', in E. Campos and S. Pradhan (eds), *The Many Faces of Corruption* (Washington, DC: World Bank, 2007).

6 A. Shah and M. Schacter, 'Look before You Leap', *Finance and Development*, vol. 41, no. 4 (2004).

7 P. Stålgren, 2006.

committed leadership helped turn around the public water utility of Phnom Penh (page 48). But this is only one side of the coin. Bottom-up approaches are important to add checks and balances. They help monitor flows of money (e.g. social audits of infrastructure projects – page 51) and water (e.g. the creation of irrigation user associations – from page 75), benchmark performance (e.g. report cards for water users – page 51) and disclose failure (e.g. water pollution mapping – page 27). Relying on grassroots support helps make corruption and policy capture at all levels more difficult.

Sequence anti-corruption reforms and responses to ensure that recommended actions have been appropriately tailored to the context. The general school of thought on how to combat corruption in water is that certain measures can prove extremely effective: user associations, citizen report cards, legal entitlements to access and community-managed irrigation programmes, among others. Each of these will have to be tailored to the needs of users and the specific characteristics of corruption in the community. But adapting anti-corruption policies to local contexts also entails rethinking the sequencing of reforms. For example, privatising a city's water services requires having a strong regulator in place to prevent and manage corruption at every step in the process. Establishing water rights for citizens will be successful only if effective judicial institutions exist to uphold the laws. Pushing transparency and civil society involvement without developing matching capabilities or creating the space for their engagement threatens to create public cynicism or apathy about anti-corruption initiatives.

Leverage existing commitments to make water governance more accountable; there is no need to reinvent the wheel. Chapter after chapter in this report lists existing legal frameworks, conventions and declarations that outline the responsibilities of governments and other stakeholders on managing water resources and addressing corruption. They cover everything from respecting transboundary waters and environmental sustainability to guaranteeing drinking water, access to environmental information and corruption-free practices. Both the United Nations Convention Against Corruption (UNCAC) and the OECD Anti-bribery Convention – as well as various regional agreements – contain articles that clearly stipulate the obligation of signatories to prevent and punish many of the abuses that currently plague the water sector. If they are serious about turning pledges into concrete commitments, governments can find ready-made templates in these and other frameworks to tailor and use. Several governments have already ratified similar agreements. Civil society can leverage international pressure to encourage the country in question to adopt the same measures and honour the many elements in these frameworks that are useful for rolling back corruption in the water sector, including participatory structures for governing and sharing water, access to water-related information, the transparent procurement of water services and measures to protect wetlands and water resources.

Taking action: recommendations for tackling corruption in water

The *Global Corruption Report 2008* has presented a number of promising strategies and tools to tackle corruption in water resources management, drinking water and sanitation, irrigation and hydropower. As has been emphasised throughout the report, a particular country's dynamics determines the right mix and sequence of anti-corruption reforms. The following recommendations summarise the most promising strands of reform. If implemented, they should foster changes in the current context of corruption in the water sector.

Scale up and refine the diagnosis of corruption in water; the momentum and effectiveness of reform depend on it

Much work remains to be done on studying the scope and nature of corruption so as to allow a deeper understanding of its drivers. Such knowledge is needed to tailor anti-corruption responses to specific contexts and determine how best to prioritise resource spending, sequence interventions and monitor progress. Tools such as corruption impact assessments, public expenditure tracking or poverty and corruption risk-mapping help to shed valuable light on different aspects of the puzzle. These tools need to be refined, adopted widely and adapted to specific local contexts to lay the foundations for targeted reform.

One promising diagnostic tool for sketching an overview of the problem is a **water integrity national survey (WINS)**. This survey can cover all the components, actors, practices and institutions that make up the water sector and can be used to help to capture the issues affecting performance. In addition, the conclusions and recommendations of tools such as the WINS could be used by governments in developing time-bound, monitorable action plans with concrete indicators. To help secure buy-in to its recommendations, the WINS should be carried out by an independent reputable organisation or a group of organisations (such as a university or a research centre) skilled in both water sector and governance issues. As experience with similar studies shows, the resulting analysis can serve as a starting point to prioritise, strategise and promote reform.[8]

Strengthen the regulatory oversight of water management and use

Governments and the public sector continue to play the most prominent role in water governance. As the entrusted executors of citizen will, they are responsible for the allocation of water resources, protecting the environment, representing the interests of future generations and overseeing the different dimensions of the sector. They are empowered to negotiate transboundary water-sharing, set sectoral policies and manage investments. Governments are also the principal shareholders that own and oversee the infrastructures in place for a country's drinking water, sanitation, irrigation and hydropower needs.

Governments' broad authority on matters of water must be leveraged as part of any strategy to tackle corruption. A central task for states is to establish effective regulatory oversight, whether for the environment, water and sanitation, agriculture or energy. In the age of public–private partnerships, regulators must take on additional roles and ensure that ventures are transparent,

---

8 See, for example, the National Integrity System country study for Bosnia-Herzegovina (BH), carried out by TI BH and adopted by the government of BH as the model for its national anti-corruption plan; available at www.transparency.org/content/download/15693/169907/file/NIS_bosnia_herzegov.pdf.

particularly in relation to power purchase agreements. Where relevant, regulators also need to pay special attention to addressing potential corruption risks deriving from decentralisation.

But setting up regulatory mechanisms presents a dilemma: in a high-corruption environment, regulatory bodies are likely to fall prey to capture and face multiple conflicts of interest, especially when a government department assumes the roles of water service provider and regulator at the same time. If the means to combat corruption also become the mechanism that spreads it, countries are left with the conundrum of figuring out where to start. There are institutional reforms that can make regulatory capture less likely and therefore should be prioritised: capacity-building and training for regulatory staff, adequate resources (human, financial, technical and administrative), creating a clear institutional mandate and power, transparent operating principles and a public consultation and appeals process. In addition, existing benchmarking tools such as the International Benchmarking Network for Water Utilities (IBNET) can assist regulators in fulfilling their mandate.[9]

There are global examples of regulatory and administrative authorities that have been able to establish the oversight, insight and integrity needed to counteract corruption in water. World-class organisations such as the Public Utilities Board (PUB) of Singapore and the Panama Canal Authority (or ACP in Spanish) have taken active measures to inculcate a culture of integrity within their organisations. For example, the PUB has developed codes of governance and conduct, set up effective internal control processes and established mechanisms to prevent and punish corruption. The ACP also promotes integrity and oversight through regulations that it has passed regarding staff ethics and behaviour. These codes deal with conflicts of interest, abuse of position and acceptance of gifts. As these examples show, strengthening regulatory oversight requires a focus on two interrelated objectives: it means putting in place the mechanisms that strengthen the *mandate* and *independence* of the regulator and at the same time establishing internal structures and incentive systems that ensure the *integrity and accountability* of its employees.

Improve the management of water utilities to reduce corruption and help deliver in the water and sanitation sector

Water utilities play an important role in delivering water and sanitation services. To lower corruption risks, water utilities should be autonomous, financially viable, well staffed and accountable for performance and delivery.[10] They can improve service delivery to the poor and directly combat corruption by subsidising connection fees and tariffs for low-income households, setting up inspection teams to find leaks and illegal connections, reducing the manipulation of billing and collection through installing meters for all connections, computerising billing systems and maintaining an up-to-date customer database. Management

9 International Benchmarking Network for Water Utilities; see www.ib-net.org.

10 H. Elshorst and D. O'Leary, 'Corruption in the Water Sector: Opportunities for Addressing a Pervasive Problem', presentation at World Water Week, Stockholm, August 2005; A. Baeitti *et al.*, 'Characteristics of Well-performing Public Water Utilities', Water Supply and Sanitation Working Note no. 9 (Washington, DC: World Bank, 2006).

contracts and performance-based service contracts can help utilities significantly improve performance and reduce 'petty' corruption. This needs to be supported by strong political will and determined leadership by top management from the utility. The experience of the Phnom Penh Water Supply Authority described in chapter 3 shows that this can be done.

Ensure fair competition for and accountable implementation of water contracts

Contractual agreements are used when the government bids out parts of its water service responsibilities to the private sector. These can include the expansion and running of a city's water supply, the construction of a rural irrigation system or the management of a country's hydropower dams. Designing, tendering and monitoring such contracts comes with major corruption risks.

In some countries, the private sector has embraced basic anti-corruption measures as part of its standard operating procedures, often within the rubric of strengthened corporate governance practices. These tend to focus on promoting sound financial management, regular company reporting, effective internal performance monitoring and other initiatives to account to investors and shareholders, as well as to stakeholders. TI's **Business Principles for Countering Bribery**,[11] for example, can offer guidance and benchmarks specifically for corporate anti-bribery programmes.

While private enterprises in the water sector may enforce a level of compliance that assists anti-corruption efforts, additional actions are necessary, often by government, to address the areas that fall outside their control. The urgent need for action is inspired by the fact that future business opportunities are expected to be concentrated in corruption-plagued countries. Nine out of the ten largest growth markets for private sector involvement in water services are in nations that score below 3.8 on a scale between 0 (highly corrupt) and 10 (clean) on TI's Corruption Perceptions Index, marking them as countries with high levels of corruption.[12]

To help foster clean contracts and fair competition, different tools exist that rely on promoting stakeholder collaboration and buy-in. Since the mid-1990s TI has been using **integrity pacts**. These pacts are typically developed for public procurement processes and include a signed promise between the government and all interested bidders that neither side will offer, demand or accept bribes during the bidding and execution of contracts. IPs have been applied successfully in many countries and sectors.[13] In Pakistan, an IP that was used as part of the Greater Karachi Water Supply Scheme led to an 18 per cent reduction in costs compared to the original estimates.[14] In Mexico, a similar pact for a hydropower project

---

11 See www.transparency.org/global_priorities/private_sector/business_principles.

12 Global Water Intelligence, *Global Water Market 2008: Opportunities in Scarcity and Environmental Regulation* (Oxford: Global Water Intelligence, 2007); TI, Corruption Perceptions Index 2007, in *Global Corruption Report 2007* (Cambridge: Cambridge University Press, 2007).

13 TI, *Curbing Corruption in Public Procurement Handbook* (Berlin: TI, 2006).

14 TI Pakistan, 'A Pakistan Success Story: Application of an Integrity Pact to the Greater Karachi Water Supply Scheme, Phase V, Stage II, 2nd 100 MGD, KIII Project' (Karachi: TI Pakistan, 2003).

helped achieve a saving of more than 8 per cent. An IP can also be signed for an entire sector. In Argentina, water pipe manufacturers – accounting for 80 per cent of the market – struck an agreement based on the IP principles to ensure fair bidding for public contracts.[15] Stiff fines for bribe-takers and strong rules for debarring bribe-givers can further reduce incentives for corruption. The publication of performance criteria and contract terms is another indispensable measure for public trust and public oversight, but it is not yet common practice in many countries.

Mainstream due diligence in the financing of private sector water projects

Corrupt practices in the form of bribery abroad underscore the need for export credit agencies, commercial banks, international financial institutions and donors to take action as part of their fiduciary responsibilities.[16] When supporting investments, including processes that involve procurement, they must ensure that mechanisms are put in place that create the right incentives – to discourage firms from engaging in corrupt activities.

ECAs, commercial banks and the private sector lending wings of IFIs, such as the World Bank's International Finance Corporation, should expand their due diligence requirements to include anti-bribery provisions. These measures can apply to each interested developer and should cover the entirety of a company's global operations.

Prior to making an application for funding or a guarantee, the applicant(s) should be required to disclose if they are under investigation, have ever been convicted of violating anti-corruption laws (such as the United States' Foreign Corrupt Practices Act) or have been debarred by any IFI.[17] For example, the Overseas Private Investment Corporation (OPIC), an ECA based in Washington, DC, requires companies seeking OPIC financing or 'cover' for a project to have anti-bribery programmes in place, such as TI's BPCB.

Donors can also contribute in important ways to promoting the right incentives and signals for private companies interested in doing business with them. They can strengthen the anti-corruption components of water projects and support initiatives that promote civil society capacity-building and media development. Such measures will help to put in place the institutional building blocks necessary to create an environment that fosters greater accountability. Internally, donors can take steps to improve their own accountability by strengthening public disclosure practices and penalties for misdoings. Specific measures include: public consultations of project documents, stiffer sanctions against corrupt staff, the blacklisting of corrupt project partners and an unambiguous and coordinated no-bribes commitment by the donor community.

---

15 L. Haas *et al.*, 'Setting Standards for Communications & Governance: The Example of Infrastructure Projects', Working Paper no. 121 (Washington, DC: World Bank, 2007).

16 P. Stålgren, 'Corruption in the Water Sector: Causes, Consequences and Potential Reform', Policy Brief no. 4 (Stockholm: Swedish Water House, 2006).

17 TI, 'Using the Right to Information as an Anti-Corruption Tool' (Berlin: TI, 2006).

Step up citizen monitoring of water service delivery: civil society has a pivotal task to complete the accountability circle

When it comes to combating corruption in the water sector, civil society organisations (CSOs) can help monitor the commitment and effectiveness of government and private contractors at all levels. CSOs have the capacity to mobilise communities and express their demands for change. **Citizen report cards** are an example of a community-level monitoring tool that helps to channel community needs into action. Report cards are survey questionnaires that citizens complete to assess the quality of service delivery and whether service providers have fulfilled their obligations (in terms of budgets, resources and promises).

In the water sector, the tool has proved very successful for getting users to interact directly with utilities and air their concerns. The experiences of Bangalore in India, where citizen report cards were first adopted in 1994, are impressive: since the surveys began, the percentage of people 'satisfied' with water and sanitation services has skyrocketed from 4 per cent to 73 per cent (2003).[18]

Monitoring the satisfaction with water services is not an add-on measure with populist appeal. It is important, because it makes the water provider more directly answerable to the citizenry. It shifts attention to outputs and outcomes, turns individual dissatisfaction into public pressure and thereby complements the recommendations that focus on accountability for inputs (budgets, staffing) and integrity of processes (fair tendering and effective regulation) outlined earlier.

Adopt transparency and participation as guiding principles for all water governance

Adding up the elements needed to tackle corruption in the water sector, two central elements stand out: transparency and participation.

Transparency must come to characterise how both public and private stakeholders conduct water sector activities. Water budgets, resettlement funds and the rules of procurement need to be carried out in a transparent manner and disseminated to the public. Measures must be put in place requiring public officials and sector managers to disclose their assets publicly as a means to ensure that resources are not being siphoned off from the sector and into their bank accounts. The public shaming of water polluters and debarred contractors should be encouraged as a way to add a social cost to any legal and financial penalties incurred.

Transparency is also encouraged by more research and information-sharing. Analysis is needed to show who the major beneficiaries are of subsidies targeting rural wells, irrigation networks and drinking water systems.[19] Tendered bids should be read aloud in community meetings,

18 G. K. Thampi, 'Community Voice as an Aid to Accountability: Experiences with Citizen Report Cards in Bangalore', presentation at the seminar 'Can We Meet International Targets without Fighting Corruption?', World Water Week, Stockholm, August 2005.

19 See articles starting on pages 40 and 67.

planning blueprints publicly posted, donor documents and water quality indicators uploaded to websites, and materials produced in a simple and accessible language – from service contracts to audit reports. All these measures should help to shift behaviour in the sector and create an environment in which transparency is expected and valued. Even when projects are technical or the matters require expertise, citizens should have the opportunity and voice to demand basic information and explanations (e.g. about infrastructure specifications, experts hired, contractors selected and prices set). Strong freedom of information (FOI) laws that create enforceable entitlements for citizens to inspect public records provide the foundation for transparency in the water sector.

Increased participation has been documented throughout this report as a mechanism for reducing undue influence and capture of the sector. When effective, citizen engagement forces public and private sector counterparts to be more transparent and accountable in their actions. Participation by marginalised and vulnerable groups in water budgeting and policy development can provide a means for adding a pro-poor focus to spending. Community involvement in selecting the sites of rural wells and managing irrigation systems helps to make certain that small landholders and poor villagers are not last in line when it comes to accessing water. Engagement in infrastructure planning or environmental impact assessments gives civil society stakeholders a platform for holding decision-makers accountable for extending the benefits of new water mains or dams to everyone. Participation in auditing, environmental pollution-mapping and performance-monitoring of water utilities creates a system of checks and balances to see whether contracts have been fulfilled and violators of water regulations punished.

Transparency and participation build the very trust and confidence that accountable water governance demands. They are essential elements for keeping the lure of corruption low and the system functional. Transparency and participation help to reassure small landholders and poor people that they are heard and need not bribe to get their fair share of water. Private companies are given greater confidence that they do not have to sweeten their bids for water contracts. Industry is reassured that competitors are not gaining an unfair advantage by bribing their way around environmental rules. Neighbouring countries are provided with assurances that water-sharing arrangements will not be violated.

Of course, transparency and participation are no magic cure. They work in tandem with other measures, such as clear legal entitlements to water and strict sanctions against corrupt behaviour. They depend on having the necessary capacity in place to use the information made available and participate effectively in decision-making.

These challenges notwithstanding, transparency and participation are prerequisites for ensuring that water governance is less corrupt and more accountable, democratic and equitable. They are indispensable elements for tackling corruption in the context of the global water crisis today. And they are important principles for reforming governance frameworks and laying the foundation for anti-corruption strategies in the future.

A critical crossroads has been reached that mandates a radical shift in the status quo of how water and corruption are addressed. Climate change, the search for fossil fuel alternatives, the expansion of commercial agriculture and continuing demographic trends (in terms of lifestyles, urbanisation and population growth) have made the need for a response urgent. The stakes in the global water crisis could hardly be higher. The lives and livelihoods of billions of people, the sustainability of our ecosystems and energy footprint, the prospects for equitable human development and international political stability are all interlinked with solving the global water crisis. Fighting corruption in water is an important dimension of working towards a solution. As the *Global Corruption Report 2008* shows, this fight against corruption in water is very challenging, but it is feasible and rewarding and it is more urgent than ever.

Part two

Country reports

7 Corruption through a national lens

Introduction

Rebecca Dobson[1]

Transparency International's fight against corruption is carried forward by a truly global movement of national chapters and contact groups in all regions of the world. This section of the *Global Corruption Report 2008* draws on this unique breadth and diversity of in-country expertise and experience. Importantly, it includes views from both developed and developing countries.

The thirty-five contributions that follow offer a glimpse of major corruption-related events and a review of progress in institutional anti-corruption reforms during the reporting period from July 2006 to July 2007, as seen at national and local levels.[2] In so doing, the reports provide a sense of the corruption issues that are most prevalent and of common concern across countries, from political corruption to corruption in the water sector. A few of the main themes that emerge across the country reports are as follows.

Corruption in both politics and the judiciary have appeared as recurring concerns in reports from all regions. Political corruption is revealed in relation to public procurement, access to information and, in particular, around elections and political financing. While there have been attempts to improve political integrity, in some cases this has led to the exploitation of new loopholes. This is particularly the case in Armenia, Latvia, Kenya, the United Kingdom and Austria, where, despite legislative revisions in relation to elections and party financing, corruption either persists or has re-emerged in new forms. In Latvia, for example, amendments made to party financing rules were circumvented by unregulated third parties that campaigned on behalf of leading political parties. Questionable practices of party financing also feature in the 'loans for peerages' scandal reported from the United Kingdom, where some of the individuals who made undeclared loans to political parties were later nominated for titles of nobility.

Improvements in the judiciary are evident in some countries, notably in Mexico, with the introduction of oral trials, and in India, where the Supreme Court continues to be outspoken about corruption. Establishing an independent and accountable judiciary still presents great

1 Rebecca Dobson is the contributing editor to the *Global Corruption Report*.

2 Each of the country reports begins with the country's ranking in the Corruption Perceptions Index 2007 and a list of anti-corruption conventions signed and ratified by that country. The reports then focus on key corruption issues in each country during the period under review.

challenges, however. The insecurity of tenure for judges in Argentina and Senegal or the inadequacy of funding for the judiciary reported from Sierra Leone highlight some of the most basic challenges to judicial independence. As reports from Romania and Bangladesh indicate, judicial reforms are neither simple nor always effective. In Romania, for example, significant steps to reform the judiciary have taken place, but conflicts of interest persist and fewer than a half of the magistrates believe that the newly empowered Superior Council of the Magistracy can effectively ensure their independence.

The *international reach of corruption* is another central theme that emerges from the country reports. Several contributions present incidences of corruption that play out at national or local level, but also have an important international dimension. In Germany and Switzerland, for example, the importance of anti-corruption laws addressing transnational corruption is highlighted in the Siemens and Swissair cases. While both countries have instituted laws banning the bribery of foreign officials, neither country appears to have been successful in systematically preventing this corrupt practice. These cases confirm a rather dissatisfactory tenth anniversary assessment of the OECD Anti-Bribery Convention, one of the most important legal cornerstones for fighting corruption across borders: by November 2007 only fourteen of the thirty-seven signatories had complied substantially with the convention.

On the plus side, a number of country reports document the fact that bilateral collaboration in fighting corruption continues to expand. A new agreement between Indonesia and Singapore, while leaving loopholes, has enabled the extradition of corrupt individuals back to Indonesia to face trial. The former Zambian president, Frederich Chiluba, and his associates were convicted of corruption on civil charges in a London court. The Swiss government has extended the freeze on assets belonging to former Haitian dictator Jean-Claude Duvalier, in order to allow the Haitian government to launch a case of mutual legal assistance.

Institutional anti-corruption reform continues to be high on the agenda of governments, according to many country reports. In Chile, Nicaragua, Georgia, Romania, Cameroon and Zambia, the development of new anti-corruption institutions is flourishing. In Chile, a bill on access to information has been favourably received by the government and a proposal to create an autonomous body for access to information has been accepted. A new integrity system in Zambia is impressive in scope. It establishes integrity committees charged with preventing corruption in each government agency and department. Misgivings about the true autonomy of such institutions remain, however, and experience elsewhere justifies caution. In Indonesia, the phenomenon dubbed 'corruptors fight back' describes a situation in which, despite progress in the early years of this decade, more recently the fight against corruption has been undermined, culminating in a series of challenges to the legitimacy of anti-corruption institutions.

Corruption in the water sector, the focus of the analytical section of the *Global Corruption Report 2008*, has been addressed by almost a half of the country reports. Corruption in the water sector is multifaceted, and the approaches to preventing or rectifying it are equally diverse. An initiative in Bangladesh collected and analysed the different types of corruption in the water sector as reported by the media, indicating that asset-stripping and negligence of duty are prevalent problems. In Kenya, interviews with water utility customers in 2005 indicated the significant scale of corruption in the sector, with 62 per cent of respondents claiming to have witnessed petty corruption relating to water service provision. A survey reported from India identified the supply of water tankers and meter installations to be perceived as particularly corrupt. Evidence of successful reform is also provided, from examples in India and Mexico. Grass-root projects engendered better transparency via toll-free helplines in Bangalore and Hyderabad for use by the poor. An initiative of the National Water Commission in Mexico reintegrated delinquent water consumers into the payment system, collecting approximately US$121 million in unpaid fees.

Taken together, these reports illustrate the pervasiveness of corruption and its ability to distort all types of political, economic and cultural context. Global efforts to draw attention to the corruption curse, to create a normative framework for preventing corruption and realising practical cross-border mechanisms for combating corruption continue to be crucial. At the same time, as this collection of country reports from TI national chapters around the world shows, national and local efforts by all stakeholders are crucial for anti-corruption reform to take hold and be effective – and for people around the world to feel its positive effects.

7.1 Africa and the Middle East

Cameroon

Corruption Perceptions Index 2007: 2.4 (138th out of 180 countries)

Conventions

UN Convention against Corruption (signed December 2003; ratified February 2006)

UN Convention against Transnational Organized Crime (signed December 2000; ratified February 2006)

Legal and institutional changes

- The adoption on 29 December 2006 of **the 'ELECAM' law created a new, independent body, Elections Cameroon** (ELECAM), which will be responsible for the organisation and supervision of electoral operations and referendums. The new body will draw up, manage, update and maintain a national voters' register, revise voters' lists, issue voter cards, organise electoral materials, train electoral staff, supervise electoral budgets, and so on. It takes over duties previously assigned to the Ministry of Territorial Administration and Decentralisation, and the National Elections Observatory. ELECAM was set up in response to demands for a more independent body to organise elections, but many critics regard it as an empty gesture. They claim there are inadequate guarantees concerning the impartiality of members of the directorate general and the electoral council, the two bodies set up to administer ELECAM. All are appointed by the president.

- **The Code of Penal Procedures adopted on 27 July 2005 came into force on 1 January 2007**. The code lays particular emphasis on reinforcing the rights of individuals prosecuted under criminal law. It includes three major innovations that will help limit abuses of the criminal justice system by magistrates and police. The first extends the lawyer's role to include the preliminary phase of the penal lawsuit. All suspects taken to a police station now have the right to assistance from a lawyer. The second change is the introduction of an examining magistrate. This puts an end to the joint prosecution and investigation functions of the public prosecutor, which have previously been a source of corruption. The third innovation, which might appear trivial at first sight, is the obligation of judges to write down their rulings before they are delivered. Delay in drafting rulings before delivery was identified as one of the principal causes of legal delays, providing a number of opportunities for corruption. This obligation was reiterated in the law of 29 December 2006 relating to judicial acts.[1]
- **The CHOC-Cameroon programme (CHOC stands for 'Change Habits, Oppose Corruption') was launched in February 2007**. Initiated by the government and the '8+6 Group',[2] and supported by international donors,[3] the three-year programme is intended to reduce corruption by creating a national governance programme, enforced by the recently created National Anti-Corruption Commission (CONAC), anti-corruption cells within ministries and the new Financial Investigation Agency (ANIF). Established by decree in March 2006, CONAC was supposed to be an independent, public agency, but it is dominated by President Paul Biya, who appointed its president, vice-president and membership on 15 March 2007.[4] Paul Tessa, CONAC's new head, is a stalwart of the ruling party with no particular experience of fighting corruption, but outside observers were encouraged by the appointment of several other members noted for their integrity. CHOC-Cameroon must now draw up a national anti-corruption strategy, implement the UN Convention against Corruption, operationalise CONAC and ANIF and strengthen the role of civil society. The last of these entails drawing up an anti-corruption charter for civil society organisations, establishing a national committee to coordinate and train member organisations, determining a network action plan and providing financial support for a national awareness-raising anti-corruption campaign.
- The **government published its second conciliation report on oil revenues and volumes within the framework of the Extractive Industries Transparency Initiative** (EITI) on 2 April 2007 (see below). Unlike the first report, it includes an explanatory note from the Initiative Monitoring and

1 Law 2006/16.

2 The 8+6 Group in Cameroon is made up of ambassadors from Canada, France, Germany, Greece, Italy, Japan, the Netherlands, the United Kingdom and the United States; and delegates from the European Commission, IMF, World Bank and United Nations. The group works on governance issues.

3 Canada, France, Germany, the Netherlands, the United Kingdom, the United States, the European Commission, the World Bank, the African Development Bank, the OECD Development Assistance Committee (DAC) and the UNDP.

4 *Cameroon Tribune* (Cameroon), 16 March 2007.

Implementation Committee, which explains certain differences between the corporate figures provided and the Treasury figures as noted by the conciliator. Although this note does not cover all the differences identified, it provides a response to one of the complaints from the CSOs sitting on the tripartite committee of public authorities, oil companies and civil society.

Limitations of EITI monitoring

The extractive industry sector in Cameroon is dominated by oil, its third largest source of revenue after taxes, and customs and excise. Industrial exploitation of natural gas and minerals, such as iron, bauxite, nickel and cobalt, is in its early stages, but looks extremely promising.

There have been three major changes with respect to the management of oil revenues. They were successively managed through an off-budget account, through a so-called 'operations' account and, following pressure from the structural adjustment programmes set up in 1988, under 'budget guidelines' that advocate the inclusion of oil proceeds in the annual state budget.

Access to information entered a new phase after the first audit in 1991 of the Société National des Hydrocarbures, the state-controlled oil company. The subsequent involvement of civil society after Cameroon adhered to the Extractive Industries Transparency Initiative in March 2005 has meant that the government has published two conciliation reports, the first on 28 December 2006 and the second on 2 April 2007.[5] These reports, which have not yet been submitted for validation under EITI regulations,[6] were drawn up in accordance with a process that presents some weaknesses, particularly with regard to the conciliator's terms of reference, the composition of the Initiative Monitoring Committee (IMC) and Technical Secretariat (TS), the roles of civil society and international institutions, and the interests of Cameroonians themselves.

An EITI process comprises four phases: adhesion to the initiative; appointment of a tripartite committee composed of delegates from government, the oil industry and civil society; recruitment of an independent auditor; and publication of a report and an appraisal of the process, for which the tripartite committee can request the opinion of a 'validator'.

In Cameroon, the Mazars and Hart Group Consortium was selected as conciliator following a call to tender for the periods 2001/4 and 2005.[7] The terms of reference drawn up by the committee presented two weaknesses. First, they restricted the scope of study to oil alone, while the industries covered by the EITI also include gas and mining. Industrial nickel and cobalt mining is carried out in the region of Lomie in east Cameroon by Geovic Cameroun, which paid 'superficiary tax' and extraction royalties for 2004, 2005 and 2006.[8] These taxes are not included in the two conciliation reports, in breach of the first criterion of the EITI Source Book.[9] Second, the conciliator's role is limited to data collection, removing all possibility of criticism or any chance to formulate recommendations regarding the process to which he is supposedly a key contributor.[10]

The government wings of the IMC and TS of the EITI are headed by senior executives from the public administration. The minister of economy and finance chairs the IMC and the president

5 Available at www.eitransparency.org/Cameroon.
6 Ibid.
7 See www.spm.gov.cm/detail_art.php?iddocument=451&id_art=1273&type=doc&lang=en.
8 See conciliator's terms of reference on the government website, www.spm.gov.cm.
9 Geovic Cameroun paid US$116,764 in land royalties for 2004, and US$259,600 in land royalties for 2005 and 2006.
10 See conciliator's terms of reference.

of the Technical Committee for Industry Rehabilitation chairs the TS. Qualified though they may be, the status and responsibilities of these high-ranking drivers of the EITI mean that they have little time to monitor the initiative.

The type of civil society working in the IMC and TS is twofold. The first category, appointed by decree[11] to monitor the EITI, is composed of retired public sector executives. Although they are members of religious congregations and representative of local communities, they tend to fall in line with the steamrolling administrative reasoning. This category 'facilitates' the EITI management process in the name of civil society, while reducing its impact and relevance.

The second category is essentially made up of organisations in the Cameroonian 'Publish What You Pay' Coalition,[12] whose militancy concerning the extractive sector was fired by the construction of the Chad–Cameroon transit pipeline. This project enabled them to unite and to reinforce their position with regard to the extractive sector in general. They were working for the EITI before Cameroon joined the process and provided constructive criticism. But their antagonism towards government, resulting partly from their own intractability and partly from the authorities' divide-and-rule approach, may be detrimental to the image of civil society, impeding its wider growth.

Cameroon joined the EITI at a time when being admitted to the club of Heavily Indebted Poor Countries (HIPC) was a key government priority. The World Bank supported the EITI process from adhesion to report publication, while the IMF's *Guide on Resource Revenue Transparency*[13] was duly noted in Cameroon. While neither institution made accession to the EITI one of the five HIPC goal activators, it became a tacit condition.

Among the three economic structures shared by countries bordering the Gulf of Guinea – oil-dependent, oil-dominant, and non-oil-dependent and diversified – Cameroon enjoys the third, a combination stemming from its diversity of national resources and a political determination to divert popular focus away from oil. As a consequence, the population has been galvanised into respect for the management of forestry revenues to a greater degree than revenues from oil, though the latter are of more budgetary importance. The EITI inevitably suffers, as it remains an elitist initiative remote from the everyday concerns of the people. The process, which fits into the broader framework of budgetary monitoring, would certainly gain from greater local and community-level support, but this is a role for civil society, not government.

Operation Sparrow Hawk

Long seen as ineffective due to its focus on building institutions rather than clear, dissuasive sanctions, the fight against corruption has entered a new phase since the start of President Biya's fifth term in office, this time for seven years. When he took office in October 2004 he promised that 'corruption will no longer be tolerated', and since then he has committed himself to stepping up the fight against it. Meanwhile, influenced by newspapers that are only too ready to publish lists of 'presumed embezzlers' and radio stations that air debates on corruption throughout the day, public disgust with the phenomenon is growing.

11 See www.spm.gov.cm/detail_art.php?iddocument=451&id_art=1273&type=doc&lang=en.

12 Decree no. 2005/2176/PM of 16 June 2005 pertaining to the creation, organisation and running of the EITI Monitoring Committee in Cameroon. Created on 10 December 2005, the coalition comprises eight civil society organisations: FOCARFE, CED, ERA, SNJP, SeP, TI Cameroon, AGAGES and RELUFA.

13 IMF, *Guide on Resource Revenue Transparency* (Washington, DC: IMF, 2005).

This was the context in which 'Opération Epervier' (Operation Sparrow Hawk) was launched in early 2006 to track down the most notable embezzlers of public money. Its first target was state-controlled companies, many of which had been transformed into private banks by their bosses and managers.[14]

Opération Epervier, primarily involving police, gendarmerie and the justice system, led to the arrest and trial of the general managers of the real estate company Crédit Foncier du Cameroun, the Port Authority of Douala and the Inter-Communal Mutual Aid Fund (FEICOM). Many of their staff and some chairmen faced corruption charges. At least fifty people were brought to court in Yaoundé and Douala, charged with diverting public funds and related offences.

The most important case, because of the amounts involved and the personality of its defendant, was the trial of FEICOM's former top executive, Gerard Ondo Ndong, which led to record-breaking verdicts in Cameroon's legal history.[15] After seven months in court fourteen of the thirty-one defendants were convicted by the Yaoundé High Court on 28 June 2007 for looting millions of dollars, receiving prison sentences ranging from ten to fifty years. Ondo Ndong was sentenced to fifty years in prison for misappropriating CFA13 billion (US$26 million).[16]

The money was primarily stolen from additional local taxes FEICOM was responsible for using to upgrade local investments in accordance with directives from the Ministry for Local Administration and Decentralisation.[17] In some cases, money was diverted for fictitious overseas missions, at other times through unjustified financial assistance and other payments to members of the board, who were expected to work for free.[18]

Ondo Ndong and his co-defendants built up incredible personal wealth with the misappropriated money. The court identified Ondo Ndong's assets so far as: a BICEC bank account (CFA6 million, US$13,400); a bank account in Monaco (CFA34 million, US$76,000); six cars and one lorry; an unfinished 7,000 m^2 building in Simbock, Yaoundé; a three-apartment concession in Biyem Assi, Yaoundé; a duplex for his wife's nephews in Yaoundé; duplex residences for his children in Ngousso, Yaoundé; a duplex second home in Fouda, Yaoundé; a duplex in Maetur Golf, Yaoundé; two residences in Ambam; two concessions in Assandjick, his native village; a shopping mall in Ambam; the Hôtel la Couronne; a forty-room rental accommodation in Soa; a residence in Nsiméyong, Yaoundé; an 8,000 m^2 palatial residence; a block of eight luxury apartments in Dragage, Yaoundé; and the Chapel of Assandjick.[19]

Within the context of public disgust with government lethargy and, in particular, the apparent crackdown since the Mounchipou case (see *Global Corruption Report 2005*), the trial was a showcase to demonstrate a change in attitude and a complete break with the past. The public prosecutor was quick to spell this out. In his summing up, he invited the Cameroonians to see in the Ondo Ndong case 'the resounding echo of a new era for those who may be tempted to divert public funds'.

Was this case truly a symbolic act marking the end of impunity? While it appears to reflect the intention of the legal system to become an effective weapon in the fight against corruption, it lies with the government to confirm its determination by bringing to justice all those suspected of misappropriation. Over and above the ultimate sentences, emphasis needs to be

14 *Cameroon Online* (Cameroon), 25 October 2006; *Cameroon Online*, 27 June 2007.
15 *PostNewsLine* (Cameroon), 18 November 2005.
16 *Reuters* (UK), 29 June 2007.
17 See www.dibussi.com/2007/01/100_ways_to_pil.html.
18 Ibid.
19 *Cameroon Online* (Cameroon), 30 June 2007.

laid on the restitution of all misappropriated funds.

Raymond Dou'a and Maurice Nguefack (TI Cameroon)

Further reading

F. Foka, 'La Corruption, les Infractions Assimilées et les Droits Économiques et Sociaux au Cameroun', master's thesis, Université catholique d'Afrique Centrale, Yaoundé, 2007.

A. Voufack, 'Légalité et Égimité de l'Action de TI', doctoral thesis, Institut des Relations International du Cameroun, Yaoundé, 2003.

TI Cameroon: www.ti-cameroon.org.

Kenya

Corruption Perceptions Index 2007: 2.1 (150th out of 180 countries)

Conventions

African Union Convention on Preventing and Combating Corruption (signed December 2003; ratified February 2007)

UN Convention against Corruption (signed December 2003; ratified December 2003)

UN Convention against Transnational Organized Crime (acceded June 2004)

Legal and institutional changes

- In December 2006 parliament **approved a witness protection programme**, coordinated by the attorney general on behalf of the police and other law enforcement agencies. The act provides for the establishment of new identities, the relocation of witnesses and financial assistance,[1] but affords only limited cover outside the criminal legal arena. By limiting disclosure to the state, law enforcement agencies, courts and tribunals, the law fails to protect witnesses appearing before quasi-judicial hearings, such as commissions of inquiry and parliamentary committees, nor does it apply to whistleblowers in private corporations. Notably, the law would not have protected the late David Munyakei, the whistleblower employed by the Central Bank of Kenya (CBK) who received TI's 2004 Integrity Award.[2]

1 For more on the Witness Protection Act 2006, see www.kenyalaw.org.

2 In 1993 David Munyakei, a clerk at the Central Bank of Kenya, provided opposition MPs with documents detailing how S24 billion (US$342 million)was siphoned off to a company called Goldenberg International under the bank's pre-shipment finance scheme. Following his actions, Munyakei faced police harassment and was sacked by the bank. He died destitute on 18 July 2006, despite efforts to pursue restitution from the CBK through TI Kenya and the International Commission of Jurists.

- The Finance Act 2006, approved on 30 December, addresses measures to be taken against tax refund fraud and lays out guidelines on tax administration insofar as value added tax (VAT), customs and excise, and income tax are concerned. The legislation provides **sanctions on corrupt practices and expands the tax bracket to capture a wider tax base**, thereby reducing opportunities for tax evasion.
- Revisions to the Public Procurement and Disposal Act 2005 and the accompanying 2001 regulations, passed at the end of December 2006, established a legislative framework that focuses on enhancing **good governance in public procurement**. Under the act, a procurement oversight authority will oversee and coordinate public procurement and disposal. Its success will depend on the efficiency of the public procurement complaints review and appeal board empowered to investigate and resolve procurement-related disputes.
- In March 2007 the Kenya Anti-Corruption Commission (KACC) **implemented an internationally certified, webbased, anonymous reporting system**. The Business Keeper Monitoring System (BKMS®) is the only whistleblower system in the world whose anonymity has been certified by forensic investigators in Germany.[3]
- On 24 May 2007 Chief Justice Evans Gicheru appointed an ethics and governance committee of the judiciary whose terms of reference are, *inter alia*, to **collect information on and determine the levels of corruption in the judiciary**, report on individual cases and recommend remedial measures (see *Global Corruption Report 2007*). The chief justice appoints these committees every two years.
- The KACC is one of the highest-funded institutions of its kind in the country, with combined revenues in 2005 and 2006 of US$26 million. Critics have warned that the disparity in resource allocation between the investigative KACC and the office of the public prosecutions director could result in meticulously investigated corruption cases failing to lead to convictions because of **weaknesses in prosecution caused by resource constraints**. Efforts to empower KACC to prosecute suspects have been pursued through the Statute Law Miscellaneous Amendments Bill 2007, presented to parliament in July 2007. The bill was initially shelved, however, following a public outcry over proposed increases to MPs' perks attached to the same legislation.[4]

The slow retreat from secrecy

Like many countries in Africa, secrecy surrounding state operations has always been an entrenched component of Kenyan bureaucracy – so much so that it is often said the Swahili name for government, *serikali*, is shorthand for the phrase *siri kali*, or 'big secret'.

The constitution grants limited rights to communicate information,[5] but fails to provide clear guarantees on access. Meanwhile, the colonial-era Official Secrets Act provides restrictions on the use of information, giving the government powers to impede the dissemination of information if it is deemed 'prejudicial to the safety or interests of the republic'.[6] Furthermore, arrest without warrant, wide prosecutorial powers to exclude the public from proceedings, presumptions in favour of allegations without express proof of commission, and restrictions on citizens' freedom of association are some of the provisions that contradict prevailing notions of due process.

Recent events have demonstrated a drive by the administration towards greater freedom of information through the inclusion of provisions in

3 See www.kacc.go.ke/default.asp?pageid=62.
4 *Daily Nation* (Kenya), 26 July 2007.
5 Article 79.
6 Official Secrets Act, section 3, cap. 187.

the draft constitution[7] – though this was rejected in a 2005 referendum – and the issuance of a draft bill on FOI in early 2007. As these developments progress, however, public oversight of government performance remains impeded by opaqueness in transactions and bureaucratic barriers to accessing them.[8]

Some of the most glaring examples can be seen through the public's engagement with local government. The Local Government Reform Programme was crafted with the aim of improving the service delivery of local authorities. Significant amounts of money were allocated through the Local Authority Transfer Fund (LATF), which caters for the improvement of service delivery, financial management and reducing public debt. Although citizens are expected to participate in selecting projects, little has been done to improve access to information for those who actively seek it.

In a parallel process, resources are devolved to constituency level through the Constituency Development Fund (CDF), one of eight decentralised funds.[9] The fund aims to control imbalances in development brought about by partisan politics. Under the CDF Act, at least 2.5 per cent of government revenue is channelled towards the fund each year, of which 75 per cent is allocated to the country's 210 constituencies.

A 2006 study by the Kenya Institute for Public Policy Research and Analysis[10] identified several operational, institutional and legal weaknesses that could form avenues for corruption. In many parts of the country, management committees are either incapable of keeping records or unwilling to disseminate information about their activities. This leads to inadequate involvement by stakeholders, who are discouraged from taking a more active monitoring role. Improving the transparency of CDF management will require enhancements in a number of areas. Stringent evaluation is required to follow up the utilisation of CDF resources; civic education is needed to make communities understand the CDF is not a gift from the local MP; and the CDF's supervisory systems should be reviewed to place management in the hands of community representatives, and not the local MP.

As a result of privatisation and liberalisation, more opportunities for corruption are found in public procurement and the public administration. In 2004 the infamous Anglo-Leasing and Finance Limited scandal rocked the country. According to some reports,[11] the government issued promissory notes worth more than S50 billion (US$757.5 million) to companies including Anglo-Leasing, which reportedly received notes totalling S7 billion (around US$106 million). An audit in June 2006 revealed government commitments of approximately S56 billion in the Anglo-Leasing style of contract, of which S16.37 billion had been paid by June 2005.

The Anglo-Leasing deal was only one of eighteen sham contracts entered into with different companies, most of which were non-existent entities paid for supplying fictional or price-inflated security services. A 2006 report by the Public

7 Sections 49–51, Kenya Gazette Supplement no. 63, 2005, draft constitution.

8 See E. Ojiambo, 'Participatory Governance and Access to Information: Holding Government to Account,' TI Kenya, *Adili* newsletter, no. 58, 26 July 2004.

9 The other decentralised funds are the Secondary Education Bursary Fund; Roads Maintenance Levy Fund; Rural Electrification Programme Levy Fund; Local Authority Transfer Fund; HIV/AIDS Fund; Youth Enterprise Development Fund; and Women's Enterprise Fund.

10 The Kenya Institute for Public Policy Research and Analysis is an autonomous public think tank established by act of parliament; see www.kippra.org.

11 *Daily Nation* (Kenya), 29 April 2007. For more detailed analysis of this and other recent corruption cases in Kenya, see 'Illegally Binding: The Missing Anglo-Leasing Scandal Promissory Notes', at www.marsgroupkenya.org.

Accounts Committee found that all eighteen companies were connected, as evidenced by shared addresses and directorships, and because the contracts were structured so similarly as to suggest a centrally controlled conspiracy. After the scandal was uncovered the government claimed the contracts had been cancelled and the promissory notes returned. Meanwhile, the physical existence and whereabouts of Anglo-Leasing and Finance have yet to be ascertained.[12]

Increasing the state's obligation to reveal information to the public provides some restraints to the impunity with which such fraud takes place. An effective freedom of information law would force the minister of finance, for example, to explain how apparently irrevocable promissory notes amounting to S56 billion are now claimed to have been revoked. Some of the allegedly cancelled notes have been displayed to the public, but they have not been submitted to parliamentary scrutiny. Active freedom of information legislation would compel the minister to open up the issue to greater scrutiny and alert the public as to the size of the debt it would face in the future if the notes are upheld. This was the case with Zambia, which lost a high court case relating to promissory notes in London in April 2007.[13]

The government has recently taken more positive steps to develop FOI legislation. After issuing a draft policy for public consultation, it drafted a freedom of information bill for tabling in parliament that provides for both proactive disclosure and repeal of the Official Secrets Act. While these developments demonstrate a commitment to greater accountability, some key provisions require further attention for such a law to be effective, including:

- greater clarity concerning exemptions on the right to access information;
- proper systems for record-keeping and information retrieval;
- the comprehensive elaboration of penalties for public officials who deny access to information; and
- the creation of an autonomous and independent commission to develop the provisions of the proposed law.[14]

Reforming party finance: self-help or help yourself?

Kenya is a signatory to the African Union Protocol to Prevent Corruption, which calls on members to adopt legislation to regulate private funding to political parties. There is no coherent body of laws governing political parties in Kenya, however: they operate under the Societies Act 1961, which is also responsible for non-political associations.

The existing environment has conditioned party law reform. Decades of single-party rule have blurred the distinction between the ruling Kenya African National Union (KANU) and the state, with KANU plundering government revenues to perpetuate its hold on power. This resulted in the entrenchment of patronage politics, which, coupled with increasing poverty, created a culture of dependency in which citizens expected leaders to give them money or goods in exchange for political support.

Kenya has over 152 registered parties, of which ten are represented in parliament.[15] Multi-party politics in Kenya have been characterised by parties based on ethnic or class interests that

12 Public Accounts Committee, 'Special Audit on Procurement of Passport-issuing Equipment by the Department of Immigration', Office of the Vice-President and Ministry of Home Affairs, National Assembly, Ninth Parliament, Fifth Session, 2006.

13 *Daily Nation* (Kenya), 29 April 2007.

14 The government FOI draft bill was not tabled in parliament, but instead a private member's bill was tabled, which incorporates these and other provisions.

15 Figures obtained from the Electoral Commission of Kenya, 30 May 2007.

focus on factional rather than ideological concerns, are devoid of internal democracy and are marred by inter- and intra-party feuds. In addition, a nexus between corruption and political financing has emerged over the years. The *harambee*, or self-help, system of raising funds was intended to enhance village participation in national development, but it has seen gross abuse, with monies diverted to personal use and local elites. It has evolved into a key platform for buying local votes.[16]

The connection between party financing and corruption was clearly displayed during the commission of inquiry into the Goldenberg affair (see *Global Corruption Report 2005*) in February 2003, amid revelations that KANU used some of the stolen money to finance the 1992 general election. The scandal involved fictitious claims for compensation on gold re-exports to third countries. A total of about S58 billion (US$879 million) was claimed and paid out by the Treasury. The abuse of state corporations as a channel for illegal campaign finance was also raised in the Anglo-Leasing scandal. A number of 'white elephant' projects were reportedly used to amass public money for political activities.[17]

Any campaign against corruption that does not address the issue of political financing is destined to fail. The absence of a transparent regulatory framework for party funding has cost Kenya billions of shillings. Indeed, grand corruption persists in spite of all the measures taken against it, such as the enactment of elaborate anti-corruption legislation,[18] the establishment of anti-corruption institutions[19] and resort to judicial redress.

Despite these interventions party financing is still under-regulated, being limited to a ban on vote-buying and monetary deposits to discourage frivolous candidates. There are no disclosure rules, no ceilings on campaign expenditure and no restrictions on the amount or source of political contributions. The absence of a suitable legislative framework undermines the oversight of parties and encourages the formation of weak institutional structures. Despite laws relating to the electoral process and collateral laws relating to constitutional office remuneration, blatant abuse of process continues.

The draft constitution of 2005 made detailed provision regarding parties, including registration under the electoral commission, public funding entitlement and greater public scrutiny. The draft constitution has yet to become law, however, and the Societies Act does not provide in detail for party operations. Furthermore, the registrar of societies – who has no security of tenure and is directly appointed by the president – has far-reaching powers of discretionary de-registration which have been widely abused over the years.[20]

---

16 See TI Kenya, 'Harambee: Pooling Together or Pulling Apart' (Nairobi: TI Kenya/Friedrich Ebert Stiftung, 2001). Prevailing legislation now seeks to regulate the public collection of money and property in accordance with the 2003 Report of the Task Force on Public Collections.

17 The Turkwell Gorge Project lost over S7.5 billion in 1992, according to the Centre for Governance and Development: *CGD Policy Brief*, March 2005; available at www.accessdemocracy.org/library/1881_ke_cgdpolicybrief.pdf.

18 These include the Anti-Corruption and Economic Crimes Act 2003, Public Officer Ethics Act 2003, Government Financial Management Act 2004 and Public Audit Act 2003.

19 In May 2004 the government set up the National Anti-Corruption Campaign Steering Committee (NACCSC) to spearhead a national campaign to ensure zero tolerance on corruption. The public prosecution's department also reorganised itself into the anti-corruption, economic crime, serious fraud prosecution and asset forfeiture section; the counter-terrorism, narcotics, organised crime and money laundering prosecution section; and the general prosecution and appeals section.

20 For example, it refused to register the opposition party SAFINA in 1997. See *CGD Policy Brief*, March 2005; www.accessdemocracy.org/library/1881_ke_cgdpolicybrief.pdf.

The Political Parties Bill 2007 is the latest effort to impose transparent governance on political parties. Parliament passed the bill in September 2007 and it received presidential assent in October 2007. Its scope has moved beyond party funding to issues of registration and regulation. Areas covered by the legislation include: the creation of a fund administered by the registrar; provisions on alternative funding sources; disclosure, audit and record-keeping; the winding up of political parties; and making regulations.

The legislation also attempts to address the problem of defections. By December 2006 more than 74 per cent of the 222 MPs elected to the ninth parliament in 2002 had defected to other parties.[21] The act would force MPs to remain in their original parties or face by-election.

Unfortunately, subsequent events following the 27 December 2007 presidential election could not be covered by this report.

Government addresses inefficiency in the provision of water

The government's National Water Policy of 1999 envisages universal access to safe water by 2010: the current figure is a little over 50 per cent. Though huge investments were made in the 1980s and 1990s, they failed to produce an efficient water service, and the majority of schemes collapsed due to underinvestment in maintenance, poor management and a confusing array of institutional frameworks.

To address the almost total collapse in the water sector, the government approved the Water Act 2002 as a vehicle for addressing inefficiency. A central tenet is the separation of policy formulation, regulation, asset ownership and control. Formalising relationships between these functions is expected to reduce conflicts of interest and increase transparency in service provision.

Although the Ministry of Water and Irrigation remains at the helm, the act created new bodies with explicit roles. The most important change was to bar local authorities from running water and sewerage services. To conform to these requirements, water providers, modelled on commercial principles, sprang up in every corner of the country. The major challenge facing providers is to stem corrupt practices that migrated with the operational structures and staff inherited from the local authorities. In the case of the Nairobi Water Company, a 2005 survey showed that 62 per cent of consumers had witnessed petty corruption in relation to water service provision.[22]

A Citizens' Report Card in May 2007, based on a survey of almost 3,000 households in Kenya's three largest cities (Nairobi, Mombasa and Kisumu) showed mixed results.[23] Although few households reported paying 'incentives' outside official billing, the report indicated increased reliance on landlords to pay water bills, opening the prospect of bribery at the interface between them and water companies.

Lisa Karanja, Kennedy Masime, Fred Owegi and Lawrence Gikaru (TI Kenya)

Further reading

T. Barasa, 'Reforming the Political Market in Kenya through Public Party Funding', Discussion Paper no. 088/2006 (Nairobi: Institute of Policy Analysis and Research [IPAR], 2006).

21 *Sunday Nation* (Kenya), 3 December 2006.

22 TI Kenya, 'Nairobi Water and Sewerage Company: A Survey' (Nairobi: TI Kenya, 2005); available at www.tikenya.org/publications.asp?DocumentTypeID=3&ID=12.

23 Ministry of Water and Irrigation *et al.*, 'Citizens' Report Card on Urban Water Sanitation and Solid Waste Management in Kenya' (Nairobi: Republic of Kenya, 2007).

P. M. Lewa, 'Management and Organisation of Public Procurement in Kenya: A Review of the Proposed Changes', Discussion Paper no. 092/2007, (Nairobi: IPAR, 2007).
Ministry of Justice and Constitutional Affairs, 'Government Law Justice and Order Sector Reform Programme: National Integrated Household Baseline Survey Report' (Nairobi: Republic of Kenya, 2006).
National Anti-Corruption Campaign Steering Committee, 'The State of Corruption in Kenya' (Nairobi: NACCSC, 2006).
TI Kenya, 'Kenya Bribery Index 2001–06' (Nairobi: TI Kenya, 2001–6).
'Paying the Public of Caring for Constituents, Preliminary Findings from a Pilot Survey of Seven Volunteer MPs' (Nairobi: TI Kenya, 2003).
'Ufisadi Jijini: Corruption in Services and Electoral Processes in Urban Kenya' (Nairobi: TI Kenya, 2004).
'Living Large: Counting the Cost of Official Extravagance in Kenya' (Nairobi: TI Kenya, 2006).
TI Kenya: www.tikenya.org.

Niger

Corruption Perceptions Index 2007: 2.6 (123rd out of 180 countries)

Conventions

Africa Union Convention on Combating and Preventing Corruption (signed July 2004; ratified February 2006)

UN Convention against Transnational Organized Crime (signed August 2001; ratified September 2004)

Legal and institutional changes

- The **Minister for Health, Ari Ibrahim, and the Minister of Education, Harouna Hamani, were dismissed from office** on 27 June 2006, following allegations of corruption by the European Union during the course of the ten-year Education Development Programme (PDDE; see below). In October the former ministers were imprisoned after an audit showed that CFA4 billion (US$8.8 million) in EU aid had gone missing between 2002 and 2006. Partly as a result of the scandal, the government led by Hama Amadou, the prime minister, lost four no-confidence votes on 31 May 2007.[1] On 3 June President Mamadou Tandia named the former

1 *BBC News* (UK), 1 June 2007.

trade minister, Seyni Oumarou, as Amadou's successor with a mandate to 'promote good governance, and to struggle against corruption and the embezzlement of public monies'.[2]

- On 18 August 2006 the new government amended articles 15, 126 and 127 of the **Public Procurement Code**, and Prime Ministerial Orders 113 and 114 on 10 October elaborated the requirements with respect to the composition and powers of public procurement evaluation committees on the one hand, and the public procurement opening sessions on the other. All ministers, institutions, state-controlled companies and public–private partnerships are required to use these committees to authorise their procurement needs. A further Prime Ministerial Decree on 11 October reactivated the public procurement regulation agency, a measure confirmed by Presidential Decrees 2007/038 and 2007/076 on 13 January and 31 March 2007, respectively.
- A number of recent press reports have thrown the spotlight on **continuing corruption in Niger's Customs Department**. Under customs regulations established in 1961, only 40 per cent of the income from fines and sanctions at customs goes to the public Treasury, with the remainder split between customs inspectors according to a set formula.

Shooting the messenger

On 4 August 2006 Maman Abou, publisher of the privately owned weekly *Le Républicain*, and his editor, Oumarou Keïta, were detained and charged with spreading false information, officially in response to an article on 28 July criticising Hama Amadou for abandoning the West in favour of closer ties with Iran, partially due to its need for Nigerien uranium.[3] In fact, it was thought by some that the arrests may have also been in response to a special issue of *Le Républicain*.

The special issue explained in detail how two former ministers had embezzled CFA4 billion in donor funds for the PDDE education programme; how the expenditure was justified by overbilling to the tune of 239 per cent; how payments were made for undelivered goods; how orders for school and building equipment had been placed with companies belonging to the family of the former education minister, and his cronies; and it even leaked the findings of the internal audit of the PDDE and the Ministry of Basic Education and Literacy (MEBA) by the European Union and other technical donors. It included photocopies of correspondence between MEBA and its contractors. According to *Le Républicain*, corrupt handling of the programme by the prime minister's team led to the blocking of CFA26 billion and other technical support, jeopardising the reputation of primary school pupils for the 2006/7 academic year and Niger's reputation as a reliable aid partner.[4]

On 1 November 2006 a Niamey court sentenced Abou and Keïta each to eighteen months in prison, CFA5 million (US$11,000) in damages and CFA300,000 (US$660) in fines. 'Everything suggests that Abou and Keïta are now the prime minister's personal prisoners,' said the journalists' defence group, Reporters without Borders.[5] 'President Mamadou Tandja should realise that these heavy sentences will not benefit either Niger or his prime minister and constitute a serious breach of press freedom.' The two journalists were finally released following an appeal hearing on 27 November.[6]

2 *Jeune Afrique* (France), 3 June 2007.
3 *Le Républicain* (Niger), 4, 5 August 2006.
4 Ibid.
5 Reporters without Borders, 27 November 2006; available at www.rsf.org/article.php3?id_article=18746.
6 Ibid.

Meanwhile, at a meeting on 16 August Niger's largest opposition party, the Parti Nigérien pour la Democratie et le Socialisme (PNDS), denounced four businessmen close to the government for embezzling CFA117 billion (US$37.3 million) from public Treasury accounts held in the Central Bank of West African States (BCEAO). According to the PNDS, the money had been divided four ways: CFA70 billion (US$79 million) for one conspirator, CFA36 billion for a second, CFA6 billion for a third and CFA5 billion for a fourth.[7]

To reassure the public that all was well, the prime minister published a list of the public procurement contracts approved by the Cabinet during the period in question in a government magazine.[8] Unfortunately, malpractices came to light after publication, which were later picked up by other newspapers.

The consequences of the parallel scandals were twofold. First, official reluctance to charge the two ministers incriminated in the PDDE case – while persecuting their accusers above and beyond the letter of the law – undoubtedly contributed to the vote of no confidence in Amadou's government in May 2007, fully nine months after the allegations of ministerial corruption first came to light. By that point, *Le Républicain*'s journalists had become global figures. In addition, the World Bank, the French Development Agency, Belgium and Danida had frozen their allocations to the educational programme; and France, the European Union, the World Bank, Belgium and Germany said they were waiting for the government to 're-establish confidence' before resuming long-term aid.[9]

The second consequence was a renewed focus on loopholes in the Public Procurement Code in a belated attempt to restore the confidence of a donor community that provides Niger, the world's poorest nation,[10] with most of its investment budget and famine relief.

Idrissa Alichina Kourgueni (Association Nigérienne de Lutte contre la Corruption [ANLC], TI Niger)

Further reading

G. Blundo *et al.*, 'La Corruption au Quotidien en Afrique de l'Ouest: Approche Socio-anthropologique Comparative: Bénin, Niger et Sénégal: Rapport Final October 2001', study financed by the Commission of European Communities and Direction du Développement et de la Coopération Suisse (Marseilles: 2002).

OECD, '2006 Survey on Monitoring the Paris Declaration' (Paris: OECD, 2007); see Niger chapter at www.oecd.org/dataoecd/45/9/38949577.pdf.

TI Niger, Association Nigérienne de Lutte contre la Corruption, 'Etat de la Corruption' (Niamey: TI Niger, 2006).

7 *La Roue de l'Histoire* (Niger), nos. 313, 314, 315, 316, 317, 318 and 322, 2006; *Opinions* (Niger), 16 August 2006; *La Nouvelle Tribune du Peuple* (Niger), 22 August 2006.

8 *Sahel Dimanche* (Niger), 15 September 2006.

9 *Le Républicain* (Niger), 4 August 2006; available at www.planeteafrique.com/republicain-niger/files/special7aout2006.pdf.

10 Niger ranks 177th out of 177 on the Human Development Index 2006.

Palestinian Authority

Legal and institutional changes

- On 22 April 2006 **the government froze the registration of all NGOs**, in contravention of article 26 of the Basic Law and the Charitable and Non-Governmental Organisations Law, which both guarantee the right of civil institutions to work without harassment. The Coalition for Accountability and Integrity (AMAN), the TI chapter in Palestine, requested clarification of the decision on 3 June and, though it was forthcoming, it contained no legal foundation. On 22 June AMAN submitted a case to the high court, which threw out the decision to freeze NGO registration.[1]
- On 14 June 2006 the **Palestinian Legislative Council (PLC) called on the governing Palestinian Authority (PA) to inform it of all future loan agreements it draws up**. This became effective on 29–31 August, when the PLC approved three loan agreements with the Islamic Development Bank.[2] If enacted regularly, the decision is expected to enhance the transparency of government finances, though admittedly it has not been recent PA 'culture' to publish any more information than the loan amount and the name of the granting agency.
- On 9 July 2006 the attorney general created the **PA's first department for combating corruption**, to take charge of prosecuting crimes committed against public finances.[3] Once established, it will be responsible for bringing to justice all those tarnished by allegations.
- The PLC has created an **internal affairs committee to deal with administrative reforms.** Chaired by PLC head Dr Aziz Dwaik and comprising members of all political parties, its meetings began on 1 August 2006. It was designed to examine the PLC's general policy directions, its mandate, its financial status and ways to activate its legislative committees and enhance its administrative competence, thereby contributing to the elimination of systemic corruption.[4] However, it is important to mention that the new committee was formed before the infighting between Fatah and Hamas, Palestine's two leading parties, and it may consequently have lost both support and jurisdiction (see below).
- On 1 April 2007 **PLC members submitted disclosures of their bank accounts to the head of the Supreme Court** in the presence of the chairperson of the Control and Financial Bureau.[5] This was the first step of its kind in the PLC's history and was in compliance with the Law of Illicit Enrichment, which it ratified in January 2005 (see *Global Corruption Report 2005*). The law aimed to register and monitor the income of senior officials, including ministers, PLC members and members of their immediate families. There are no exemptions, insofar as the law is concerned, but it has not yet been enforced and will be subject to the same limitations imposed by the collapse in relations between Fatah and Hamas.[6]
- On 14 April 2007 **the Hamas-led unity government ratified the Security Reform Plan** with a view to ending the violence within the Palestinian territories. A day later PA President

1 Copies of all correspondence and the case submitted to the high court are available from AMAN.
2 Palestinian Independent Commission for Citizens' Rights, *Annual Report* no. 12, 2006; www.piccr.org/index.php?option=com_content&task=view&id=76&Itemid=99&lang=en.
3 Palestinian Independent Commission for Citizens' Rights, 2006.
4 'Report on the Work of the PLC in its First Round', PLC Report (2007).
5 *Al-Quds Al-Arabi* (UK), 3 April 2007.
6 Ibid.

Mahmoud Abbas issued a decree creating the Palestinian National Security Council with himself as chairman and the prime minister as his deputy. The council was intended to take responsibility for unifying the country's various security bodies in accordance with the Mecca agreement of 8 February 2007, signed by both Fatah and Hamas. On 14 May the interior minister, Hani Kawasmi, who is responsible for leading the strategy, resigned, claiming he had inadequate authority and resources to deal with the deteriorating security situation.[7]

- AMAN organised **the first Transparency Festival** in December 2006 to raise awareness in the Palestinian community on the need to curb corruption and encourage institutions to work towards a national integrity system.[8] The festival included the signature of anti-corruption codes of conduct for private and public sector organisations, and for local authorities. Three integrity awards were granted to three employees from the public, the media and local authorities. This encouraged the signature of similar documents in the non-governmental sector on 10 July 2007. AMAN is now considering expanding the process by training trainers on the content of the codes and educating them on transparency, accountability and integrity.

Change in government halts anti-corruption advancement

The political system in the PA is dominated by Fatah, a liberal party organisation created by former President Yasser Arafat in 1965, and Hamas, a Sunni resistance movement established in 1987 by the Gaza branch of the Muslim Brotherhood. Fatah supporters, led by current President Abbas, have dominated the PA's ministries and security forces since 1994.

The legislative election of January 2006 caused a seismic shift in this equilibrium when Hamas decisively defeated Fatah for the first time. Though the election was perceived as transparent, the United States, European Union and Canada had previously listed Hamas as a terrorist organisation. As a consequence, the international community imposed a boycott on the new government in March 2006, freezing financial transfers to the Finance Ministry. Israel halted the monthly transfer of US$55 million in customs and tax receipts,[9] and a number of banks that had maintained the PA's treasury accounts halted all transactions in order to abide by the US Office of Foreign Assets Control ban.[10] Institutions including the World Bank warned that such measures would have negative consequences for transparency and accountability in the PA. Indeed, the new prime minister, Ismail Haniyeh, was prevented from entering the Rafah border-crossing after he was discovered to be carrying US$30 million in donations from sympathisers in the Gulf and Iran.[11]

At the time of writing, the PA was subject to an intense power struggle between newly elected Hamas and Fatah. Conflicted interests between the two sides had escalated into fighting, creating more fertile soil for corruption, particularly in the area of public appointments.

An employees' strike over lack of wages in September 2006 and the imprisonment of forty PLC members by the Israeli Defence Forces,

7 *CRI News* (China), 15 May 2007.

8 See www.aman-palestine.org/English/activitiesE.html#2006.

9 Q. Hadeel *et al.*, 'Reconstruction National Integrity System Survey' (Ramallah: AMAN, 2007).

10 World Bank, 'Coping with the Crisis: Palestinian Authority Institutional Performance' (Washington, DC: World Bank, 2006).

11 *Ha'aretz* (Israel), 16 December 2006.

combined with movement restrictions and the destruction of ministry buildings in Gaza by Israeli shelling, further aggravated the internal situation. Governorates in the West Bank are now isolated, badly affecting national identity and making it impossible to have a unified approach to combating corruption.

This isolation facilitated the capture of cities and towns by small militant groups, which impose practices that lack accountability. More than 70,000 men were employed by six different security forces in 2006, and 345 murder cases were reported that year, compared to 176 in 2005.[12] There were 273 attacks on public institutions in 2006, compared to forty a year earlier, and 150 kidnappings, compared to thirty-six in 2005. According to a recent report by Al-Mezan Center for Human Rights, the number of murders in the first quarter of 2007 reached 147 – double the number in the same period in 2006.[13]

With increasing rivalry between the government and the president, the number of men under arms rose from 3,500 to 6,500 members in the new executive force, with an intention to double the force to 12,000 in early 2007.[14] In the ministries at this time of crisis, the International Monetary Fund noted a 17 per cent increase in the government's wage bill between 2005 and 2006.[15] AMAN noted that these appointments were not in accordance with the Civil Service Law, which stipulates that no one is to be appointed without meeting a job's requirements.[16] By the end of 2006 the Ministry of Education appointed 1,000 new religious teachers at the expense of investment in civic and secular education.[17]

Most of the PA's institutions are paralysed. The Finance Ministry's website has been shut down since November 2006, denying public access to information about tax and procurement procedures. With PLC support the government managed to make expenditures in the absence of a legal framework for the 2006 budget, but the 2007 budget was not submitted on time. As far as legislation is concerned, the PLC ratified two decisions: approving an extension period for submitting the 2006 Budget Law, and a Presidential Decree concerning PLC voting dates for police and security officers.

The PLC was unable to play its monitoring role in checking government performance in relation to the programme on which it had gained electoral victory; it failed to hold question-and-answer sessions with the Hamas interior minister on three separate occasions in December due to movement restrictions, the abduction of forty PLC MPs and the general chaos; it did not question the government on its inability to submit a proper budget on time; it did not query the government's inability to submit final accounts for 2005; it did not question the late or non-payment of public sector employees' salaries; and, finally, it did not request an account of the US$30 million in cash that Haniyeh brought over the Rafah border-crossing.[18]

12 Palestinian Independent Commission for Citizens' Rights, 2006.
13 Al-Mezan Center for Human Rights, press release, 24 April 2007.
14 Palestinian Independent Commission for Citizens' Rights, 2006.
15 IMF, 'West Bank and Gaza: Fiscal Performance in 2006' (Washington, DC: IMF, 2007).
16 AMAN, 'Recruitments in the 10th Palestinian Government, Coalition for Accountability and Integrity' (Ramallah: AMAN, 2006).
17 Jerusalem Media and Communications Center (JMCC), *Good Governance Monitoring Report* no. 1 (Jerusalem: JMCC, 2006).
18 An account of some of the problems faced by the PLC can be found at www.jmcc.org/goodgovern/06/eng/plc.htm.

These events set Palestine's nascent institutions back by a decade. The PLC played no role in activating anti-corruption legislation, such as the Law on Illicit Gain of January 2005, which was established to register the income of public officials and lawmakers.

The internal power struggle, the international aid boycott, the bypassing of the Finance Ministry and continued oppression by the Israeli occupation force have undermined reform efforts and nourished corruption. Public recruitment is perceived as tarnished, shaking people's trust in the system. This all shows that national and international commitments are badly needed to sustain reform efforts and eliminate corruption. Unfortunately, both are unavailable for the present.

Frosse Dabit (TI Palestine/AMAN)

Further reading

AMAN, 'Report on the Citizen's Right to Free Access to Information in the PLC' (Ramallah: AMAN, 2006); www.piccr.org/index.php?option=com_content&task=view&id=76&Itemid=99&lang=en.

'Report on the Level of Compliance of the Ministry of Finance to the Basic Principles of Freedom of Information' (Ramallah: AMAN, 2006).

'Recruitments in the 10th Palestinian Government' (Ramallah: AMAN, 2006).

'The Unbalanced Separation between the Legislative, Judicial, and Executive Powers and the Weakness in Monitoring One Another. Case of Palestinian Judicial Authority' (Ramallah: AMAN, 2007).

JMCC, *Good Governance Monitoring Report* no. 1 (Jerusalem: JMCC, 2006).

Palestinian Center for Policy and Survey Research (PCPSR), Poll no. 23, (Ramallah: PCPSR, 2007); www.pcpsr.org/survey/polls/2007/p23e.pdf.

Palestinian Independent Commission for Citizens' Rights, *Annual Report* no. 12, 2006.

Q. Hadeel *et al.*, 'Reconstruction National Integrity System Survey' (Ramallah: AMAN, 2007).

Transparency Palestine/AMAN: www.aman-palestine.org.

Senegal

Corruption Perceptions Index 2007: 3.6 (71st out of 180 countries)

Conventions

African Union Convention on Preventing and Combating Corruption (signed December 2003; ratified April 2007)
UN Convention against Corruption (signed December 2003; ratified November 2005)
UN Convention against Transnational Organized Crime (signed December 2000; ratified October 2003)

Legal and institutional changes

- On 25 April 2007 the **Council of Ministers adopted a new Public Procurement Code**[1] known as the 'code des marchés publics'. Drawn up in a participative process involving civil society, the private sector and donors, the new version maintains improvements made to date and introduces several new elements. It is considered more transparent than its predecessors.
- More than twenty-five years after Senegal set up the now dormant **Court for the Repression of the Unlawful Accumulation of Wealth** (CREI), President Abdoulaye Wade created an ad hoc committee in April 2007 in a bid to revive the law.[2] While some saw the measure as a way of forcing the former prime minister, Idrissa Seck, to surrender funds he allegedly misappropriated at Thiès in 2004 (see below),[3] the draft law is challenging on its own account. The committee proposes reversing the burden of proof, thereby obliging the accused individuals to provide evidence that the origins of their assets are, in fact, legal. While this could strengthen the fight against bribery and corruption, it raises justice issues with respect to the presumption of innocence.

New efforts to tackle procurement corruption

The benchmark was set by the 2004 Thiès construction scandal, which involved an opaque mix of poor planning and rivalry at the highest levels of power that cost unknown millions. These events prompted the OECD in 2006 to question 'the ability of the state to manage major projects with transparency'.[4]

1 Decree no. 2007-545.
2 *Walfadjri* (Senegal), 17 April 2007.
3 Idrissa Seck and his followers were members of President Wade's Senegalese Democratic Party (PDS) until the two fell out in 2004. Seck, the mayor of Thiès, was later investigated and jailed for irregularities in procurement contracts for work carried out for the independence celebrations in 2004. The whereabouts of his allegedly stolen millions are unknown. See J.-C. Fall, 'Les Chantiers de Thiès: Prétexte à une Réflexion sur les Marchés Publics', presentation at the Civil Senegal Forum for Public Governance, Dakar, December 2005.
4 OECD, 'Senegal', *Africa Economic Outlook, 2005–06*; see www.oecd.org/dataoecd/33/14/36741806.pdf.

The works at Thiès were part of an infrastructure programme timed to coincide with the 2004 independence celebrations, but which also coincided with President Wade's plans to modernise Senegal's regional capitals. Initiated in the middle of the financial year, the government found a solution to the resultant cash flow crisis by requesting interested companies to pre-finance the building work – in violation of provisions of its own Public Procurement Code, passed in May 2002.[5]

In February 2006 Idrissa Seck, a former prime minister who had fallen out of favour with President Wade, was tried and imprisoned for seven months for usurpation of title for awarding a contract to a businessman, Bara Tall, without having the authority to do so, and for colluding in overpricing public works contracts.[6] Seck was cleared of some charges and released in 2006, though the accusation of illegal enrichment remains to be answered.

Until the new code is implemented, public procurement will continue to be governed by the 2002 code. This code introduced several new elements, but a fundamental problem remains. There is contradiction between the Public Administration Code, which is a law, and the Public Procurement Code, which is a decree, resulting in lack of transparency.[7]

In 2006 the contracts giving rise to the greatest concern were those that went through 'agencies' – bodies entrusted with ministerial functions but which enjoy greater autonomy. Of Anglo-US inspiration, the 'agencies' were created in response to a perceived need to introduce private sector techniques into public services management.

The Agence chargé de la Promotion de l'Investissement et des Grands Travaux (APIX) and the Agence Nationale de l'Organisation de la Conférence Islamique (ANOCI) now handle a large share of state projects. While the two agencies' legal status is somewhat indeterminate, their accounts are separate from Treasury accounts, despite the fact that they benefit from budgetary transfers. The World Bank made its concerns clear with respect to this situation in a 2006 report.[8]

The most recent case of alleged corruption related to contracts awarded by ANOCI in preparation for the eleventh Islamic Summit in Dakar in 2006 (now scheduled for March 2008). In May 2006 Abdoulaye Baldé, ANOCI's chief executive, was accused of accepting kickbacks in a US$64.5 million, Kuwait-backed contract to widen Dakar's 10km-long western corniche.[9] The allegations came in an open letter to the coordinator of the National Programme of Good Governance from Pape Malick Ndiaye, a Senegalese student in France who represents the Collectif de Réflexion et d'Action contre la Corruption (CRAC).

The managing director of ANOCI took up the issue with the National Anti-Corruption Commission (CNLCC) in a letter on 30 May

5 See article 6 of Decree no. 2002-550 of 30 May 2002 in relation to the Public Procurement Code, which states: 'The finalisation of a public procurement contract involving financing from the state, the local authorities and public institutions, or national companies and mixed-capital limited companies with a majority public holding, is dependent on the existence of sufficient budgetary credit and adherence to the regulations governing expenditure by the aforementioned public organisations.'

6 See ruling no. 4 of the examining committee of the High Court of Justice, 7 February 2006.

7 For a description of corrupt procurement practices, see A. Fall *et al.*, 'Gouvernance et corruption dans le secteur de la santé au Sénégal' (Dakar: Civil Forum, 2005).

8 *Walfadjri* (Senegal), 17 April 2007. See also World Bank, 'Senegal: Développements Récents et les Sources de Financement du Budget de l'Etat', PREM 4 Région Afrique Rapport No. 36497-SN (Dakar: World Bank, 2006).

9 *Le Quotidien* (Senegal), nos. 1156, 1159 and 1175, November 2006.

2006.[10] The commission referred the matter to the National Financial Information Cell (CENTIF)[11] and also sent letters to various banks mentioned in the denunciation. ANOCI's managing director was cleared of all charges, on the grounds that the whistleblower had failed to provide sufficient proof.[12] The decision was criticised for its excessive haste, as it did not wait for the results of the CENTIF investigation, and because it relied on documents produced by one of the accused, Abdoulaye Baldé.

A preliminary inquiry at a Dakar regional court on 30 June 2006, however, accused Ndiaye of 'fraud, use of forgery and libel'. For reasons of health, he was placed under judicial control, contrary to a request by the public prosecutor that he be committed to prison. After an appeal by the prosecutor, Ndiaye was sent to prison, but later he was released on bail. The court case is still pending.[13]

A further case involving ANOCI concerned a contract for work on the Northern Slip Road awarded in June 2006. The conditions under which the contract was granted to the Consortium Sénégalais d'Entreprises raised questions about the technical ability of the companies bidding for the contract. It was alleged that other bidders would have been more suitable than the ultimate winner.[14] Questioned about this, ANOCI's managing director stated that the other companies lost the contract because of their technical ability.[15] No legal proceedings were taken.

The new procurement code should provide effective solutions to these issues. At an institutional level, the code provides for a regulating body whose principal functions are to:

- create a new disputes settlement committee to resolve them as they arise, especially during the procurement phase;
- propose legislative reforms required in line with economic changes;
- train procurement services and others in marketing techniques;
- centralise statistical data and assess the impact of public procurement on the economy; and
- ensure post-procurement monitoring through audits and studies, and monitor the application of the resulting recommendations.[16]

Home-grown justice

Article 92 of the Senegalese constitution states: 'Judicial power is independent of legislative power and executive power.' It specifies that judges and law officers are subject only to the law in the performance of their duties and cannot be removed from office.

In practice, the Conseil Supérieur de la Magistrature (CSM), which manages the appointment, promotion and transfer of judges, is chaired by the president, who is assisted by the justice minister. This means that the executive branch can influence how justice is conducted. This may take the form of posting a troublesome judge to a remote jurisdiction or promoting more pliable ones to senior positions.

There is evidence of systematic political interference and pervasive impunity across the judiciary.

10 See CNLCC, ca144se no.13/2006, M. Baldé vs. M. Ndiaye, in CNLCC 2006 Activity Report.
11 The CENTIF was set up on 18 August in accordance with article 16 of Law 2004-09. Its mission is to receive and process information related to the fight against money-laundering.
12 CNLCC, case no.13/2006.
13 US Department of State, '2006 Country Report on Human Rights Practices: Senegal' (Washington, DC: US Department of State, 2007).
14 *Le Quotidien* (Senegal), 4 November 2006.
15 *Walfadjri* (Senegal), 24 March 2007.
16 See article 2 of Decree no. 2007-546 of 25 April 2007, concerning the organisation and management of the Public Procurement Regulations Authority.

Use of the law for political ends was demonstrated explicitly in the Thiès affair, when proceedings were based on the conclusions of the state inspector general, who accused both the former prime minister, Idrissa Seck, and the then finance minister of systematic contract overpricing.[17] Seck alone faced legal action.[18]

The scale of corruption in the judiciary led to a public outcry in July 2006, after a recording of an attempt to bribe the public prosecutor was aired on private radio and printed in some newspapers. Aminata Mbaye was heard accepting CFA15 million (US$32,000) to divide with fellow judges Cheikh Bamba Niang, Jean Louis Turpin and Ibrahima, and two court clerks, to rule in favour of Momar War Seck, who allegedly embezzled CFA100 million (US$211,500) from Mohamed Guèye in a 1995 property deal.[19] The justice minister referred the case to the inspector of judicial administration, whose investigation led to early retirement for Mbaye, and transfer and other sanctions for her colleagues.

Against this background the government launched a programme of judicial reform. Proposed improvements include doubling the judicial budget from CFA7.4 billion in 2000 to CFA15.7 billion (US$33.2 million) and more than doubling monthly allowances for judges and law officers to CFA800,000 (US$1,700) in October 2006 ('allowances' are taxable supplements to judges' salaries).

The continued existence of virtuous circles within the judiciary must be stressed. Far from retreating into institutional secrecy, the Union of Judges requested that full light be shed on judicial corruption.[20] The CNLCC, operational since 2004, launched an inquiry into an appeal by a Swiss national, Dame Schluep, on 11 July 2006 against seizure of her property with the alleged collusion of a magistrate and other law officers.[21]

In the Aminata Mbaye case, it is striking to note that, while the corrupters went to prison, the corrupted were merely disciplined. Following a complaint on 5 September 2006 by one sanctioned law officer, the CNLCC denounced the 'two-tier justice system' that allowed for two procedures for the same offence.[22] The procedure for disciplining judges and law officers is set out in Organic Law 92-27, while the CSM sits in judgment when a case is brought against a judge.

In its 2006 activity report, the CNLCC said that it would refer a reform proposal in this regard to the Council of State, indicating that, while sensitive to the legal arguments, it did not want to upset the CSM.

Therefore, the CNLCC is not entirely blame-free. While it claimed not to be qualified to question the disciplinary measures – which is true – it could have urged the president to bring proceedings against the judges convicted of corruption under article 3 of Law 2003-35, which laid its own legal foundations. Article 3 reads: 'When the Commission considers that it holds information that can justify the opening of legal proceedings, it shall transmit a detailed note and recommendations to the President of the Republic, specifying the identity of the people or organisations susceptible to legal proceedings.'

The CNLCC's faint-hearted attitude in a case such as this, like its performance in the corruption case against ANOCI's handling of

17 J.-C. Fall, 2005.
18 *IRIN News* (Kenya), 5 April 2006. See ruling no. 4 of the examining committee of the High Court of Justice, 7 February 2007.
19 See www.seneweb.com/news/engine/print_article.php?artid=4744.
20 *Weekend* (Senegal), 24–30 May 2007.
21 See CNLCC, case no. 17/2006, Dame Schluep vs. Dianka, in CNLCC 2006 Activity Report.
22 Recommendation no. 18, CNLCC 2006 Activity Report.

Dakar's western corniche enlargement, shows little evidence of institutional determination to stamp out corruption either in the justice sector or procurement policy.

Semou Ndiaye (Forum Civil/TI Senegal)

Further reading

G. Blundo and J.-P. de Sardan, 'La Corruption au Quotidien', *Politique africaine*, no. 83 (2001).

CNLCC, *2006 Activity Report* (Dakar: CNLCC, 2006).

A. Fall, 'Gouvernance et Corruption dans le Secteur de la Santé au Sénégal' (Dakar: Civil Forum, 2005).

'Gouvernance et Corruption dans le Domaine des Ressources Naturelles et de l'Environnement au Sénégal' (Dakar: Civil Forum, 2006).

M. Ndoye, 'Traitement Judiciaire de la Corruption', master's dissertation, Cheikh Anta Diop University, Dakar, 2001.

Forum Civil/TI Senegal: www.forumcivil.sn.

Sierra Leone

Corruption Perceptions Index 2007: 2.1 (150th out of 180 countries)

Conventions

African Union Convention on Preventing and Combating Corruption (signed December 2003; not yet ratified)

UN Convention against Corruption (signed December 2003; ratified September 2004)

UN Convention against Transnational Organized Crime (signed November 2001; not yet ratified)

Legal and institutional changes

- On 18 July 2006, with support from the World Bank, the European Commission, the African Development Bank and the United Kingdom's Department for International Development, the government published an **Improved Governance and Accountability Pact**, which promised to 'take forward' and implement ten key reforms by July 2007.[1] The pact's corruption-related objectives included: the agreement by all stakeholders on an implementation plan for the anti-corruption strategy by the end of 2006; an increase in the number of significant 'public interest' prosecutions by the Anti-Corruption Commission (ACC); the establishment of a law reform task force to review the Anti-Corruption Act 2000

1 See www.dfid.gov.uk/pubs/files/sierra-leone-igap.pdf.

by the end of 2006; the introduction of legislation for declaration of assets by public officials; the strengthening of the auditor general's office and the parliamentary accounts committees, including the timely publication of their reports; and the implementation of the Public Procurement Act in all ministries by the end of 2006. Plans were also in hand to introduce the regulations of the Money Laundering Act 2005 by the end of 2006, and to increase the capacity of the Bank of Sierra Leone to oversee and monitor it.

- On 23 September 2006 NGOs with a focus on mining met in Freetown to form the **National Advocacy Coalition on the Extractives** (NACE), with the aims of enhancing transparency in the extractive industries and sensitising the public on the Extractive Industry Transparency Initiative process, which has broadly been led by DfID since Tony Blair launched it in 2002.[2] The minister of mines wrote to the World Bank in June 2004 to request technical assistance in implementing an EITI, and a framework and work plan are now in place for diamonds, bauxite, rutile and gold. The first phase is to verify data with the four large-scale mining companies currently working in Sierra Leone.
- On 25 January **DfID published a thirty-five-page review of its support for the Anti-Corruption Committee**, which found it had made no substantial impact on any of the four key indicators during 2006.[3] According to the review, the ACC had made no progress on the overall goal of reducing corruption, had made no impact on reducing real or perceived levels of corruption, had suffered a fall in institutional capacity since the previous year and could not provide clear evidence of community mobilisation. There were seven prosecutions of nine public officials for corruption in 2006, of which only two had actually originated in that year. Some 30 per cent of cases were completed in good time, but the average length of investigation was 146 days, raising questions about cost effectiveness, given the generally small amounts embezzled. There was no progress on processing high-level prosecutions. The review team found an 'under-spend' of £750,000 (US$1.5 million), suggesting that 'work relevant to the achievement of the operational plan in investigations, preventions and community relations is not being carried out'. 'Without the functioning support institutions,' it concluded, 'the ACC cannot operate effectively.'[4] The report recommended that DfID no longer support the ACC, but integrate its anti-corruption efforts into existing programmes and develop alternative initiatives.
- In June 2007 **three gender laws passed into law**: the Domestic Violence Act; the Registration of Customary Marriage and Divorce Act; and the Devolution of Estates Act.[5] While the first of these has no corruption dimension, the other two definitely do. The Registration of Customary Marriage and Divorce Act prohibits children under the age of eighteen from marrying. It stipulates that both parties must consent to marriage; women are entitled to own and dispose of property in their own right; and, in the event of separation or divorce, dowries do not have to be repaid. Prior to the law, women whose husbands died or divorced them found it difficult to lay claim to shared property. The Devolution of Estates Act directs that wives should automatically inherit their husbands' estates if they die without a will. The implementation of these laws will be complicated, however, given the levels of corruption in local police forces and the judiciary.

2 See eitidev.forumone.com/section/countries/_sierraleone.

3 DfID, 'Annual Review 2006 of DfID Support to the Anti-Corruption Commission Phase 2 in Sierra Leone' (London: DfID, 2006).

4 Ibid.

5 For more details, see www.unifem.org/news_events/story_detail.php?StoryID=606.

Politics sways chieftaincy election

Commentators note that one cause of the Sierra Leone civil war (1991–2002) was the government's practice of imposing 'paramount chiefs' who did not hail from local 'ruling houses'. Paramount chiefs are responsible for the day-to-day administration of a 'chiefdom', or district, while a ruling house denotes a family eligible by tradition to put up a candidate for the chieftaincy elections. Elections for the paramount chieftaincy are hotly contested, and often end in violence or in court.[6]

In recent years the lists of those qualified to stand have been revised to exclude those considered cronies of the central government, and regulations were introduced to ensure the proper conduct of elections. These steps include the creation of electoral colleges and the appointment of assessor chiefs to vet lists in accordance with tradition, although this varies from district to district.

At the end of the war there were more than sixty vacancies for paramount chieftaincies across the country. One was for Biriwa in the Bombali district, Northern Region. Biriwa is inhabited by the Limba and Madingo ethnic groups, who have coexisted amicably since 1890, although the former is dominant. There was controversy as to who was eligible to stand for chieftaincy, however. The Limbas argued that the Madingo contestant was not from one of Biriwa's four ruling houses and, as settlers, the Madingo should not be entitled to rule over them. The Madingo countered that they had participated in chieftaincy elections since 1952 and, under the constitution, had the right to put up a candidate.

Prior to the election several aspirants had vied for the vacancy, and the Limba raised objections. This triggered pre-election violence in the chiefdom's capital, Kamabai, and on 26 May 2006, the day fixed for the candidates' declaration of intention, proceedings were again marred by violent conduct. On 3 June the declaration proceedings were disrupted by a group of young Limbas wielding sticks and cutlasses. On 12 August 2006 the Ministry of Local Government conducted the election, but the six Limba candidates refused to declare, leaving Issa Mohamed Sheriff, the only Madingo, as the sole candidate qualified to contest the election.[7] Out of the 473 tribal authorities in Biriwa, only 133 voted.

Sources said the Limbas refused to present candidates or vote on the grounds that the entire electoral process had been contaminated by political undertones and that Sheriff was preferred by the central government (President Ahmad Tejan Kabbah is also Madingo).[8] In apparent protest, on 19 August 2006 the Limbas elected a former chiefdom speaker, Pa Alimamy Conteh, as their own paramount chief.

This added up to a serious controversy between government and the Limbas as to whose responsibility it is to conduct paramount chief elections. The Limbas argued that the provincial secretary's office had no mandate to conduct the Biriwa election, and had therefore abrogated the constitution and undermined the legitimacy of the National Electoral Commission, which had rejected the call for an election on the grounds that there was insecurity in the chiefdom and there were procedural issues.[9] The government

6 This issue was taken to the Supreme Court in November 2006: 'In the matter of the elections for the office of Paramount Chief of Biriwa Cheifdom, Bombali District in the Northern Province of the Republic of Sierra Leone, held on 12 August 2007.' See *Awareness Times Newspaper* (Sierra Leone), 13 November 2006.

7 UN Security Council, 'Second Report of the Secretary-General on the United Nations Integrated Office in Sierra Leone (S/2006/695)', 29 August 2006.

8 US Department of State, Bureau of Democracy, Human Rights and Labor, 'Sierra Leone: Country Reports on Human Rights Practices' (Washington, DC: US Department of State, 2007); *Standard Times Press* (Sierra Leone), 30 December 2006.

9 UN Security Council, 2006.

maintains, however, that the conduct of chieftaincy elections falls within the purview of the Ministry of Local Government.[10]

Chieftaincy elections are a live issue in Sierra Leone. Paramount chiefs command huge respect in their districts and are the first point of contact with the chieftaincy. Most politicians are keenly aware that the first step in winning the hearts and minds of people in a chiefdom is to have the backing of the paramount chief.

Threadbare justice in a recovering state

An effective judiciary is a cornerstone of a stable and successful state, and judges must be free from gratuitous influences, whether real or imagined, to decide a case on the basis of the facts and in accordance with the law. The judiciary in Sierra Leone is perceived by some as corrupt, inefficient and symbolic of injustice because of extended delays to cases, the incompetence of some judges and poor conditions of service.[11] Though there have been some improvements compared to ten years ago, the judicial system is in such a dilapidated state that a change of direction is urgently needed.

Justices of the peace have not been paid their sitting fees for two years, while magistrates have reportedly not been paid for months. Poor conditions of service mean that few new practitioners seek to take up appointment, with the result that the majority of judges are past retirement age.[12] The attorney general and minister of justice, Frederick Carew, who served under the government led by Tejan Kabba (2002–7), has done little to redress these problems to date.

On the eve of the civil war, in 1991, Sierra Leone's judiciary could protect only the 'haves' or those in power, because its independence was compromised. As a result many citizens resorted to extrajudicial means to resolve their disputes. The law still does not guarantee judges' independence. Subsections 3 and 5 of article 136 of the 1991 constitution empower the president to terminate their contracts, while subsections 2 and 4 allow the president to recruit judges from the High Court of Appeal and Supreme Court even though they may have reached retirement. A judiciary that serves only at the pleasure of the president cannot be impartial, since there will be no true separation of powers.[13]

Since 2005 the five-year, government-led Justice Sector Development Programme, funded by DfID and managed by the British Council, has launched pilot projects in the Western Region and Moyamba district to restore the rule of law, prevent further conflict and improve access to affordable justice for the poor. Among its objectives are to revise outdated laws, speed up case resolution, improve police responses to community needs and reduce prison congestion.

But there is a long way to go. The Law Society building in Freetown has no electricity, so judges have to light their chambers with candles, and they must take turns being transported to court.[14] Nor is there any certainty that available resources will be used wisely. Frederick Carew, as attorney general, recently recruited three new judges on six-month contracts to serve in the High Court and Appeal Court. This decision has been criticised because the three new judges were paid between £5,000 and £8,000 (US$10,000 and US$16,000) per month, while other judges in the

10 Ibid.
11 *The Monitor*, vol. 18, 2006. This is the official newsletter of the Sierra Leone Court Monitoring Programme; see www.slcmp.org.
12 Ibid.
13 Ibid.
14 Ibid.

hard-pressed judiciary were earning less than £1,000 per month.

Nothing for free: corruption in education

The US$40 million Sababu education project was launched in 2002 by the government, World Bank and African Development Bank with the goals of restoring basic education and providing vocational skills. The project encompassed the construction and rehabilitation of classrooms, the purchase and distribution of textbooks, and skills training, including teacher training. Due to end in December 2007, Sababu covers all 4,328 primary schools in Sierra Leone's chiefdoms and districts.

Legislation mandates free primary education for all, and primary textbooks are also supposedly free of charge. Sadly, however, this is not always the case. Commentators have called into question many of the decisions of the Ministry of Education, Science and Technology throughout the years – for example, the trial and conviction in 2001 of the former director general of education, Soluku Bockarie, for misappropriating US$1 million from the 'salaries' of 26,000 'ghost' teachers.[15] This came to light only after a Public Expenditure and Tracking Survey (PETS) was set up to trace the flow of resources downwards from ministries to ascertain any points of leakage.

These problems continued in the Sababu project and were again uncovered due to a PETS. In December 2006 twenty-three contractors building schools in Kailahun and Kenema districts were dismissed for using substandard materials, following a PETS inspection visit to the sites two months previously.[16] Similarly, primary textbooks procured for the project have found their way into the black market and have been sold.

According to research from 2006,[17] 32 and 23 per cent of pupils in Kenema and Kailahun districts, respectively, were not provided with free textbooks. In four districts surveyed, only 10 per cent of children received textbooks for mathematics, 8 per cent for English and 6 per cent for general science. As for the government practice of supplying school fees for primary pupils, the 2004 PETS revealed that, of a total allocation of L980.8 million (US$332,000), only 45 per cent was received by schools, and an estimated L587.9 million (US$200,000) worth of teaching and learning materials were not accounted for.[18]

Teachers augment their salaries by charging for private classes that children are forced to take if they hope to pass exams. It is in these classes that pupils actually learn what they should be learning in school. One parent said he had to pay L5,000 (US$1.50) per subject per month for extra lessons, and that applied to each of seven subjects.[19] According to Samuel Brima, lecturer at the University of Sierra Leone, admission to junior secondary school amounts to US$66 per child, against the annual average income of US$150–200.[20]

Yusuf Umaru Dalhatu (National Accountability Group/TI local partner, Sierra Leone)

Further reading

Campaign for Good Governance, 'Report on Basic Education in Sierra Leone', March 2006; see

15 S. Jabbi, 'The Sababu Education Project: A Negative Study of Post-war Reconstruction' (Freetown: National Accountability Group [NAG], 2007).

16 'Report on PETS Inspection Team' (Freetown: Ministry of Finance, 2006).

17 'Report on Basic Education in Sierra Leone' (Freetown: Campaign for Good Governance, 2006); see www.slcgg.org/Basic%20Education%20Report.doc.

18 Ibid.

19 *IRIN* (Kenya), 18 September 2007.

20 Ibid.

www.slcgg.org/Basic%20Education%20Report.doc.
Centre for Economic and Social Policy Analysis (CESPA), 'Service Delivery Perception Survey' (Freetown: CESPA, 2007).
DfID, 'Annual Review 2006 of Support to the Anti-Corruption Commission Phase 2 in Sierra Leone' (London: DfID, 2006).
Ministry of Finance, 'Public Expenditure Tracking Survey' (Freetown: Ministry of Finance, 2002–5).
NAG, 'Participatory Service Delivery Assessment of the Activities of 19 Local Councils in the Health and Sanitation, Agricultural and Educational Sectors: Citizen Report Card' (Freetown: NAG, 2006).
'Dissemination of the National Anti-Corruption Strategy: Final Report' (Freetown: NAG, 2007).
National Accountability Group, Sierra Leone: www.accountability-sl.org.

Zambia

Corruption Perceptions Index 2007: 2.6 (123rd out of 180 countries)

Conventions

African Union Convention on Preventing and Combating Corruption (signed August 2005; ratified March 2007)
UN Convention against Corruption (signed December 2003; not yet ratified)
UN Convention against Transnational Organized Crime (accession April 2005)

Legal and Institutional changes

- To enhance transparency and accountability during elections, the **Electoral Act was passed in July 2006**, repealing previous legislation. In effect, corruption in the electoral process is criminalised. The law criminalises influencing voters, either directly or indirectly, by any means such as offering to give, lend or procure money.[1] It is also illegal to offer any inducement or reward for joining a party, attending a political event or influencing candidates or nominations. It is a crime to abuse a position of power, privilege or influence for political purposes, to use government facilities for campaign purposes (this does not apply to the president or vice-president) or to use government transportation or facilities for assisting voters to polling stations.[2]
- In September 2007 the **Anti-Corruption Commission Bill** was in its final stages, having been debated since 2004. It seeks to

1 Electoral Act, no. 12 of 2006.
2 Electoral (Conduct) Regulations 2006.

strengthen the existing Anti-Corruption Commission Act, proposing to offer protection for whistleblowers and criminalise further acts of corruption associated with elections. The bill has been widely consulted, with input from many stakeholders, including those in the private sector, parliamentarians and civil society.

- In 2006 the **Zambia Development Agency Act** was passed. This act effectively merges five institutions: the Zambia Privatization Agency, the Zambia Investment Centre, the Export Board of Zambia, the Zambia Export Processing Zones Authority and the Small Enterprises Development Board. The act will enable the government to channel resources more easily to one agency rather than five, thereby reducing bureaucracy. The aim of the agency is to 'foster economic growth and development'. It is structured to ensure efficiency in the delivery of services and accounting, and will help the agency bring an end to the bureaucratic problems that have been frustrating potential investors by regulating the investment industry.[3]
- In May 2006 the **Judicial Code of Conduct Act** made amendments to a 1999 act of the same name. These amendments strengthened and renamed the Judicial Complaints Committee as the Judicial Complaints Authority. This enhanced the notion that the public can seek remedial redress in cases of alleged misconduct by judicial officers. The Judicial Code of Conduct encourages judicial officers to uphold the integrity, independence and impartiality of the judicature in accordance with the constitution, code of conduct or any other law.[4]

Zambian elections improve but still marred by malpractice

Although the September 2006 elections were generally held to have been an improvement on 2001, they were still marred by malpractice. The new Electoral (Amendment) Act 2006[5] was in place, outlining illegal practices relating to elections, including bribery and various forms of vote-buying. The act was intended to meet the demands of stakeholders for minimum standards of 'free and fair' elections.

The 2006 tripartite elections were monitored by several civil society and international organisations.[6] The general opinion was that the 2006 election campaigns were flawed by malpractice and corruption, beginning with the issuance of National Registration Cards (NRCs). All Zambian citizens aged eighteen and above and in possession of an NRC are eligible to register as voters.[7] According to the Anti-Voter Apathy Project (AVAP), the issuance of NRCs was politicised, as some candidates directed registration officers to concentrate on their constituencies, thus depriving other areas and disenfranchising thousands of people.[8]

The registration process supervised by the Electoral Commission was successful initially, but resulted in confusion at the voters' verification stage. There were numerous anomalies on the commission's records, mainly due to poorly

3 *Times of Zambia* (Zambia), 14–21 March 2006.

4 Judicial Code of Conduct Act, no. 13 of 1999.

5 The Electoral (Amendment) Act and its code of conduct were enacted in May 2006, only four months before the elections.

6 Including TI Zambia, the Southern African Centre for Constructive Resolution of Disputes (SACCORD), the Anti-Voter Apathy Project (AVAP) and the Foundation for Democratic Process (FODEP). The European Union Election Observation Mission (EU EOM) was among the major foreign observers.

7 Electoral (Amendment) Act, no. 12 of 2006.

8 D. Phiri and D. Mumba (eds.), 'The Anti-Voter Apathy Project: Zambia's 2006 Tripartite Elections Report' (Lusaka: AVAP, 2006).

trained electoral staff. AVAP observed double and/or multiple voter registration, misspelled details of registered voters, missing or incomplete details of registered voters, portraits of registered voters missing or swapped, and inappropriate recording of dates of birth and other details.[9]

The most common form of corruption was vote-buying. Practices included collecting NRCs from unsuspecting registered voters, donating goods to traditional authorities and 'treating' would-be voters.[10] For example, in Chipata Central, the ruling Movement for Multi-party Democracy (MMD), led by Sinoya Mwale and Getrude Sakala, was caught collecting cards from registered voters, and, in Lusaka Central, MMD candidate Rose Zimba was reportedly distributing beer to voters.[11] TI Zambia collected and documented over thirty cases of electoral malpractices, which were reported to the Anti-Corruption Commission (ACC).[12] By September 2007, however, no feedback from the commission had been received.

EU observers also noted bias in the use of the public media during campaigns.[13] Despite each contesting party having the opportunity to participate in a television interview called *Elections 2006*, in general the coverage of the Zambia National Broadcasting Corporation (ZNBC) TV and radio clearly gave more time to the ruling MMD compared to opposition parties. There were also several examples of development projects being announced by the government just two months before the elections. As a means to encourage votes, some thirty projects were suddenly to be funded to the tune of K518.6 billion (US$156 million).[14]

During the elections there were further examples of irregularities. At one polling station in Chilenje, a woman was seen by fellow voters casting two presidential ballot papers.[15] This led to confusion, as other voters demanded two presidential ballot papers as well. In many cases there were inadequate electoral materials: at one polling station in Bangweulu constituency, too few local government ballot papers were delivered, disenfranchising a number of voters.[16]

Once the votes were cast, major discrepancies were found between the number of ballots for parliamentary and presidential candidates announced at polling stations, and those announced for the same candidates at collation centres. For example, in Munali constituency the total sum of votes recorded for presidential candidates and parliamentary total votes varied by a significant margin (some 20,000 votes).[17]

Considering this evidence, it is not surprising that, within eight months of the elections, three parliamentary seats (Kapoche, Nalolo and Mbala) were nullified by the High Court, citing overwhelming evidence of such illegal practices as vote-buying, bribery, false statements and voter intimidation.[18]

Despite these issues, there is some cause for optimism. Positive steps were made, such as involving civil society in the reception of ballot papers

9 Ibid.
10 Ibid.
11 Ibid.
12 TI Zambia, press statement on the 2006 tripartite elections, 10 October 2006.
13 'European Union Election Observation Mission Zambia 2006' (European Union, November 2006).
14 C. Sikanyika, '2006 Election Monitoring Report Zambia Tripartite Elections 28th September 2006' (Lusaka: SACCORD, 2006).
15 TI Zambia, 2006.
16 D. Phiri and D. Mumba, 2006.
17 Ibid.
18 *Zambia Daily Mail* (Zambia), 21 April 2007; *The Post* (Zambia), 22 May 2007.

to verify the numbers received and working with civil society to inform people, especially those in rural areas, about changes to the electoral process. Furthermore, this progress was made in a relatively short time, from the passing of the new electoral act in May 2006 to the elections in September 2006. It is now up to the present government and the Electoral Commission to ensure that all stakeholders, the public and political parties are sensitised to the new law in time for the next elections, due in 2011.

Instilling integrity in the Zambian public sector

In May 2006 the Zambian government signed a two-year Threshold Programme worth US$22.7 million with the Millennium Challenge Corporation (MCC). The programme is intended to reduce corruption and improve government effectiveness through three components: preventing corruption in government institutions, improving the effectiveness of public services, and improving the management of trade at the borders.[19]

Part of this anti-corruption drive, instituted by the Anti-Corruption Commission, included the establishment of Integrity Committees within government ministries, departments and agencies. Integrity Committees have been formed in eight government institutions. The first institution to establish such a committee, as well as a code of ethics, was the Zambia Revenue Authority, in May 2007.[20] This was closely followed by the Ministry of Lands, the Immigration Department, Lusaka City Council, Ndola City Council, the Public Service Pensions Fund and the Zambia Police Service. In order to lead by example, the ACC itself has also formed an Integrity Committee.[21] The public sector is seen as an essential focus for anti-corruption initiatives. The Zambia National Governance Baseline Survey found that almost 40 per cent of people surveyed had experienced being asked to pay bribes in order to obtain public services.[22]

The Integrity Committees are premised on the belief that all organisations should be responsible for preventing corruption. Integrity Committee members are appointed from within organisations and should normally include at least four members at a senior level with a direct reporting line to the chief executive. They should be 'tasked with taking steps to prevent corruption within their organization's sphere of control'.[23]

Steps to prevent corruption include producing an institutional corruption prevention plan, reports of which will be made to the Cabinet Office on a quarterly basis; participating in training senior and middle management on service delivery, ethics and integrity, and measures of transparency; enlisting support and understanding of anti-corruption work; ensuring that codes of conduct and ethics are understood and compliance is enforced; receiving and considering complaints from within and outside the institution; and recommending administrative action to management in response to complaints.[24] Integrity Committees should also provide an interface between the public and the ACC, as they will receive and collate complaints from the public and forward them to the ACC for further investigation and prosecution, reducing the ACC's workload.[25]

19 For more information, see MCC, 'Zambia Fact Sheet' (Washington, DC: MCC, 2006).
20 *People's Daily Online* (China), 8 May 2007. For more information, see www.zra.org.zm/Committee.php.
21 The Anti-Corruption Commission of Zambia, Department for Corruption Prevention.
22 For more information, see www.worldbank.org/wbi/governance/zambia/pdf/GBS%20RPT%20PRESENTN.pdf.
23 www.zambiaimmigration.gov.zm/zims/The_Integrity_Committee_Initiative.aspx.
24 Ibid.
25 Ibid.

In essence, the idea is widely supported. Because of these opportunities for improving the interface between the public and the ACC, the Integrity Committees, as mediator between the two, will be in a position to follow up on matters they forward to the ACC on behalf of their clients. Despite the benefits, Integrity Committees are faced with challenges. The organisations in which many of them operate are very large and complex, posing difficulties for what is a small group of individuals (four committee members), who will also be involved in substantive organisational responsibilities. In addition, as the members will be embedded within the system, there is the danger that some individuals will be reluctant to report cases of corruption for fear of victimisation.

If not properly implemented, Integrity Committees are likely to suffer from rigidity and possible rejection from within institutions. The committees' legitimacy will depend on a number of factors, including ensuring adequate oversight of activities by the Cabinet Office, shielding the committees from superfluous bureaucracy and potential conflicts of interest, and ensuring that committee appointments are made in a transparent way. It is essential for the initiative's success that Integrity Committees retain their own internal integrity by being independent of their organisations' internal management structures.

The trials and tribulations of former President Chiluba

On 4 May 2007 the former Zambian president, Frederick Jacob Titus Chiluba, was found liable by the High Court in London for defrauding US$46 million of public money. The civil case in London was conducted simultaneously with an ongoing criminal case against the former president and his associates in Zambia.

Chiluba was elected president in 1991. He was replaced in January 2002, following defeat in the presidential election, by Levy Patrick Mwanawasa. On 11 July 2002 President Mwanawasa addressed the Zambian parliament, detailing a number of allegations of corruption against his predecessor and making the case for revoking the former president's immunity to prosecution. This statement was broadly welcomed, and followed demonstrations from civil society organisations calling for Chiluba's immunity to be removed so that he could face prosecution for widely publicised corruption allegations (see *Global Corruption Report 2004*).

The government formed a Task Force on Corruption in July 2002 to deal with alleged cases of corruption committed during Chiluba's regime. In December 2003 Chiluba was charged with 168 counts of theft totalling more than US$40 million of public money.[26] He was accused of conspiring with senior officials to divert public funds to a London bank account for private use. Chiluba denied the charges and insisted that the investigation was a political witch-hunt.

The government also began civil proceedings in the London High Court in December 2003 with a view to recovering some of the alleged stolen public funds. The case was brought to London for a number of reasons, not least because the money had passed through London bank accounts before being dispersed. There were also claims against defendants from many different countries, including the United Kingdom, Belgium, Switzerland, the United States and the British Virgin Islands, so the United Kingdom was considered the appropriate jurisdiction in which the various claims could be properly determined.[27]

26 *BBC News*, (UK), 9 December 2003.
27 *The Lawyer.com* (UK), 23 July 2007.

The defendants, Chiluba, Stella Chibanda, Aaron Chungu, Faustin Kabwe and Francis Kaunda, issued applications challenging the decision to commence the proceedings in the UK court. They argued they should have had the right to attend their trial in person, but, as they were simultaneously on trial in Zambia on criminal charges, their passports had been taken from them. Moreover, they argued that a fair trial could not take place in London without jeopardising the criminal proceedings in Zambia.[28]

The application was rejected, however, with the concession that the judge in the case, Justice Peter Smith, would be willing to go to Zambia as a special examiner to hear the defendants' evidence; the remainder of the proceedings would take place via video link; the trial and all interlocutory applications would take place in private; and materials produced during the civil proceedings would be disqualified from being used in criminal proceedings without the court's permission.[29]

The London High Court ruling found Chiluba and nineteen others liable for defrauding the Zambian government and ordered them to pay back 85 per cent of the funds. Chiluba dismissed the judgment as political, however. The registration of the London judgment is being challenged in the Zambian courts and the trial on the criminal aspect of the case is still ongoing in Zambia.

Louis Bwalya, Goodwell Lungu and Kavwanga Yambayamba (TI Zambia)

Further reading

D. Phiri and D. Mumba (eds.), 'The Anti-Voter Apathy Project: Zambia's 2006 Tripartite Elections Report' (Lusaka: AVAP, 2006).

C. Sikanyika, '2006 Election Monitoring Report Zambia Tripartite Elections 28th September 2006' (Lusaka: SACCORD, 2006).

TI Zambia: www.tizambia.org.zm.

28 Ibid.
29 Ibid.

7.2 Americas

Argentina

Corruption Perceptions Index 2007: 2.9 (105th out of 180 countries)

Conventions

OAS Inter-American Convention against Corruption (signed March 1996; ratified August 1997)
OECD Convention on Combating Bribery of Foreign Public Officials (signed December 1997; ratified February 2001)
UN Convention against Corruption (signed December 2003; ratified August 2006)
UN Convention against Transnational Organized Crime (signed December 2000; ratified November 2002)

Legal and institutional changes

- In December 2006 Congress passed the **law on political parties' financing**, which modifies the law on political party and electoral campaign finance. The maximum permitted expenditure for political parties in electoral campaigns increases from P1 (US$0.32) to P1.50 per citizen. The law also stipulates that the bank accounts in which political parties deposit funds designated to sustain their campaign activities should be unified. The old law obliged parties to maintain two different accounts, one for current expenditure and another for expenditure during electoral campaigns. The new law also prohibits contracting publicity services from third parties. Given that the law is important only if put into practice during elections, it is too early to comment on its value.
- A 1992 law on financial administration regulated the reach and method of **modifications to budgetary law**. Decisions that affected changes in expenses, financial applications and the purposes of the general budget were reserved to Congress. Since August 2006, however, the head of the Cabinet has had the power to modify the total amount approved by each law of budget.
- At the end of April 2007 a draft law was introduced **reforming the law on financial administration**. The constitution regards the National Auditing Office (AGN) as an autonomous public body responsible for providing the legislative branch with technical assistance. The draft sought to strip the AGN of its autonomy by granting highly discretionary power to the commission responsible for coordinating the AGN's activities with Congress to make decisions on the AGN's behalf. Civil society organisations pronounced against the draft law, which is now suspended.[1]

1 *La Nación* (Argentina), 21 May 2007.

Putting the brakes on government by decree?

In July 2006, after a long period of non-compliance, Congress sanctioned the Decrees of Need and Urgency Regime, which calls for the creation of a permanent, bicameral committee to validate decrees of need and urgency (DNUs) ordered by the president.

These decrees became common during the presidency of Carlos Menem (1989–99) and were heavily used by his successors, Fernando De La Rua, Eduardo Duhalde and Nestor Kirchner. Decrees of need and urgency enabled presidents to issue orders that were legislative in character, though they clearly undermined the role of Congress. In a Congress controlled by the executive, as now, DNUs mean that the president can rule unchecked.

According to the law, the new committee:

- is composed of eight deputies and eight senators, in proportion to their parties' representation in each chamber;
- operates even when Congress is not in session;
- decides its rulings on an absolute majority of votes;
- issues a statement on a decree's validity or invalidity at plenary sessions of each chamber of Congress; and
- has ten working days to arrive at a decision and to raise it in plenary sessions of each chamber of Congress.

Once a decision has been raised at the plenary sessions, both chambers must give it immediate treatment. If ten days pass without an official statement, both chambers will see to the express treatment of the decree.

There were two issues that mainly concerned members of Congress. The first involved the period of time allotted for Congress's consideration of DNUs. In the law finally sanctioned, time periods were not specified; the law states only that both chambers ought to 'give immediate treatment' to the committee's judgment.

The second issue concerns what becomes of DNUs that Congress does not review, a situation that may arise given that no specific time frame has been prescribed. The laws regulating DNUs state that they are valid until abolished, by express will, in both chambers of Congress. This would imply that, when Congress fails to review issues raised by the committee, there would be an artificial sanctioning of the decree – a practice that the constitution expressly prohibits.

The passing of the law provoked numerous questions among legislators, as well as among constitutional specialists. A case was brought before the Supreme Court in a bid to declare the unconstitutionality of the law creating the bicameral committee. It argued that the law violates the separation of powers, and establishes a system of 'tacitly sanctioning laws'.[2]

It is unlikely that the bicameral committee will succeed in becoming an effective mechanism for regulating DNUs. This appears to be even more so the case considering that the Senate constitutional affairs committee charged with the treatment of the law is chaired by Senator Cristina Kirchner, the president's wife, who was herself standing for office in October 2007. As an opposition senator in the 1990s, she opposed such a solution. As a possible president, she is likely to be even more convinced of the necessity to retain the option of decrees of need and urgency.

2 Asociación por los Derechos Civiles, 'Acción de Amparo contra la Ley que Reglamenta los Decretos de Necesidad y Urgencia', 15 December 2006; see www.haciendocumbre.org.ar.

Corruption and a gas pipeline

The case of the Skanska pipeline is important, because it reflects on the integrity of the Kirchner presidency, which, from its beginnings in 2003, has sought to project the image of a political culture free from corruption.

The case concerns the construction of a natural gas pipeline from Bolivia in order to make up shortfalls in gas supply caused by the devaluation of the peso in 2002 and the consequent freezing of tariffs.[3] Transportadora de Gas del Norte (TNG), the pipeline operator, organised competitive tenders. Some were won by Skanska, a Swedish company, despite complaints by TNG that the bids were coming in at too high a price. Nevertheless, the regulator instructed TNG to pay the excess amount, and still maintains that the operator's estimates were too low.[4]

Following an anonymous tip-off and an internal audit, evidence emerged that Skanska may have paid bribes to win the contract and been involved in tax evasion. Evidence was found that the company received receipts from Infiniti, which is considered a 'phantom company', and that Skanska executives had paid out P13.4 million (US$4.3 million) in bribes.[5] Skanska sacked seven of its executives – who were subsequently arrested – and paid the tax authorities US$5 million.[6]

The president tried to dissociate the public authorities from the scandal, maintaining that it was purely private sector corruption. In May he acknowledged that government officials may have been involved, however, and two officials under investigation were fired.[7]

There have been other developments that indicate a political interest in ensuring that the investigation is not conducted with the necessary independence or rigour. First, the case has been split between two different courts, on the grounds that two separate offences had been committed: bribery and tax evasion. As a result, the judge who initiated the investigation, and is therefore the most qualified to lead inquiries, has been prevented from working on the part of the investigation pertaining to bribery. Meanwhile, the judge investigating the bribery offence has been offered the post of security minister in the city of Buenos Aires, an offer he promptly accepted.[8] Similarly, the public prosecutor has been offered a Cabinet post in the province of Buenos Aires. This signals a serious threat to the investigation, because these kinds of cases need time, patience and strong levels of detailed knowledge relevant to the trial. If any jury backs off, the replacement will need a long period of adaptation, further delaying resolution of the case.

When acting judges are a threat to judicial independence

Federal judges of lower courts are elected by the Judicial Council, a multi-sector organ with representatives from the three branches of state. The selection commission makes an open call for the submission of résumés and objections, and issues a report that the plenary body uses to select the three most qualified candidates. That list is sent to the national executive branch, which convenes a public audience and presents observations about the candidates. The president then makes the choice and requests Senate approval. The lack of judicial appointments in accordance with the above

3 *The Economist* (UK), 10 May 2007.
4 Ibid.
5 *Offnews* (Argentina), 16 May 2007; *The Economist* (UK), 10 May 2007.
6 *The Economist* (UK), 10 May 2007.
7 *Reuters* (UK), 30 May 2007.
8 *La Nación* (Argentina), 29 September 2007.

process has led to the temporary status of acting judges being extended. At present, acting judges account for nearly 20 per cent of all working judges.[9]

The problem is that, once the list of qualified candidates has been sent to the president, it does not necessarily proceed to appointment. The Judicial Council passed Resolution 76/04, so that, in the case of prolonged vacancies (more than sixty days), it can designate a member from the list of the three most qualified candidates from the court that produced the vacancy. A judge so designated continues his or her duties until the definitive cessation of the vacancy, or until twelve months have passed, which can be extended by six more months by a well-founded decision.[10] As a result, these judges, who provisionally assume the role of doling out justice, do not enjoy the privilege of tenure.

In view of the discretion the council can exert in its nominations and removals, this puts temporary judges in an extremely vulnerable situation and has an impact on judicial independence.[11] Additionally, in cases in which acting judges become part of the list of the most qualified candidates – after having successfully passed the selection process – the delay aggravates their vulnerability, as they are already waiting for appointment and Senate approval.

This way of appointing judges is contrary to the standards and principles designed to measure and guarantee the independence of the judiciary. The Basic Principles on the Independence of the Judiciary state that '[a]ny method of judicial selection shall safeguard against judicial appointments for improper motives'. Moreover, the Inter-American Convention against Corruption presupposes publicity as a requirement for selection procedures. None of these requirements is heeded when the Judicial Council choose judges from a list of lawyers or secretaries of lower courts.

On 24 May 2007 the Supreme Court declared the judicial substitutions system unconstitutional and ordered Congress and the executive to establish a definitive system for regulating the replacement of judicial vacancies in accordance with judicial resolutions within a year.[12]

Water failures down to poor regulation and inadequate sanctions

According to World Bank data for 2006, an index displaying failures in water supply ranks Argentina lower than the region and the world (6.09, 13.32 and 13.57, respectively).[13] A report by the general auditor indicated that, as of July 2004, the company responsible for providing water services in Argentina, along with the corresponding regulatory body, had not adequately fulfilled the terms of its mandate.[14] It had not maintained adequate water quality and had failed to meet deadlines for the development of network infrastructure. In addition, the company failed to make significant investments despite this obligation being stipulated in its contract.

9 Ministry of Justice response to an access to information request by Poder Ciudadano. See the Report on Acting Judges, available at www.poderciudadano.org/up_downloads/news/272_1.pdf.

10 Law 25.876, article 1.

11 It should be emphasised that, from November 2006, the composition of the Judicial Council has been reformed. The dominant political party has broken the equilibrium that the constitution established for this multi-sectoral body.

12 National Supreme Court of Justice in *Rosza, Carlos Alberto y otro s/ recurso de casación,* R.1309 XLII; available at www.mpf.gov.ar/Novedades/R%201309%20L%20XLII%20ROSZA.pdf.

13 World Bank and IFC, 'Encuesta de Empresas' (Washington, DC: World Bank and IFC, 2007).

14 See www.agn.gov.ar/informes/Aguas.pdf.

The report also suggested that penalties imposed by the regulatory body for such breaches proved ineffectual. For example, the company promised to invest US$2.5 million in a treatment plan in the south-east in 2001. When the company failed to make the investment, the regulatory body fined it a mere US$8,740 – less than 1 per cent of the planned investment. Furthermore, some systems and item provisions did not follow contractual technical specifications. For six years of the company's contract excessive levels of nitrate and chromium were found in the water network, and for four consecutive years studies cited an increase in the number of wells containing arsenic. In 2003, while the company was transporting the sewage of 5.7 million inhabitants, it was treating only 12 per cent of the waste, allowing the rest to spill into Rio de la Plata without adequate treatment.[15]

Although there are no direct allegations of corruption, it is possible to identify a number of irregularities related to non-fulfilment of contracts, neglect of regulations and an absence of appropriate sanctions for non-compliance.

Federico Arenoso, Gastón Rosenberg, Martín Astarita, Pablo Secchi, Varina Suleiman and Lucila Polzinetti (Poder Ciudadano/TI Argentina)

Further reading

P. Arcidiácono *et al.*, *Vulnerable Public Contracting* (Buenos Aires: Fundación Poder Ciudadano and Talleres Gráficos Manchita, 2006).

Centro de Análisis e Investigación-México, *Latin American Index of Budget Transparency* (Managua: CISAS, 2005).

Fundación Poder Ciudadano, in collaboration with Centro de Implementación de Políticas Públicas para la Equidad y el Crecimiento (CIPPEC), 'Informe de Sociedad Civil en el Marco de la Segunda Ronda del Mecanismos de Seguimiento de la Implementación de la Convención Interamericana Contra la Corrupción' (Buenos Aires: Poder Ciudadano and CIPPEC, 2006).

M. J. Pérez Tort and C. Ribeiro dos Santos, *Una Mirada Atenta sobre el Consejo de la Magistratura* (Buenos Aires: Fundación Poder Ciudadano and Talleres Gráficos Manchita, 2006).

S. Rose-Ackerman, *Corruption and Government: Causes, Consequences and Reform* (Cambridge: Cambridge University Press, 1999).

Poder Ciudadano: www.poderciudadano.org.

15 Ibid.

Chile

Corruption Perceptions Index 2007: 7.0 (22nd out of 180 countries)

Conventions

OAS Inter-American Convention against Corruption (signed March 1996; ratified September 1998)

OECD Convention on Combating Bribery of Foreign Public Officials (signed December 1997; ratified April 2001)

UN Convention against Corruption (signed December 2003; ratified September 2006)

UN Convention against Transnational Organized Crime (signed December 2000; ratified November 2004)

Legal and institutional changes

- **Eight bills were introduced** in December 2006 addressing transparency, anti-corruption measures and a new era of constitutional reform; some of the most important are discussed below.
- An **Access to Information Bill** was intended to bring Chile into line with global standards of transparency of information (see below). The bill considers the creation of an Access to Information Commission, composed of four members, to act as an appeal body for citizens whose requests for information have been refused. It was expected to pass Congress by the end of 2007.
- A **Constitutional Reform Bill** aims to resolve a number of issues related to congressional and political party matters. The bill seeks to establish the public nature of the declarations of assets and interests required of MPs, government officials and judges.[1] This was made necessary by a Constitutional Court ruling in 2005 that a conflict existed between the right to privacy and the public's right to know. The bill would also authorise the holding of primaries for the selection of party candidates in congressional elections, with public funding for those campaigns; partial state funding for party campaigns; increased power to dismiss members of Congress who have contravened regulations on political campaign finance;[2] and tighter conflict of interest regulations. Members of Congress who are lawyers would not be allowed to participate in court cases, since senators approve nominations to the Supreme Court; nor would they be able to vote in legislation affecting their personal interests.

1 The declaration of private interests by members of Congress, mayors, judges, high-ranking government officials – including ministers – and other authorities was established under Law 19.653 in 1999 and the assets statement requirement was required under Law 20.088 in 2005.

2 The financing of political party campaigns was regulated for the first time by Law 19.884 in 2003. The legislation set limits on political party contributions from private, anonymous and corporate sources, and also introduced the principle of public funding being allocated according to the number of votes obtained in the previous election.

Finally, the bill aims to widen the oversight of the state comptroller to state-owned companies and not-for-profit institutions that receive public funding.

- A **Revolving-Door Bill** would disqualify for one year some senior officials from accepting jobs in companies supervised by the bodies for which they once worked, and provides financial compensation for the ban. The same bill would prevent civil servants from working as lobbyists for two years after leaving office. Until now the law determined a six-month ban and for only a limited number of authorities.[3]
- Another bill seeks to improve the system for **depoliticising the civil service**. Created in 2003, the *Sistema de Alta Dirección Pública*, or High Public Management System, established a more objective human resources policy aimed at limiting the power of the presidency to appoint over 3,000 political supporters to positions in the administration. The new system intends to reduce this traditional network of patronage significantly, to about 600.[4] The bill would eliminate weaknesses in the selection procedure and tighten the time frame for filling posts currently occupied by political appointees.
- **A bill for whistleblowers** was intended to protect public officials who have exposed institutional corruption. It passed through Congress in July 2007. Although a general improvement in the legal framework, its powers to protect whistleblowers are actually quite weak. Few public officials will be encouraged to risk careers and reputations under such conditions.
- While presenting this legislative agenda, the government enacted a **presidential decree on transparency and access to information** that obliges all twenty ministries and 240 agencies of the executive branch to publish a host of information relevant to the public on their websites (see below).

Back to the transparency agenda

Though corruption is far less entrenched in Chile than in other Latin American countries, it remains a recurrent phenomenon that is difficult to uproot. The most serious cases in the past decade have tended to be associated with the use of public funds for political campaign purposes.[5]

Cases of corruption revealed in 2002 led to a political agreement between the centre-left government of Ricardo Lagos and the opposition.[6] The agreement included elements of reform that promoted greater transparency and better instruments for controlling corruption. For the first time, legislation was passed to regulate campaign financing, establish formulas for transparency and introduce a degree of public funding that enabled greater political competition and reduced pressure on candidates to solicit campaign funds. It also created a more open system of public procurement and contracting through internet publication. Declarations of interests and assets were also introduced for public officials.

Significant as these steps were, their implementation showed up important flaws. In political finance regulation, the implementation of three

3 Contained in article 56 of Act 18.575, as amended through Act 19.653 of 1999.

4 See www.lyd.com/LYD/Controls/Neochannels/Neo_CH3915/deploy/exp%2007%20ADP%20uaiLyDres.pdf.

5 According to the World Economic Forum Executive Opinion Survey, only 3 per cent of businesspeople pointed to corruption as one of three main problems for those doing business in Chile. Around a half, however, expressed concern about irregularities in political campaign funding and the difficulty of gaining access to information. See A. Bellver and D. Kaufmann, 'Transparenting Transparency: Initial Empirics and Policy Applications', preliminary draft discussion paper, presented at the IMF Conference on Transparency and Integrity, Washington, DC, July 2005.

6 'Acuerdos Politico-Legislativo para la Modernización del Estado, la Transparencia y la Promoción del Crecimiento' (2003); available at www.modernizacion.cl/1350/articles-47984_Acuerdo.pdf.

ways of designating private money (anonymous, secret and public) has proved problematic. In the case of public contributions, the purpose was to ensure that when the donation surpassed a certain amount the donor should be known to both the candidate and the public. In some cases it has proved hard to verify the real identity of the donor, since false names may be used; in some cases this means that only the candidate can identify the donor. In the case of anonymous contributions – allowed for small amounts – the fragmentation of the real contribution makes it difficult to prove that the threshold for each donor has not been exceeded.[7]

In November 2006 President Michelle Bachelet formed an Experts' Commission to draw up proposals to control corruption and improve probity in public affairs. The commission consisted of economy minister Alejandro Ferreiro Yazigi, under-secretary for finance María Olivia Recart Herrera, Carlos Carmona from the legal division of the Ministry of the Secretary General of the Presidency, and a group of experts, including Davor Harasic, president of Chile Transparente, TI's national chapter.

The group was brought together in reaction to a series of scandals related to campaign funding during the legislative election of December 2005. Several investigations revealed the diversion of public funds from the Sports Ministry agency Chiledeportes, and several emergency employment programmes. The Chiledeportes case was uncovered during an audit of dozens of small programmes conducted by the state comptroller. A considerable number of the programmes could not justify their use of resources. Later it came to light that part of the funds had ended up supporting political campaigns.[8]

Further irregularities, involving false invoicing, were discovered in the campaign financing reports of a few MPs. Using this technique, candidates could justify requests for more public funding while avoiding the obligation to return unspent private contributions. The investigation found that officials of the tax office had worked with a paper company that organised fraudulent invoices to evade taxes. In November 2006 the head of the electoral service filed a suit at the prosecutor's office to investigate irregularities in the accounting reports of some candidates.

The Experts' Commission made over thirty separate recommendations for changes in the institutional structure,[9] which President Bachelet announced on the same day that Chile ratified the UN Convention against Corruption.[10] The legislative agenda was broadly put into practice through the package of proposals sent to Congress in December 2006. Though these moves in favour of greater transparency are significant, whether they will have any observable impact on opportunistic behaviour will depend on their final design and President Bachelet's political commitment.

Law on access to information and presidential decree

On the Experts' Commission's advice, the government produced a bill on access to public information in December 2006. It also accepted a proposal to create a specialised autonomous body for access to information, with powers of control, supervision and discipline. Mexico's Federal Institute of Access to Public Information and the United Kingdom's Information Commissioner were the preferred models for similar institutions in Chile.

7 See www.cepchile.cl/dms/lang_1/buscar.html?tipologica=or&textobuscar=salvador%20valdes&pagina=4.
8 Report of the special investigatory committee of Chiledeportes, available at www.camara.cl/comis/docINF.aspx?prmID=50.
9 See www.chiletransparente.cl/doc/InformeFinal06.pdf.
10 Available at www.modernizacion.cl/1350/article-137949.html.

The previous government had made a timid attempt in 1999 to make information more transparent, by passing a regulation known as the Probity Law.[11] This established a set of obligatory principles for officials, including a high level of transparency in their everyday activities. An administrative resolution authorised services to decree the secrecy criteria autonomously, however.[12] By 2005 some 100 resolutions had been passed that limited access to information on internal investigations, disciplinary proceedings, tenders, correspondence and pay, among other procedures.[13] Both the intention and the spirit of the law were clearly contravened.

The current bill does not grant Chile's access to information agency sufficient autonomy, proposing instead that its governing body should be appointed by the presidency. During the first parliamentary debate on the proposal it was conceded that Senate approval would be required for the appointment of the most senior board officials, but the legislative process had not ended at the time of writing.

The aim of the bill was to institutionalise the principles of transparency and access to public information incorporated in the 2005 constitutional reform. Shortly after introducing the bill to Congress the government proactively promulgated a presidential decree that obliged all ministries and departments to publish on the internet most of the information that the law, once passed, would have required. This includes complete lists of officials and advisers, current and concluded tender and purchasing processes, expenditures and fund transfers, information on the ownership of contractors and bidders, and normative frameworks and all resolutions affecting third parties.

This presidential decree brings about a significant change of direction, although past experience tends to show that any move towards greater transparency can be swiftly reversed. For this reason it is essential to establish an institution whose principal functions are to monitor full compliance with the law, apply sanctions, offer technical assistance and recommend legal, regulatory and procedural improvements.[14]

Civil society monitoring is essential to this end, but it is not sufficient. A *Corporación Humanas* study in 2006 highlighted restrictions in access to the assets and interests statements of members of Congress,[15] negating the very objectives that transparency policies aimed to achieve. These surreptitious forms of limitation are common and require a specialised independent body, able to guarantee that existing obligations are met. If access to this kind of information is denied or is accessible only with great difficulty, there can be no true accountability in government.

Felipe de Solar (TI Chile)

Further reading

A. Bellver and D. Kaufmann, 'Transparenting Transparency: Initial Empirics and Policy Applications'; available at worldbank.org/wbi/governance/pdf/Transparenting_Transparency171005.pdf.

A. Fung *et al.*, 'The Political Economy of Transparency: What Makes Disclosure Policies Sustainable?', Faculty Research Working Paper

11 Law 19.653.

12 Act 18.575, article 13, paragraph 11.

13 M. Sánchez, 'Secretismo de Chile: Revisión de la Práctica Administrativa 2001–05' (Santiago: Agrupación Defendamos la Ciudad, 2005).

14 D. Banisar, *Freedom of Information around the World: A Global Survey of Access to Government Information Laws* (Washington, DC: Privacy International, 2006).

15 See www.humanas.cl/documentos/RESUMEN%20ESTUDIO%20DECLARACION%20PATR%20abril07.pdf.

OP-03-04 (Harvard, MA: John F. Kennedy School of Government, Harvard University, 2004).
T. Mendel, *Freedom of Information: A Comparative Legal Survey* (Paris: UNESCO, 2003).
'Parliament and Access to Information: Working for Transparent Governance', working paper (Washington, DC: World Bank Institute, 2005).
J. Stiglitz, 'Transparency in Government', in *The Right to Tell: The Role of Mass Media in Economic Development* (Washington, DC: World Bank Institute, 2002).
Corporación Chile Transparente/TI Chile: www.chiletransparente.cl.

Mexico

Corruption Perceptions Index 2007: 3.5 (72nd out of 180 countries)

Conventions

OAS Inter-American Convention against Corruption (signed March 1996; ratified May 1997)
OECD Convention on Combating Bribery of Foreign Public Officials (signed December 1997; ratified May 1999)
UN Convention against Corruption (signed December 2003; ratified July 2004)
UN Convention against Transnational Organized Crime (signed December 2000; ratified March 2003)

Legal and institutional changes

- In August 2006 the Ministry of Public Administration released the **Annual Operative Programme of the Professional Career Service (SPC)**, a comprehensive document that establishes guidelines for the service's operation.[1] Twenty-four indicators are combined to evaluate the SPC of the federal public administration and identify measures to correct its functions. Of particular note are the indicators geared towards increasing competitiveness in public service hiring procedures and improving the evaluation of public servants' performance. These indicators help to eliminate spaces for discretional authority and other irregular practices in hiring processes, and confer transparency on the admission, evaluation and separation of public servants from their posts. The law for the SPC came into effect in 2003 and comprises close to 40,000 government posts subject to an open application process.[2]

1 See www.funcionpublica.gob.mx/pt/difusion_disposiciones_juridicas/doctos/POA_SISTEMA_2006_publicacion%2006.pdf.
2 Secretaría de la Función Pública (SFA), 'Informe de Labores' (Mexico City: SFA, 2007).

- In December 2006 Congress approved an amendment to article 73 of the constitution, granting it the autonomous power to set up **administrative courts to discipline public servants who harm the public interest** in the exercise of their duties. Such acts include the improper use of public funds and the illegal acquisition of goods. Although reporting such crimes is primarily incumbent on their managers, channels exist through which citizens can inform the authorities of wrongdoing by public servants. The reform will strengthen the independence of administrative courts, given that it grants a broad legislative framework that amplifies their scope of action and their powers.
- During its first year in government the administration of President Felipe Calderón presented its **Manifesto of Anti-Corruption Efforts and measures to promote accountability, transparency and access to information** on International Anti-Corruption Day on 9 December 2006. The manifesto focuses on six priority areas: accounting and fiscal reform; educational content; regulatory simplification; the promotion of transparency; institutional accountability; and the professionalisation of public service. To increase the efficiency of auditing, public accounting systems will be harmonised across the administration and auditing procedures in the federal executive branch will be intensified. The public administration ministry is seeking to strengthen the transparency institutions established by the outgoing Fox administration, including the Federal Institute of Access to Information (IFAI) and the SPC.
- The **National Network in Favour of Oral Trials** (*Red Nacional a Favor de los Juicios Orales*), a civil society organisation, submitted a constitutional reform bill in November 2006 to establish oral trials. The presidents of the Justice Commission and the Commission on Constitutional Issues have agreed to present it as a formal proposal to modify the federal judicial system. The proposal was in its final round of discussion when this text was submitted (see below).
- In April 2007 Congress passed an amendment to article 6 of the constitution, which refers to the **right of access to information**. The amendment obliges all levels of government – federal, state and municipal – to standardise their access to information laws according to international practices. States used to make their own provisions regarding access to information, and the procedures and content vary significantly. The objective is to oblige states to adopt practices that have proved successful both nationally and internationally, such as the use of electronic portals to request information and to submit an appeal when a request is denied or when the user is dissatisfied with the information provided. The amendment came into effect in July 2007.

Towards trial by spoken testimony

A major concern regarding the administration of justice is the high incidence of corruption (see the *Global Corruption Report 2007*). The justice system is founded on the creation of a written file, which serves as the basis for all cases and which, once composed, is submitted to a judge, who passes the appropriate sentence according to the arguments presented. The system draws out the process of resolving cases significantly. Together with the large number of documents handled when a judge is absent, it creates numerous opportunities for corruption and contributes to the public's alarmingly low level of confidence in judges and the justice system.

These concerns have generated vigorous debate over the years. It was only in November 2006, however, that a formal proposal to introduce oral trials was presented to Congress. An important aspect of the proposal was that it resulted from the work of the National Network in Favour of Oral Trials, which brings together representatives from academia, CSOs, the media, business groups and lawyers' associations. Several laws, such as the federal Law for Transparency and Access to Information, approved in 2002, were similarly crafted by fusing a citizens' proposal with an executive policy.

The network's intention is to reform five constitutional articles so as to incorporate oral trials into federal criminal cases. These include: restrictions on the use of preventive imprisonment; the prohibition of illegally obtained evidence; the incorporation of presumptive innocence into the constitution; principles of orality and publicity; the expansion of rights for victims of crimes; limitation on the district attorney's current monopoly of penal action; and a multiplication of the available channels for citizens to take direct recourse to justice. Some observers expect the bill, if passed, to unblock the courts' workload, rationalise the penal process and eliminate the opaque areas where corruption flourishes. By conferring more transparency on the many judicial decisions made in private, those involved should be better able to monitor the progress of their legal cases.

The bill seeks to construct a dual justice system, similar to that in Chile, in which minor cases are settled by reaching agreement on damage caused and the payment of a fine, while more serious cases are resolved through oral trials. It remains to be seen which courts will oversee each process. Trial exercises are currently under way.

The presidents of the Justice Commission and the Commission on Constitutional Issues received the bill in November 2006 and it is still being analysed and modified. President Calderón has declared himself in favour and encouraged legislators to put the bill to the vote as soon as possible. The climate appears to bode well.

There have been other projects to introduce oral trials at local level, some more ambitious than others. The most advanced state is Chihuahua, which instated a new penal code that includes oral trials in January 2007. In Nuevo Leon, the state of Mexico and Oaxaca, the implementation of similar provisions is under way, while oral trials are in the design or approval phase in six other states.

The quality and results of the first experiments have been mixed, as they depend to a great extent on state legislation. There has been a high degree of heterogeneity among the legal provisions, but the arguments in favour of oral trials are still little understood by the primary beneficiaries, the general population.

Access to information and what the 2006 election revealed

On 2 July 2006 Mexico witnessed the most contentious presidential election in its history. Never before had the count produced such a small margin between votes for the opposing candidates, in this case Andrés Manuel López Obrador of the left-leaning Coalition for the Good of All (*Coalición por el Bien de Todos*) and Felipe Calderón Hinojosa of the National Action Party (*Partido Acción Nacional*).

The Federal Electoral Institute (IFE) had to wait several weeks before announcing the official result, finally naming Calderón the winner with a majority of 0.56 per cent. López Obrador appealed for a recount to the Supreme Court in Electoral Matters (TEPJF), which approved a partial but representative recount of the votes cast in booths across the country. The recount repeated the result of the original and did little to resolve the controversy. On 5 August, the very day the verdict was reached, López Obrador reiterated his demand for a second, comprehensive recount of every vote cast. The TEPJF turned down the petition on technical grounds, later ruling that the electoral process had not been tainted by pervasive irregularities.[3]

That same month a group of CSOs and media outlets petitioned the IFE to grant them access to the electoral ballots through an online information request under the Federal Law for Transparency and Access to Information. The group was poised to conduct a 'citizens' recount' of ballots and had issued calls inviting interested parties to participate. A total of 16,806 people signed up.[4]

The initiative awoke the interest of several international organisations, including the National Security Archive and Global Exchange, which agreed to act as foreign observers. The IFE denied the group access to the ballots, on the grounds that they were not public documents and that releasing them would violate the confidentiality of the vote.

Faced with this obstacle, the network appealed to the Federal Institute of Access to Information, the agency in charge of upholding the right to public information as guaranteed by the Law for Transparency and Access to Information. Though lacking legal authority to impose sanctions, the IFAI can make recommendations. In August 2006 it issued a press release stating that the resolution of any electoral dispute remained beyond its jurisdiction. As a constitutionally autonomous entity, the IFE is not subject to the IFAI's authority.

Towards the end of April 2007 – nearly nine months after the election – the electoral court upheld the IFE's finding that there was no legal basis for releasing the ballots. In its ruling the court argued that the denial does not limit the right to access the information in the electoral ballots, as that information was incorporated in the official election documents that constitute public information, *ipso facto*.[5]

The validity of the arguments presented by both sides is debatable. What is possibly more important, however, is that, regardless of the court's decision, this stand-off over the vote count gave a clear indication of the growing influence of the access to information institutions and how they can force institutions to reconsider arguments, procedures and legislation, and to draw them into trials lasting several months.

Success for the National Water Commission

Although no major water-related scandals came to light in the past year, there have been several persistent problems.[6] The large number of delinquent users, opacity in concessions and public bidding, and the illegal siphoning of water have the most significant financial consequences.[7] Another problem is environmental crimes by organised groups that profit from, and exploit, the lack of enforcement of the National Water Law (*Ley de Aguas Nacionales*). It is not uncommon to find illegal drainage or wastewater dumped on beaches and tourist areas where the concessions granted to some hotels and industries do not comply with legal provisions.[8]

In 2001 the National Water Commission (*Comisión Nacional del Agua*), the body responsible for supervising the provision of water, began

3 TEPJF press releases 074/2006 and 081/2006; see www.trife.gob.mx.
4 *El Universal* (Mexico), 17 August 2006.
5 TEPJF press releases 074/2006 and 081/2006; see www.trife.gob.mx.
6 Transparencia Mexicana, 'National Index of Corruption and Good Government, Results 2001, 2003, 2005' (Mexico City: Transparencia Mexicana, 2006).
7 The National Water Commission publishes an annual report that features statistics based on water that has been recovered from irregular activities, available at www.cna.gob.mx.
8 Comisión Nacional del Agua (CNA), 'Logros en Materia de Transparencia y Combate a la Corrupción, Enero-Junio 2006' (Mexico City: CNA, 2006).

implementation of the Operative Programme for Transparency and Combating Corruption (POTCC), aimed at improving accountability and minimising opportunities for corruption in water provision and the internal administration. The POTCC is directed by the Ministry of Public Administration's Inter-ministerial Commission for Transparency and the Fight against Corruption. The POTCC submits an annual report to the ministry concerning the CNA's progress. The 2006 report points out that, between 2001 and December 2006, close to 29,000 delinquent consumers were 'reintegrated' into the payment system and approximately P1.35 billion (US$121 million) in unpaid fees were collected.[9]

In order to address the problem of delinquent users, the CNA also created an electronic database, known as the Public Register of Water Rights, which was subsequently commended by the OECD.[10] The database includes information about water permits, users and debtors, and is publicly available on the internet. It represents an important advance in access to information regarding the provision of and payment for water. It is expected to play a critical role in helping to eliminate the discretional assignation of concessions and permits.

Transparencia Mexicana (TI Mexico)

Further reading

A. del Castillo and E. Ampudia, 'Diagnóstico sobre el Impacto del Fraude y la Corrupción en las Pymes' (Mexico City: CEI Consulting & Research, 2005).

Administración Pública Federal, 'Programa Operativo Anual del Sistema de Servicio Profesional de Carrera 2006' (Mexico City: Secretaria de Función Publica, 2007).

R. Hernández and L. Negrete, 'Opinión Jurídica sobre la Reforma Penal en México', (Mexico City: Centro de Investigación y Docencia Económicas, 2005).

Instituto Federal de Acceso a la Información Pública, 'Estudio Comparativo de Leyes de Acceso a la Información Pública' (Mexico City: IFAI, 2007).

Transparencia Mexicana: www. transparenciamexicana.org.mx.

9 CNA, 'Logros en Materia de Transparencia y Combate a la Corrupción 2006' (Mexico City: CNA, 2007).
10 OECD, 'OECD Environment Performance Reviews: Mexico' (Paris: OECD, 2003).

Nicaragua

Corruption Perceptions Index 2007: 2.6 (123rd out of 180 countries)

Conventions

OAS Inter-American Convention against Corruption (signed March 1996; ratified March 1999)
UN Convention against Corruption (signed December 2003; ratified February 2006)
UN Convention against Transnational Organized Crime (signed December 2000; ratified September 2002)

Legal and institutional changes

- **A new government, led by Daniel Ortega and the Sandinista National Liberation Front (FSLN), took power on 10 January 2007.** President Ortega's first step was to appoint his wife, Rosario Murillo, as coordinator of communications and citizenry, heading a new agency with responsibilities for advising the president, liaising with the media and disseminating information to voters. The opposition Movimiento Renovador Sandinista (MRS), Alianza Liberal Nicaragüense (ALN) and Partido Liberal Consticionalista (PLC) objected on the grounds that it flouted Law 290 by placing power in the hands of a non-elected individual. Eleven days after assuming power President Ortega ordered reforms to Law 290, but withdrew them under pressure from the opposition.[1] With only thirty-eight of the ninety-two seats in parliament, the new government will have to adopt alliances with other parties if it is to press ahead with its ambitious programme of social reform.
- The FSLN victory would not have been possible without a **tactical alliance – and a *quid pro quo* – with the former president, Arnoldo Alemán** (1997–2002), of the PLC, who, despite his continuing authority over the party, had been sentenced to twenty years' house arrest in 2003 (see *Global Corruption Report 2005*) for embezzling US$100 million. The decision to prosecute Alemán split the PLC and enabled Ortega to set one faction against the other in the elections held in November 2006. In March 2007 the administration released Alemán on parole, encouraging observers to conclude that the Ortega government was effectively the product of a power-sharing pact between the former revolutionary leader and a corrupt ex-president, boding ill for the future of accountability in Nicaragua (see also *Global Corruption Report 2006*). Alemán said he would like one day to return to the president's office.[2]
- Ex-President Bolaños, who retains a seat in the National Assembly and is a member of

1 See www.laprensa.com.ni/archivo/2007/enero/17/noticias/politica/167835_print.shtml.
2 *Associated Press* (US), 17 March 2007.

the Central American parliament, thereby enjoying double immunity from prosecution, was questioned in May 2007 about **discrepancies in his accounts amounting to US$330,000.**[3] On 23 May the attorney general, Hernán Estrada, announced investigations into sixteen corruption cases among the 340 allegedly committed on his watch. Bolaños is no stranger to corruption allegations and admitted in 2004 to having taken US$500,000 in dubious money that year to finance his election campaign (See *Global Corruption Report 2005*).

- In its first months in power **the Ortega government announced its intention to introduce Consejos del Poder Ciudadano** (Councils of Citizen Power, or CPCs) by mid-July.[4] Based on the Bolivarian Circles in Venezuela and the Popular Power Assemblies in Cuba, the CPC concept has been promoted as a medium for 'direct democracy'. Nevertheless, opposition parties, some media and the 600 plus member organisations of the civil society umbrella, Civil Coordinator, are concerned that the intentions of the CPC programme are to undermine genuine, participative democracy.[5] The twelve members of a CPC are elected by their communities to represent their interests before ministries and agencies, and to present communal problems to the authorities. The functions are currently carried out by the municipal development committees established under the Civil Participation Law, suggesting either a duplication of duties or an attempt by the new government to capture a larger proportion of local power than it won at the elections.
- On 16 May 2007 **the National Assembly approved a law on access to public information** by a majority of sixty-seven to eighteen, the latter being mainly composed of deputies from the MRS and ALN, who argued that the draft had been diluted to protect government interests.[6] The law requires civil service employees to provide information about their activities under threat of a fine equivalent to six months' salary. It also guarantees the rights of journalists to protect the identity of their sources. Public organisations will have to publish electronically details of their structure, operations, services, fees, banks, names and wages of personnel, as well as information about contracts, subcontracts and tax exemptions for companies engaged in business with the state. Access to information offices will be set up nationwide to cope with requests, while a special committee will be created to decide whether to respond to citizens' curiosity about the wealth declarations made by public officials on a case-by-case basis.[7]

Probity Commission doesn't quite clean up

The Probity Commission of the National Assembly is composed of eight congressmen – five FSLN, one ALN, one PLC and the last from Resistencia Nicaragüense. All were sworn in by President Ortega in January 2007. The commission's purpose is to analyse the cases of public officers who have been linked to corruption, the embezzlement of public wealth or illicit enrichment. The congressmen examine individual cases and, when necessary, send a request to the joint direction of the National Assembly to cancel a suspect's immunity from prosecution to enable the suspect to be tried by a court.

Two cases in particular came into focus during the period under review. The first, known as the

3 Nicaragua Network, 'Former President Bolaños Questioned on Embezzelment', 'Hotline' brief, 22 May 2007.
4 Nicaragua Network, 'Councils of Citizen Power to be Inaugurated on July 19', 'Hotline' brief, 29 May 2007.
5 Nicaragua Network, 'Controversy over Councils of Citizen Power Continues', 'Hotline' brief, 31 July 2007.
6 *La Prensa* (Nicaragua), 17 May 2007.
7 *Xinhua* (China), 17 May 2007.

Tola case, began on 27 May 2006, when TV journalist Carlos Fernando Chamorro alleged in his weekly programme, *Esta Semana*, that potential foreign investors in the tourism sector were being subjected to extortion by officials within the FSLN. The reported author of the extortion attempt was Gerardo Miranda, a former FSLN congressman for Rivas, the site of the development, but two other FSLN cadres, Vicente Chávez and Lenín Cerna, were also named.[8]

Esta Semana showed film of Nicaraguan-born US citizen Armel González being solicited for a US$4 million bribe in exchange for resolving a dispute with two local cooperatives, which have long contested the ownership of his US$88 million resort development.[9] Nicaragua's nascent tourism sector, particularly on the Pacific coast, was the source of most of the US$282 million of foreign investment received by the country in 2006.

The second case concerned Alejandro Bolaños Davis, an ALN congressman who was suspended from office after he was shown to have lied abut being born in the United States when applying to run for election to represent the town of Masaya. He apparently possessed four birth certificates, a sworn declaration and a forged identification card. The Electoral Supreme Court, the institution that has the task of generating citizens' identification cards, allowed Alejandro Davis two nationalities.[10]

While the Supreme Election Council, which is responsible for congressional ethics, retired Bolaños Davis six months after his election, the Probity Commission is having a hard time deciding whether to prosecute in either case.

The public character of water

In Nicaragua, top officials and transnational firms are alleged to have been involved in corrupt practices with regard to the payment of basic water charges, although no charges have been brought. It is suggested that bribes are paid in order to avoid paying the correct amounts relating to consumption.

The recently passed General Water Law aims to improve the regulation of the exploitation of natural resources and to prioritise human consumption in areas that now face problems of supply. It is also intended to regulate private companies and illegal connections. No agency is currently responsible for measuring the scale of the monetary losses that the country suffers each year in the water and health sectors.

The new law is also significant in that it prevents any future privatisation of water and guarantees its public character. Article 97 also vests the state with responsibility for protecting and conserving the waters of Lake Cocibolca, which has been granted the status of a natural drinking water reserve. Currently this water is heavily polluted by sewerage and chemicals from lakeside settlements and the private companies that work on its shores. The law will implement sanctions to entities that violate the regulation up to a maximum of two years in prison and C$500,000 (US$27,335) in fines. The biggest consumers, mainly producers of beers and sodas, must pay for water according to a price list issued by the National Water Authority, a new entity created by the law.

Byron López Rivera (Grupo Cívico Ética y Transparencia)

8 'Nicaragua Briefs: More on Tolagate', *Envio*, no. 312 (2007).
9 *Bloomberg News* (US), 3 October 2007.
10 *La Prensa* (Nicaragua), 30 June 2007.

Further reading

A. Acevedo Vogl, 'CENIS Bancarios: Una Exigencia Ciudadana a la Contraloria' (Managua: Coordinadora Civil, 2006).

R. García and M. García, 'Manual of Private and Public Financing monitoring and the Political Activity in Nicaragua' (Managua: Litografía Print Center, 2007).

Hagamos Democracia, 'Transparencia del Financiamiento Política: Catalogo de Buenas Prácticas' (Managua: Hagamos Democracia, 2007); available at www.hagamosdemocracia.org.ni.

Y. Losa, 'El Control de la Corrupción en Nicaragua: Instituciones y Marco Juridico' (Managua: Ardisa Impresión Comercial, 2007).

Grupo Cívico Etica y Transparencia: www.eyt.org.ni.

Paraguay

Corruption Perceptions Index 2007: 2.4 (138th out of 180 countries)

Conventions

OAS Inter-American Convention against Corruption (signed March 1996; ratified November 1996)

UN Convention against Corruption (signed December 2003; ratified June 2005)

UN Convention against Transnational Organized Crime (signed December 2000; ratified September 2004)

Legal and institutional changes

- **The Public Function Law 2000**, replaced by Law 1626, is scarcely operational because of constitutional challenges in the courts by public officials. The law establishes a test that must be taken for civil servants to be granted posts and receive promotions, and to improve the internal operations of government institutions, including the suppression of corruption. In combination with the proposed Law of Access to Public Information, Law 1626 constituted a dual-pronged attack on institutional inefficiency and malfeasance. Considerable political will will be required to resolve the stand-off. With Paraguay on the eve of elections in 2008, neither the government nor opposition will wish to alienate the civil service, which represents 200,000 direct votes and numerous indirect ones.
- **The National Integrity System (NIS) Steering Committee (CISNI)** is an autonomous body made up of public and private sector representatives that is also open to new members. CISNI's aim is to strengthen the NIS's role as the central consultative authority in relation to anti-corruption conventions. CISNI's autonomy has been undermined by the appointment of Carlos Walde as chairman, however. He is an economic adviser to President Nicanor Duarte Frutos and also a principle administrator of the US$37 million Millennium Threshold Programme, whose main focus is anti-corruption. In 2006 TI

Paraguay published reports[1] that exposed Walde's family firm as the beneficiary of public contracts, in violation of legal provisions that forbid officials and their relatives from signing contracts with the state.[2] Walde's appointment produced a crisis within CISNI, leading to the resignation of the representative of the comptroller general's office and several member organisations. As a result, CISNI has lost credibility, to the point where the media were devoting more space to Walde's family's allegedly corrupt business activities than to the CISNI initiatives.[3]

- **In 2006 a bill on political funding was presented to parliament**[4] with the aim of regulating the use of funds managed by political parties and investigating their origins. The bill has not prospered, however. While the specific reasons for this are unknown, it would not be far-fetched to suggest that support from both sides of parliament was not forthcoming because of the imminent elections. The bill on access to public information similarly failed to gain traction,[5] although a bill to regulate direct purchases by official bodies is having more success.
- **A water law**, passed on 10 July 2007, declared the public ownership of all surface and underground water, regardless of whether it is located on private or public land. Landowners will henceforth have to pay the government to use their own wells, signalling an increase in bureaucracy that is likely to usher in a lucrative new arena for corruption.

Surrender of the judiciary

With elections planned for 2008, there are serious doubts over the legitimacy of the structures required to ensure a democratic system. Shortly after taking office in 2003 President Duarte promised to 'purge the public sector and the judiciary of corruption'.[6] Many welcomed the impeachment proceedings that followed in the Supreme Court, which removed some judges for misconduct and led to others resigning. This was seen by others, however, as unacceptable interference with the judicial branch. The new members of the Supreme Court appeared to include professionals with lower-grade qualifications and experience than those they replaced, giving the impression that the president had all along intended to create a more pliable Supreme Court.

The president's control over the judiciary, as well as his apparent insincerity about addressing corruption, was exemplified by his treatment of friends and relatives discovered to be involved in corrupt practices. For example, the former interior minister, Roberto González, an inner-circle member of the president's, was implicated by one prosecutor in the smuggling of compact discs, but managed to frustrate the penal investigation with the assistance of another prosecutor.[7] The prosecutors involved were impeached for alleged formal matters relating to the case, while González was eventually appointed minister for national defence, the post he currently holds.

1 TI Paraguay, 'Actualización de los Comentarios sobre la primera Ronda' (Asunción: TI Paraguay, 2006); see also *ABC Digital* (Paraguay), 6 August 2007.
2 Article 40 of Law 2051.
3 *ABC Digital* (Paraguay), 6 August 2007; *Última Hora* (Paraguay), 7 April 2006.
4 The bill was backed by several civil organisations and aimed to make political campaign funding more transparent. Apart from subsidies provided by the state, parties receive large amounts of private funding, the origin of which needs to be clarified.
5 This bill was also backed by civil organisations, including TI Paraguay.
6 Freedom House, 'Paraguay' in *Freedom in the World 2007* (Lanham, MD: Rowman & Littlefield, 2007).
7 See www.paraguayglobal.com/noticias_efe.php?ID=2494.

Further evidence of official willingness to manipulate legal norms arose during the internal elections of the ruling Colorado Party, when Duarte stood as candidate for party president in defiance of the constitution, which prohibits the head of the executive from holding additional offices.[8]

The Superior Electoral Court interpreted this to mean that, while the president would not be able to 'exercise' the two roles simultaneously, he could 'present himself as a candidate', thus enabling him to contest, and ultimately win, the internal election. The president appealed to the Supreme Court, which suspended the initial decision, allowing him to hold the two posts simultaneously while a final decision was made. This enabled him to hand power over to the party's vice-president, José Alberto Alderete, a person of his own choosing (see *Global Corruption Report 2007*).[9] In doing so he strengthened his support base in the Colorado Party, putting him in a better position both to modify the constitution and to enable his re-election.

For good or ill, the Paraguayan constitution expressly prohibits the re-election of the president or vice-president.[10] President Duarte used his influence in an attempt to modify the constitution, despite encountering a number of obstacles. First, changing the constitution was popularly regarded as a fundamental violation of the spirit of the original document, which sought to impede the president from promoting reforms for his own benefit. Second, the principle of non-retroactivity would prevent any constitutional amendment benefiting the president who proposes it. He therefore tried to modify it through a referendum.

Colorado Party senators and deputies used every kind of weapon to obtain the votes required to support a referendum. For example, on 2 December 2006 Juan Carlos Galaverna, an influential Colorado senator, celebrated his birthday with an ostentatious party for 1,700 guests, including three of the nine Supreme Court judges, ministers, senators, deputies, military officers and business leaders. Not all parties that senior judges and politicians attend are necessarily events where influence is peddled, but the general public and media were particularly suspicious considering the political climate.

Indeed, the Judicial Ethics Tribunal, created by the Supreme Court in December 2005, took the same view. On 20 March 2007 it concluded that the three judges had violated eight articles in the judicial ethics code and should have refused the invitation.[11] The judges were unsuccessful in their appeal against the ruling,[12] but escaped impeachment proceedings that could have undermined the Colorado Party's five/four majority in the Supreme Court. Administrative proceedings were brought against Esteban Kriskovich, however, secretary and director of the Judicial Ethics Tribunal,[13] whose removal from office was considered imminent.

The consequences of the judiciary's submission to the executive – which is what this episode

8 Article 237 stipulates: 'The President of the Republic and Vice-president may not hold public or private positions, whether paid or not, while retaining their functions. Neither can they exercise trade, industry or any professional activity, and must devote themselves exclusively to their functions.'

9 P. Lambert, 'Country Report: Paraguay', in S. Tatic and C. Walker (eds.), *Countries at the Crossroads 2007* (Lanham, MD: Rowman & Littlefield, 2007).

10 Article 229 states: 'The President of the Republic and the Vice-president shall last five non-renewable years in the exercise of their functions . . . They shall not be re-elected in any situation . . .'

11 See www.pj.gov.py/etica_documentos.asp.

12 See Reconsideración Resolución no. 9/2007 del Tribunal de Ética Judicial, Caso no. 23/06.

13 *La Nación* (Paraguay), 13 April 2007.

amounts to – have not yet become clear. The president did not secure his looked-for referendum, nor will he be able to run for the presidency in the 2008 election. Significant evidence remains that he wields undue influence over the judiciary, however. The president's failed efforts to change the constitution via referendum seem inconsistent with a provision in the Penal Code barring attempts to change the constitutional order.[14] While the president's failure to change the constitution provides hope for Paraguay's fragile democracy, his actions have dealt a serious blow to the trust that the public places in its democratic institutions.

Managing the production of shared water power

Although Paraguay is landlocked and possesses no mineral or oil reserves, it does not want for water. Two hydroelectric dams, shared with Brazil and Argentina, have been built on the river Paraná. The Itaipú Dam is jointly owned with Brazil, and the Yacyretá Dam is shared with Argentina. Both dams produce an excess of energy, which the company managing Paraguayan power sells to its partners at below international market prices.

This has long been seen as an unfair situation, which can be remedied only by renegotiating clauses that allow Paraguay to sell the energy to third parties or to renegotiate prices closer to international market norms. Although these clauses do not signal corruption per se, the companies' status in Paraguay gives rise to significant corruption opportunities, which may help to explain the contracts' anti-competitive nature. There are also concerns that the dams may be operating illegally. An inspection panel appointed by the World Bank in 2003 found that the Yacyretá reservoir had been operating above the official level and, as such, might have been producing additional energy that had not been accounted for.[15]

Neither the income produced by the hydroelectric plants nor the accounts of the two companies are controlled by the office of Paraguay's comptroller general. This is because 'bi-national bodies' are not subject to the internal control or supervision of state parties.[16] This means that the dams remain beyond the supervision and reach of all three countries. Recently, Paraguay's comptroller general's office has attempted to find ways to audit the accounts of the dam's operating companies, and there have been proposals to reform the relevant legislation.[17]

Modifying the treaties between the three countries to allow control over the sums that circulate would foster transparency and reduce the discretion with which the companies currently operate. This may also provide an opportunity to discover a way of ensuring that energy produced in Paraguay is traded appropriately, for the benefit of the Paraguayan people.

Carlos Filártiga (TI Paraguay)

Further reading

S. Mesquita, 'Acceso a la Información Pública como Derecho Cuidadano' (Asunción: CISNI, 2006), available at www.pni.org.py/publicaciones.htm.

Transparencia Paraguay, 'Encuesta Nacional Sobre Corrupción' (Asunción: TI Paraguay, 2006).

14 Article 273 of the Penal Code states: 'Whoever tries to achieve or achieves changes in the constitutional order outside the procedures provided for in the constitution shall be punished with a loss of liberty of up to five years.'

15 *Environmental News Service* (US), 10 May 2007.

16 See www.transparencia.org.py/index.php?option=com_content&task=view&id=242&Itemid=245.

17 See www.ministeriopublico.gov.py/mp/menu/varios/transparencia/planes/index.php.

'Índice de Desempeño e Integridad en Contrataciones Publicas 2004/2005' (Asunción: TI Paraguay, 2006).
'Conflicto de Intereses 2006' (Asunción: TI Paraguay, 2007).
TI Paraguay: www.transparencia.org.py.

United States of America

Corruption Perceptions Index 2007: 7.2 (20th out of 180 countries)

Conventions

OAS Inter-American Convention against Corruption (signed June 1996; ratified September 2000)
OECD Convention on Combating Bribery of Foreign Public Officials (signed December 1997; ratified December 1998)
UN Convention against Corruption (signed December 2003; ratified October 2006)
UN Convention against Transnational Organized Crime (signed December 2000, ratified November 2005)

Legal and institutional changes

- The **Honest Leadership and Open Government Act of 2007** is widely considered a major step forward in reducing corruption in US politics. Enacted in response to public pressure following major congressional corruption scandals, it significantly strengthens congressional ethics and lobbying rules. It does not, however, create an independent ethics office to administer the rules free from partisan influence.
- The US Department of Justice (DOJ) and the Securities and Exchange Commission (SEC) announced on 26 April 2007 **the largest monetary sanctions imposed for violations of the Foreign Corrupt Practices Act (FCPA).** They fined Baker Hughes Inc. and a related subsidiary US$21 million and required the disgorgement of over US$23 million in profits related to approximately US$4 million in bribes over a two-year period to the Kazakhstan state-owned oil company.[1] The companies also entered into agreements requiring anti-bribery compliance programmes, an independent compliance monitor for three years and further cooperation in ongoing investigations.
- The Civilian Agency Acquisition Council and the Defense Acquisition Regulation Council have agreed on a final rule amending the **Federal Acquisition Regulation** (FAR) to require companies that win contracts with the US government to adopt written codes of

1 SEC, press release, 26 April 2007; Department of Justice, press release, 26 April 2007.

business ethics and conduct; to institute training programmes and an internal control system; and to display Federal Agency Office of the Inspector General (OIG) fraud hotline posters or institute other mechanisms to encourage the reporting of suspected improper conduct. While there are exceptions to the new rule, it imposes significantly greater requirements than in current FAR regulations.

US lobbying disclosure and ethics rules strengthened

Corruption and ethics in government was a central issue in the 2006 mid-term elections, helping the Democrats regain control of both Houses of Congress. According to national exit polls, over 40 per cent of respondents identified corruption and ethics in government as more important than any other issue, including the war in Iraq.[2] As widely reported, the apparent catalyst for the public reaction was a series of high-profile ethics scandals involving prominent lobbyists and Congress members.

When they took office in January 2007 the new majority in the House of Representatives initiated important changes to internal ethics rules. In May the House passed landmark legislation to strengthen congressional ethics and lobbying requirements. The legislation, entitled the Honest Leadership and Open Government Act of 2007, was subsequently enacted and signed into law by President George W. Bush in October 2007.

The new law mandated changes to current congressional ethics and lobbying rules that are widely regarded as the most significant in a generation. The provisions included increasing transparency in Congress by requiring lawmakers to disclose 'earmarks' (spending measures for favoured projects); lengthening the 'cooling-off' period from one to two years before former senators can engage in lobbying (in the House, the cooling-off period remains one year); prohibiting members of Congress and their staff from influencing hiring decisions by lobby firms on the sole basis of partisan political affiliation; requiring lawmakers to disclose small campaign contributions from numerous donors that are 'bundled' into large packages by lobbyists; and banning most lobbyist-paid gifts and travel to members of Congress and requiring lobbyists to certify that they did not provide or direct prohibited gifts or travel to members or staff.[3] The lobbyist certifications are part of a package of amendments to the current Lobbying Disclosure Act (LDA) that will require more frequent (quarterly) filings, additional information (especially on political contributions activity) and timely public access over the internet (currently provided only by the Senate).

The new legislation dramatically increased penalties for LDA violations, raising the maximum civil fine fourfold to US$200,000 and adding a new criminal penalty (up to five years' imprisonment) for 'knowingly and corruptly' violating the act. Other provisions mandate the spot auditing of lobbyists' filings by the comptroller general and semi-annual reporting by the Justice Department on its enforcement activity. The new lobbyist certification requirement is expected to facilitate enforcement against companies and individual lobbyists violating the rules.

One important reform *not* included in the legislation was a proposal to establish an independent ethics office to strengthen enforcement of the new rules. Many in Congress and among non-governmental organisations from across the political spectrum believe that this office is necessary, because congressional ethics committees, which are self-governing and subject to par-

2 *CNN.com* (US), 8 November 2006.
3 Public Law no. 110-81.

tisan influence, have, for some time, been viewed as ineffective and secretive. Although both chambers have codes of conduct, enforcement has been uneven and sanctions often have been considered insufficient. An independent ethics office, with investigative and subpoena power, could be more transparent and could act independently of partisan influence.

Attention is now turning to the implementation phase, with congressional staff and lobbyists adapting to the new rules, and the House again considering whether to create an independent ethics office. To realise the full potential of the new and noteworthy ban on gifts, meals and free travel from lobbyists, Congress needs to ensure that there are strong enforcement mechanisms.

US response to corporate corruption abroad intensifies

In 2006/7 the United States intensified its enforcement of the Foreign Corrupt Practices Act, prohibiting bribery of foreign public officials. In 2006 the Department of Justice instituted twelve FCPA prosecutions, a record number for a single year; it was on track to exceed that number in 2007. Fines have also increased. On 26 April 2007 the DOJ and the SEC announced the largest monetary fine for FCPA violations against Baker Hughes and a related subsidiary, including fines of US$21 million and the disgorgement of over US$23 million in profits related to the payment of approximately US$4 million in bribes over a two-year period to an official of Kazakhoil, the Kazakhstan state-owned oil company.[4]

The increase in FCPA prosecutions is in part attributable to the impact of the Sarbanes–Oxley Act of 2002 (Sarbanes–Oxley), enacted in the wake of a series of corporate scandals. Within the Sarbanes–Oxley, sweeping corporate governance and accounting reforms applicable to publicly traded companies are requirements for corporate chief executives and financial officers to certify to the accuracy of financial statements. Section 404 of the act mandates regular management assessment of corporate internal financial controls and requires external auditors to test and evaluate the systems.[5] This has led to the voluntary disclosure to the government of violations found in the course of such reviews.

The decision whether and when a company should make disclosure of an actual or potential violation to the US government is often difficult. The government has emphasised that voluntary disclosures, when combined with other forms of cooperation, including, in some cases, waiver of the work product doctrine (protecting materials prepared in anticipation of *litigation* from *discovery* by opposing counsel), and the attorney–client privilege, may substantially mitigate or even eliminate penalties that could apply if the government discovered FCPA violations in the first instance.

Recently, and particularly in the context of voluntary disclosures, the DOJ and the SEC have begun to make more frequent use of 'deferred prosecution' agreements, under which the government agrees not to prosecute a company for a period of time, usually eighteen months to several years, in exchange for the company's admission of liability and its agreement to comply with certain conditions, including the appointment of an independent monitor to ensure FCPA compliance. A compliance monitor was required, for example, as part of the Baker Hughes settlement referred to above. If the company can demonstrate reform at the end of the probationary period, the government will dismiss all charges against the company.

Through the use of compliance monitors, the US government hopes to create incentives for remediation and to deter future misconduct through

4 Department of Justice press release, 26 April 2007.
5 Public Law no. 107-204, sec. 104.

changes to corporate management and culture. It will take time to determine whether such changes, mandated through settlements, translate into permanent changes in corporate culture and practice.

Corruption an obstacle to progress in Iraq reconstruction

Corruption in Iraq, which is ranked 178 out of 180 on the 2007 Transparency International Corruption Perceptions Index, has been identified as one of the main obstacles to progress in the reconstruction process. With billions of dollars committed for Iraq reconstruction, the opportunities for corruption are substantial, particularly in public contracting.

The Special Inspector General for Iraq Reconstruction (SIGIR) is at the forefront of ensuring that procurement rules are followed and breaches detected. SIGIR is working in conjunction with other agencies involved in oversight in Iraq, including the Office of the Inspector General of the Department of Defense, the Department of Homeland Security (DHS), the Federal Bureau of Investigation (FBI) and the Department of Justice National Procurement Fraud Task Force, to coordinate and enhance procurement fraud investigations.[6] This work has yielded positive results. SIGIR reported that, as of 19 June 2007, it had opened 300 cases.[7]

While estimates of losses due to contractor fraud are difficult to gauge, Special Inspector General Stuart Bowen, Jr., testified before the US House of Representatives that 'the corruption SIGIR has uncovered to date within the US reconstruction program, while egregious in nature, amounts to a relatively small proportion of the overall US investment in Iraq'.[8] Some attorneys knowledgeable about procurement also believe that the majority of contracts related to Iraq reconstruction are performed lawfully, and that, while corruption is of great concern, the system seems to be working.

One area of concern, however, is the adequacy of oversight for contracts and contractors that plan, define, procure and supervise the performance of all acquisition contracts. Experts believe that the number of acquisition personnel is insufficient, resulting in inadequate oversight of contracts and contractors.[9] SIGIR has published numerous reports that underscore the risks of fraud and abuse associated with weak contract oversight.[10] It has also noted differences among implementing agencies' contracting procedures as a source of concern.[11]

TI USA

6 Statement of S. Bowen, Jr., Special Inspector General for Iraq Reconstruction before the United States House of Representatives Committee on the Judiciary, Subcommittee on Crime, Terrorism and Homeland Security, 19 June 2007; see www.sigir.mil/reports/pdf/testimony/SIGIR_Testimony_07-012T.pdf.

7 Ibid.

8 Statement of S. Bowen, Jr., Special Inspector General for Iraq Reconstruction, 'Assessing the State of Iraqi Corruption,' House of Representatives Committee on Oversight and Government Reform, 4 October 2007; see www.sigir.mil/reports/pdf/testimony/SIGIR_Testimony_07-015T.pdf.

9 *New York Times* (US), 24 October 2007.

10 See www.sigir.mil/reports/Default.aspx; statement of S. Bowen, Jr., Special Inspector General for Iraq Reconstruction before the Subcommittee on State, Foreign Operations and Related Programs, Committee on Appropriations, United States House of Representatives, 30 October 2007; see www.sigir.mil/reports/pdf/testimony/SIGIR_Testimony_07-017T.pdf.

11 Statement of J. McDermott, Assistant Inspector General – Audit, Special Inspector General for Iraq Reconstruction before the United States House of Representatives Appropriations Committee, Subcommittee on Defense, 10 May 2007; see www.sigir.mil/reports/pdf/testimony/SIGIR_Testimony_07-010T.pdf; Iraq Reconstruction, Lessons in Contracting and Procurement, www.sigir.mil/reports/pdf/Lessons_Learned_July21.pdf.

Further reading

A. J. Heidenheimer and M. Johnston, *Political Corruption: Concepts and Contexts* (New Brunswick, NJ: Transaction, 2002).

B. Heineman, Jr., and F. Heimann, 'Arrested Development: The Fight against International Corporate Bribery', *The National Interest*, November/December (2007).

M. Johnston, 'Understanding the Private Side of Corruption: New Kinds of Transparency, New Roles for Donors', Brief no. 6 (Bergen, Norway: U4, Chr. Michelson Institute, 2007).

S. Rose-Ackerman, *Corruption and Government: Causes, Consequences and Reform* (Cambridge: Cambridge University Press, 1999).

International Handbook on the Economics of Corruption (Cheltenham: Edward Elgar, 2006).

S. Rose-Ackerman and B. S. Billa, 'Treaties and National Security', forthcoming in the *NYU Journal of International Law and Politics*.

TI USA: www.transparency-usa.org.

7.3 Asia and the Pacific

Bangladesh

Corruption Perceptions Index 2007: 2.0 (162nd out of 180 countries)

Conventions

ADB – OECD Anti-Corruption Action Plan for Asia-Pacific (endorsed November 2001)

UN Convention against Corruption (accession February 2007)

Legal and institutional changes

- **The Public Procurement Act, passed on 6 July 2006**, provided comprehensive legal provisions to prevent corruption and promote competition on a level playing field. The Manual of Office Procedure (Purchase), inherited from the colonial era and last revised in 1977, previously laid down methods for procurement in the public sector. The caretaker government engaged various stakeholders, including TI Bangladesh, to review the new procurement rules.[1] A public–private review committee on public procurement was formed

1 TI Bangladesh recommendations were aimed at strengthening the new rules' conflict of interest dimension, stricter compliance with anti-corruption policy, and the introduction of social accountability by engaging citizens in various processes and levels. TI Bangladesh also argued for instituting a code of conduct for the public procurement authority, with clear delineation of enforcement indicators.

on 28 May 2007 to assess the impact of the new law, but had yet to start functioning at the time of writing.

- In February 2007 the **government amended the Criminal Procedure Code Ordinance**, the final step towards re-establishing the independence of the judiciary (see *Global Corruption Report 2007*).[2] The legislation had been postponed more than twenty times since 1999, when the Supreme Court first ruled in favour of greater separation between the judiciary and the executive branch of government. According to the judgment, the separation of the judiciary would be complete only when four rules came into effect.[3] On 7 May the Supreme Court accepted an amendment to the Judicial Service Pay Commission Order and directed the government to complete the process of the separation of the judiciary by 19 July. This was also the deadline for the government to create a number of courts, courtrooms and chambers for judges and magistrates.
- In a related move, the government **established a financial intelligence unit** in March 2007 to combat financial crimes, and retrieve assets and money laundered overseas. Established within the framework of the amended Money Laundering Prevention Ordinance of 2007, the financial intelligence unit will operate as part of the central bank's anti-money-laundering department. Government sources said the unit will play a key role in recovering the large sums of money siphoned off through political corruption.[4]
- The **Anti-Corruption Commission** (ACC), set up in February 2004 (see *Global Corruption Report 2006*), **remained ineffectual** for several reasons, including political bias and a lack of commitment by its three government-appointed commissioners. Its jurisdiction was seriously limited by rules that prevented access to banking, finance, money-laundering, foreign exchange records and the activities of multinational corporations. In the Anti-Corruption Act 2004 these items were not included in the schedule, and the commission's freedom was further curtailed by government control of its budget and administrative powers. On 22 February 2007 the caretaker government appointed three individuals of integrity to replace the original commissioners. With the support of a joint task force of military officers, the new ACC embarked on a high-profile anti-corruption drive, publishing on 18 February a list of fifty individuals suspected of corruption, including ministers, lawmakers, politicians and businessmen (see below).[5] On 7 March Tarique Rahman, son of Begum Khaleda Zia, the former prime minister, was arrested.[6] The most common allegations against the accused were disproportionate wealth, extortion, abuse of power and the plunder of relief goods. Twelve former ministers and MPs were allegedly involved in an extortion worth over Tk320 million (US$4.7 million) from various companies and construction projects.[7] On 29 March the ACC filed a case against the former housing and public works minister, Mirza Abbas, and eleven engineers for selling eighteen abandoned houses at lower than market prices, costing the government about Tk1.3

2 *Daily Ittefaq* (Bangladesh), 8 February 2007.
3 These rules are the Judicial Service Commission Rule 2002; the Bangladesh Judicial Service Pay Commission Rule 2002; the Bangladesh Judicial Service (Service Constitution, Composition, Recruitment, Suspension, Dismissal and Removal) Rule 2002; and the Bangladesh Judicial Service (Posting, Promotion, Leave, Control, Discipline and other Service Condition) Rule 2001.
4 *Daily Star* (Bangladesh), 23 March 2007.
5 *Daily Star* (Bangladesh), 19 February 2007.
6 *BBC News* (UK), 8 March 2007.
7 *New Nation* (Bangladesh), 31 May 2007.

billion (about US$18 million).[8] On 4 July a tribunal sentenced the former state minister for civil aviation, Mir Mohammad Nasiruddin, to ten years' imprisonment for illegally accumulating wealth and three more years for hiding assets worth Tk6.7 million (almost US$100,000) in his wealth declaration to the ACC.[9]

- The **Election Commission was reconstituted** on 5 February 2007 after the replacement of controversial commissioners with a reputation for eroding public trust.[10] It subsequently embarked on a series of consultations for the reform of election rules, aimed at creating a more equal playing field free from corruption and the influence of 'black money'. The proposals being discussed include the mandatory registration of political parties; transparency in party funding; holding local and national elections for party leadership; the reservation of one-third of leadership positions for women; disbarment of those found guilty of corruption; de-linking professional and student bodies from partisan politics; computerised voter lists; the provision of 'no votes';[11] and the barring of candidacy for election of government officials for three years.
- The Micro-Credit Regulatory Authority Act was passed in July 2006 and will facilitate the establishment of an institutional mechanism to **ensure transparency and accountability in the operation of organisations offering micro-credit**. Despite a large-scale growth in micro-financing in Bangladesh, there had previously been no regulatory framework for this sector. The act requires all micro-financing institutions, including non-profit organisations, cooperatives, societies and profit-making companies, to obtain a licence from the authority established by the act. The authority is an independent legal entity managed by a board of directors comprised of the governor of the Bank of Bangladesh and six government officials.

Taking care of corruption

Bangladesh is exploring a unique opportunity to reverse the acute failure of governance and pervasive corruption that has bedevilled it for many years. The loss to bribery in five public service delivery sectors was estimated at 7.9 per cent of household income in 2005.[12]

Amid the pre-election violence that brought the capital, Dhaka, to a standstill, President Iajuddin Ahmed appointed Fakhruddin Ahmed, a well-respected former central bank governor and World Bank economist, as 'chief adviser' of the caretaker government on 11 January 2007.

The concept of a caretaker government dates back to 1990, when, with the support of political leaders and most of the population, Chief Justice Shahabuddin Ahmed was appointed head of an interim government. The aim was to stem the growing violence between Bangladesh's two largest political alliances, led by the Bangladesh Nationalist Party (BNP) and the Awami League. Though assembled without constitutional endorsement, the 1990 caretaker government

8 *New Nation* (Bangladesh), 28 March 2007; *Daily Star* (Bangladesh), 29 March 2007.

9 *Daily Star* (Bangladesh), 5 July 2007.

10 See *Daily Star* (Bangladesh), 22 January 2007, 1 February 2007 and 5 February 2007.

11 A 'no vote' is a provision for negative voting so that a voter can express an unwillingness to accept any of the proposed candidates as a representative if he/she feels that none is eligible. In the event that the number of 'no votes' cast is a majority, a fresh election is to be held with a new set of candidates. The provision is a useful deterrent against parties nominating corrupt and unwanted candidates.

12 The sectors were education, health, justice, police and land administration. See Iftekharuzzaman, 'Corruption and Human Insecurity in Bangladesh', presentation at a seminar on International Anti-Corruption Day 2005, Dhaka.

was ratified by parliament a year later, and since 1996 has constituted a legitimate, non-partisan alternative to what has often proven a chaotic and corrupt manifestation of democracy.[13]

The current caretaker government[14] differs from its forebears by enjoying explicit support from the armed forces. The issue of military intervention in day-to-day politics came into focus when the army chief, Lieutenant General Moeen U. Ahmed, told a public seminar in Dhaka that Bangladesh needs its own 'brand of democracy'.[15] On a number of other occasions, however, he has said that the army had no specific interest in politics. In spite of such assurances, there remains concern as to whether the army will withdraw from the political environment. Elections are currently due in late 2008.

In his opening address, on 21 January 2007, Fakhruddin Ahmed made a commitment to fighting corruption and purging politics from the influence of black money.[16] The speech was followed by a series of arrests of former ministers, MPs and members of their families, allegedly involved in corruption. On 26 July a tribunal sentenced Mohiuddin Khan Alamgir, a former state minister and Awami League member, to thirteen years' imprisonment and fined him Tk10 lakh (US$14,850) for amassing Tk3.27 crore (US$485,525) through misuse of power and concealing it in his declaration of assets.[17] On the same day, a Natore court sentenced the former deputy minister for land and BNP leader, Ruhul Kuddus Talukdar Dulu, to five years in prison for arson; one year for causing damage and looting; one year for rioting with deadly weapons; and one year for abetting crimes.[18] On 27 August a court in Dhaka sentenced the former communications minister, Nazmul Huda, and his wife, Sigma, to seven and three years in prison, respectively, for involvement in a Tk2.5 crore (US$356,350) embezzlement.[19]

Reversing Bangladesh's corruption trends will prove long and hard. These institutional reforms are only the beginning of the process. Much depends on the extent to which anti-corruption legislation can be mainstreamed into public service as a whole – a process that could be beset by inertia and resistance to change. The successful prosecution of those charged by the ACC is also open to question. Although the commission has enlisted well-known attorneys, those charged can afford the very best lawyers in a country that has rarely seen the monopoly of power and influence broken in a court of law.

Sceptics will look for historical examples in which military-led anti-corruption drives have simply led to the entrenchment of the military in power. The war on corruption was greeted by unprecedented popular support, but like any other war it has brought collateral costs in terms of the erosion of human rights, public harassment and insecurity. Tens of thousands of people have reportedly been arrested since the drive started, with human rights groups alleging widespread torture and deaths in custody.[20] Political activity is banned under the state of emergency, but elections are due by the end of 2008. It is widely hoped that anti-corruption

13 See banglapedia.search.com.bd/HT/C_0041.htm.

14 The current caretaker government is composed of ten members called 'advisers' but with the rank of ministers, headed by Dr Fakhruddin Ahmed, who, as chief adviser, holds the status of prime minister. Like Dr Ahmed, most if not all advisers are non-partisan technocrats, representing various professional branches, such as business, law, economics, and retired army officers.

15 *The Economist* (UK), 6 April 2007.

16 See www.cao.gov.bd.

17 *Daily Star* (Bangladesh), 27 July 2007.

18 Bangladeshnews.com.bd, 27 July 2007.

19 *Daily Star* (Bangladesh), 28 August 2007.

20 See www.abc.net.au/news/newsitems/200706/s1945599.htm and *The Economist* (UK), 6 April 2007.

reforms will have taken root by then, though whether the new government that emerges from that ballot will adhere to them remains equally open to question. All transitions are fraught with risk. Nevertheless, considering the changes already witnessed under the caretaker government, there are reasons for guarded optimism.

Judicial independence nearly restored

According to a recent survey, two-thirds of all people who interacted with the lower judiciary said they were forced to pay bribes amounting to one-quarter of their annual income.[21] A series of constitutional amendments by military and quasi-military regimes from 1975–91 gave the chief executive authority to appoint, promote and transfer judges and magistrates, bypassing the chief justice altogether and mortally injuring the independence of the judiciary. Though democratic rule was restored in 1991, three successive governments found it convenient to retain control over the judiciary and to politicise it further.

In the historic Masdar Hossain ruling in 1999 (see *Global Corruption Report 2007*), the Supreme Court ordered the government to re-establish an independent judiciary and amend criminal procedure to meet the objective of separating it from the control of the executive branch. The government in power at the time, and the one that followed from 2001 to September 2006, made piecemeal reforms while obtaining nearly two dozen separate court extensions to legitimise their foot-dragging with regard to implementing the Supreme Court decision.

Against this backdrop the caretaker government approved the amended Criminal Procedure Code ordinance on 7 February 2007, one of four legal steps towards restoring the independence of the judiciary.[22] On 7 May the Supreme Court approved further amendments to the Judicial Service Pay Commission Order and three other orders regarding judicial service and civil procedure amendments (previously, judges and magistrates had been treated as just another branch of the civil service), and directed the government to complete the separation of judiciary by 19 July. Significant as these developments were, the judiciary's absolute independence will be realised only when article 116 of the constitution is amended in line with the above, and this will come about only after the election of a democratic government at the end of 2008.

Water corruption and land-grabbing in Dhaka

Like other sectors of the Bangladeshi economy, water is plagued by corruption and failures of integrity. An analysis of episodes of corruption reported in the print media from January to December 2006 shows the following picture.

The data show that public service officials were key actors in corruption in 84.8 per cent of revealed cases, while powerful individuals, including contractors and politicians, were key actors in the remainder (see figure 2).

Reports indicate that public service officials have flouted financial rules in tender processes, while in many cases they have been inefficient or negligent of the public interest. Engineers and other officials have been involved in corruption in major development projects, such as irrigation, river-dredging and flood prevention. In March 2007 the ACC was investigating cases of corruption in different projects run by the Ministry of Water Resources, estimated to have cost up to Tk444 crore (approximately US$1.5 billion) during 2001–6.[23]

21 TI Bangladesh, 'National Household Survey on Corruption 2005', available at www.ti-bangladesh.org.

22 *Daily Ittefaq* (Bangladesh), 8 February 2007.

23 *Daily Inquilab* (Bangladesh), 17 March 2007.

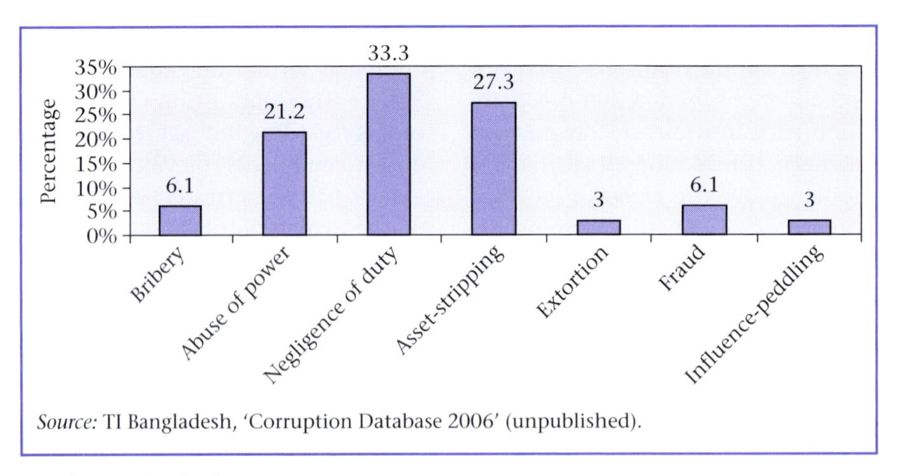

Source: TI Bangladesh, 'Corruption Database 2006' (unpublished).

Figure 2 Types of corruption in the water sector

One report[24] in 2006 concerned the illegal dredging of sand from the Monu riverbed, threatening the protective embankment of the nearby town of Moulvibazar. An influential person with local political links had removed the sand to landfill his plot in the town.

Another category of rampant corruption in urban areas involves encroachment onto the lakes and rivers flowing through cities, especially in Dhaka. Illegal occupation of the shoreline in Dhaka's Gulshan-Banani-Baridhara Lake threatens the lake's very existence. In one case of de-requisition, thirty-one acres of the lake shore were due to be reclaimed from private land-grabbers.[25] Powerful individuals with political links easily obtained court injunctions against the reclamation of their squatted land, however. In connivance with government officials, a well-organised syndicate of land-grabbers has long been active in the business of securing prime sites in the city by filling the lake shore with earth and building structures overnight.[26]

Officials in the Capital Development Authority (CDA) are often discovered to be working in tandem with land-grabbers, which is the main reason for the failure to demarcate, develop and conserve water bodies crucial for Dhaka's environment, water supply and drainage. At the time of writing, the caretaker government had launched a demolition drive against illegal structures encroaching on the lake.

Iftekhar Zaman and Tanvir Mahmud (TI Bangladesh)

Further reading

BRAC University, Centre for Governance Studies, 'The State of Governance in Bangladesh 2006' (Dhaka: BRAC University, 2006); available at www.cgs-bu.com.

R. Jahan, 'Bangladesh Country Report', in S. Repucci and C. Walker (eds.), *Countries at the Crossroads 2005* (New York: Rowman & Littlefield, 2005).

Power and Participation Research Centre (PPRC), 'Bangladesh Governance Report 2007' (Dhaka: PPRC, 2007).

R. Sobhan, 'Building a Responsible Civil Society: Challenges and Prospects', in R. Jahan (ed.), *Bangladesh: Promise and Performance* (Dhaka: University Press, 2000).

World Bank, 'Corruption in Bangladesh: Costs and Cures', (Dhaka: World Bank, 2000).

TI Bangladesh: www.ti-bangladesh.org.

24 *Daily Star* (Bangladesh), 6 May 2006.
25 *Daily Star* (Bangladesh), 10 June 2007.
26 *Daily Star* (Bangladesh), 4 July 2004.

India

Corruption Perceptions Index 2007: 3.5 (72nd out of 180 countries)

Conventions

ADB – OECD Anti-Corruption Action Plan for Asia-Pacific (endorsed November 2001)
UN Convention against Corruption (signed December 2005; not yet ratified)
UN Convention against Transnational Organized Crime (signed December 2002; not yet ratified)

Legal and institutional changes

- In September 2006 **the apex court finally granted the Indian Police Service (IPS) autonomy from political control** after twenty-five years of litigation. Police chiefs had long objected to the minister's powers of transfer and promotion over their careers, usually embodied in their subordinate relationships with the executive. In 1981 the National Police Commission (NPC) produced an eight-volume report outlining measures to modernise the police force and replace the 1861 Police Act. Two other panels, the Ribiero and Padmanabhan committees, similarly grappled with the implications of separating the IPS from the Home Department, but they also came to nothing. In 1995 Prakash Singh, a former director general of police in Uttar Pradesh, India's biggest state, and head of the border security force, filed public interest litigation in the Supreme Court seeking fundamental reforms to free police from the tutelage of political control.[1] The court ruled that henceforth there shall be: a State Security Commission to provide guidance on matters of law and order; fixed tenure for police chiefs, district-level officers, station house officers and officers who investigate crimes; state police establishment boards to ensure that officer transfers are based on merit and not political whim; and a police complaints authority to 'police the police'. The decision was initially challenged by states on the grounds that it 'violated the spirit of federalism', but the Supreme Court ordered them to implement the reforms by 31 March 2007. Small states such as Sikkim and Uttrakhand have now fully implemented the decision, but Andhra Pradesh, Gujarat, Uttar Pradesh and West Bengal have yet to comply.[2]
- TI India made a **presentation to the Second Administrative Reforms Commission in December 2006 on methods of building integrity in government operations**. In its fourth report on 'Ethics in Governance', the commission referred to a host of proposals to improve transparency, accountability and integrity in public services. These included the disqualification of criminals from public office; an ethical framework for ministers; the abolition of discretionary funds available to legislators; a code of conduct for civil servants; an ethical framework for the judiciary; the

1 *Prakash Singh v. Union of India 2006*, AIR SCW 5233.
2 *Hindustan Times* (India), 13 August 2007.

confiscation of property acquired by corrupt means; legislation for the protection of whistleblowers; the institution of an ombudsman; citizens' charters; and the introduction of integrity pacts for public procurement orders.[3]

- **Established in 2005, the Chief Information Commission came into operation in 2006/7.** It has delivered decisions instructing government, courts, universities, police, development NGOs and ministries on how to share information of public interest. State information commissions have also been opened, thus giving practical shape to the 2005 Right to Information Act, although they have not been immune to criticism. Of India's twenty-eight states, twenty-six have officially constituted information commissions to implement the act. Nine had pioneered access to information laws before the act was passed. A state report card one year on complimented the quality of law, but mourned the 'lukewarm response of a largely unaware citizenry'.[4]
- TI India has managed to persuade twenty **public sector companies to adopt integrity pacts in procurements requiring large outlays.** It convinced the Ministry of Defence to adopt the pact in all procurements of Rs300 crore (US$73 million). It is now trying to reduce that ceiling to Rs100 crore (US$24.5 million). Russia, India's largest supplier of weapons by value, has refused to sign any contract that requires an integrity pact on the grounds that it 'collides with' its own domestic laws.[5] Other Indian companies that have contacted TI India about integrity pacts are Vizag Steel, the Steel Authority of India, Hindustan Steel Construction, Gas Authority of India, Hindustan Aeronautics, Kudremukh Iron Ore, Airports Authority of India and the government of Delhi.

Supreme Court challenges states' powers

The Supreme Court takes corruption seriously in both the general and political domains. Political corruption is not confined to monetary considerations, but extends to making promises to secure votes that cannot be fulfilled, helping colleagues by granting them positions of authority, conflicts of interest and manipulating the law to help interested parties.

The Supreme Court also brings the issue of corruption into its judgments. Upholding the conviction of a person under the Prevention of Corruption Act in 2006, it observed: 'Corruption by public servants has become a gigantic problem. It has spread everywhere. No facet of public life has been left unaffected by the stink of corruption. It has a deep and pervasive impact on the functioning of the entire country. Large scale corruption retards the national building activities and everyone has to suffer on that count . . . Corruption is corroding, like cancerous lymph nodes, the vital veins of the body politic, social fabric of efficiency in public service, and demoralising honest officers.'[6]

In a case where the governor of the state of Andhra Pradesh exercised his power to pardon a defendant on the grounds that he had a 'good political record' with a prominent party, the Supreme Court held that the power exercised was arbitrary and the order had been given without any 'application of mind' and hence was *mala fide*, 'as it has been passed on the basis of extraneous or wholly irrelevant considerations'.[7]

3 Second Administrative Reforms Commission, 'Fourth Report: Ethics in Governance' (New Delhi: Government of India, 2007); see www.arc.gov.in.

4 See www.infochangeindia.org/features388.jsp.

5 See www.india-defence.com/print/2997.

6 *State of Madhya Pradesh v. Shambu Dayal Nagar 2006*, AIR SCW 5737.

7 *Epuru Sudhakar v. State of Andhra Pradesh 2006*, AIR SCW 5089.

One method that legislatures employed from 1951 to evade judicial scrutiny concerning the constitutionality of legislation has been to place such laws under the 'ninth schedule', which grants immunity from being tested in court. Remarkably, this provision enabled the legislature to prevent any judicial review of a law even when it conflicted with fundamental rights. More than 300 new laws found political sanctuary under the schedule.

The Supreme Court did away with this immunity in January 2007: 'Since the basic structure of the constitution includes some of the fundamental rights, any law granted ninth-schedule protection deserves to be tested against these principles. If the law infringes the essence of any of the fundamental rights or another aspect of the basic structure, then it shall be struck down.'[8]

In 2003 the Central Bureau of Investigation (CBI) opened a case against a former chief minister of Uttar Pradesh state, the environment minister and four other senior officials for diverting a river and reclaiming land for the construction of shopping malls, shops and amusement facilities near the Taj Mahal complex. They had not obtained planning consent, had released Rs17 crore (US$4.2 million) without sanction and had flouted other procedures.[9] It was reported that some Rs175 crore (US$42.7 million) disappeared during the so-called Taj Corridor scam.[10] After investigation, the case was recommended for prosecution. In the meantime, the CBI's director sought the attorney general's opinion of the matter without expressing his own views. On the basis of the latter's opinion, the CBI's senior public prosecutor determined that the case was not fit for prosecution.

This decision was challenged in the Supreme Court, which later argued: 'There was no question of the director of the CBI referring the matter to the attorney general. . . The superintendent of police is not legally obliged to take his opinion. In the circumstance, when there was no difference of opinion in the concerned team, the question of seeking the opinion of the attorney general did not arise.' The court accordingly directed the CBI to place its evidence before the court, which would decide the matter in accordance with the law.[11]

The politician returned to office as chief minister in Uttar Pradesh after the election of May 2007. On 3 June 2007 the competent authority refused permission to prosecute due to lack of evidence, sparking public interest litigation that challenged the decision on the grounds that it protected the corrupt from justice.[12] At the time of writing the litigation had reached the Supreme Court, and it promised to be an important test of the court's ability to resist pressure to adapt its judgments to political necessity.

Grassroots projects address corruption in water

As a scarce commodity, water is prone to exploitation. TI India's 'India Corruption Study 2005', which sampled 14,405 respondents from 151 cities and 360 villages, found that water was one of the public services most clearly identified with corrupt practices.[13] Although customer interaction with water departments was relatively low (only 12.3 per cent), the study found the most common perceived malpractices were the supply of water tankers (73 per cent), meter installation (71 per cent), bill payment (43 per

8 *I. R. Coelho v. State of Tamil Nadu 2007*, AIR SCW 611.
9 *The Hindu* (India), 19 September 2003.
10 *The Hindu* (India), 7 August 2007.
11 *Times of India*, 27 November 2006.
12 Onlypunjab.com, 8 June 2007.
13 Centre for Media Studies, 'India Corruption Study 2005: To Improve Governance' (New Delhi: TI India, 2005).

cent), and new connections or restoration of water supply (67 per cent).

Eight out of every ten respondents who claimed to have paid a bribe did so to staff, while the remainder paid agents or touts. More than 25 per cent of respondents said they had to visit the offices of the water supply department more than four times during the year. The perceptions of the respondents were as follows:

- more than a half (54 per cent) said that there was corruption in the department;
- 39 per cent felt that corruption in the department had increased; and
- nearly 25 per cent had used alternative means, such as bribery or influence, to get work done.

Concerns raised during the study included the fact that highly subsidised water can lead to considerable waste, adding to stress on limited supplies. This is exacerbated by antiquated equipment and poor infrastructure, resulting in frequent breakdowns. There is also a lack of funds available for new developments and, where projects are in place, there is poor supervision.

Various initiatives have been adopted to deal with these problems. In Gubarga district, Andhra Pradesh, a grading system has increased staff efficiency. In Mandi district, Himachal Pradesh, training camps were organised to educate officers about new technology, and increase their awareness of people's needs and how to satisfy them. Delhi Jal Board, the city's water utility, allows call centres to receive consumer complaints by SMS (Short Message Service), speeding up their registration. Toll-free help lines are available in Bangalore and Hyderabad for use by the poor.

Paramjit S. Bawa (TI India)

Further reading

Centre for Media Studies, 'India Corruption Study 2005: To Improve Governance' (New Delhi: TI India, 2005).

V. Chand (ed.), *Reinventing Public Service Delivery in India* (New Delhi: Sage, 2006).

Marketing and Development Research Associates, 'Corruption in Trucking Operations' (New Delhi: TI India, 2006).

J. Roy, 'Reforms in Criminal Justice' (New Delhi: Indian Institute of Public Administration, 2006).

Second Administrative Reforms Commission, 'Fourth Report: Ethics in Governance' (New Delhi: Government of India, 2007).

TI India, 'A Study of Citizens' Charters' (New Delhi: TI India, 2007).

'Consolidated Minutes of National Seminars and Regional Workshops on Improving Governance' (New Delhi: TI India, 2007).

TI India: www.tiindia.in.

Indonesia

Corruption Perceptions Index 2007: 2.3 (143rd out of 180 countries)

Conventions

ADB – OECD Anti-Corruption Action Plan for Asia-Pacific (endorsed November 2001)
UN Convention against Corruption (signed December 2003; ratified September 2006)
UN Convention against Transnational Organized Crime (signed December 2000; not yet ratified)

Legal and institutional changes

- On 23 August 2006 the **Constitutional Court reduced the scope of the Judicial Commission to supervise judges**, as established under Law no. 22 in 2004, due to the refusal of thirty-one Supreme Court judges to submit to supervision by what is a comparatively new institution. With no such authority, the Judicial Commission will find it hard to reduce, let alone eliminate, corruption in the court system (see below). Since it was created in August 2005 the institution has often received a withering response from the Supreme Court over its recommendations for judgeships to the parliamentary selection committee, as well as over its duty to supervise judicial integrity and conduct.
- On 19 December 2006 the Constitutional Court ordered the **drawing up within three years of new legislation authorising the Special Corruption Court**, since bundling it with the law on the Corruption Eradication Commission (KPK) had been constitutionally unsound. Civil society is concerned that the new law will require a considerable amount of debate at a time when parliament will have a full agenda ahead of the 2009 general election, leaving the authority of the KPK unclear (see below).
- The **Witness and Victim Protection Bill** passed into law in July 2006. Under article 10, witnesses, victims and whistleblowers cannot be prosecuted on charges based on testimony given in the past or future. The bill considers whistleblowers to be anybody who provides law enforcement agencies with information on illegal acts. According to Indonesia Corruption Watch (ICW), whistleblowers in twenty-four graft cases ended up being prosecuted before the law was passed, while the cases they reported on were dismissed.[1] The government is obliged to establish a body responsible for witness protection within one year of the new law's enactment.
- The signature of an **Indonesia–Singapore Extradition Pact** in April 2007 could be a significant milestone in the fight against corruption, given that the city-state, a mere hour away, has regularly been the first port of call for absconders. Although retroactive for fifteen years, the agreement has many detractors, because it guarantees only the extradition of the corruptors, not their assets, particularly if

1 *Jakarta Post* (Indonesia), 28 August 2007.

they have been invested in a 'global investment mechanism' or when the case is categorised as a civil case,[2] as with the Federal Bank of Indonesia's Liquidity Assistance (BLBI) case in 1998. Sukanto Tanoto, Samadikun Hartono, Sudjiono Timan, Adrian Kiki Ariawan, Bambang Sutrisno[3] and other big-name corruptors resided in Singapore despite Indonesian requests that they be sent back to face trial. According to Merrill Lynch and Capgemini, around a third of Singapore's 55,000 'high-net' individuals, or around 18,000 people, are Indonesians with combined assets of US$87 billion,[4] indicating that large amounts of wealth have been transferred to Singapore that would be difficult to recuperate if any of them were found to be tainted by corruption.

- The Timtas Tipikor, or **corruption eradication coordination team, was wound up** in May 2007, two years after it had been formed with a mandate to investigate and prosecute corruption in connection with sixteen state-owned enterprises, four ministries, three private companies and twelve escaped suspects. The team consisted of forty-five staff from the police, the attorney general's office and the auditor's office. Among its leading targets were Said Agil Hussein Al-Munawar, a former minister of religious affairs, and Taufik Kamil, a former director general of Islamic guidance and haj management, who were convicted in February 2006 of embezzling R750 billion (US$80.5 million).[5] The Supreme Court turned down their appeals in August 2006, sentencing them to five and four years in prison, respectively, and ordering them to pay a combined total of R300.7 billion (US$33 million) in fines and confiscations. Timtas Tipikor put five cases on trial, losing one, and recovered an estimated R3.95 trillion (US$424 million) of state money.[6] Its head, Deputy Attorney General on Special Crime Hendarman Supandji, was appointed attorney general of Indonesia in May 2007.

Corruptors fight back

The fight against corruption was showing signs of improvement in the mid-2000s. Indonesia's Corruption Perceptions Index score rose from 2.0 in 2004 to 2.2 in 2005 and to 2.4 in 2006.[7] After his election in September 2004 President Susilo Bambang Yudhoyono promised to crack down on corruption, and founded the Timtas Tipikor the following year. A survey by TI's local chapter in 2006 showed further improvement in the government's performance. This progress was set against the backdrop of an unprecedented set of corruption trials against senior officials, an ex-minister, a senior police officer, military officers, prosecutors, a judge, a governor, MPs and prominent businessmen.

Amid these developments, a phenomenon known as 'corruptors fight back' emerged in late 2006. On 3 July Mulyana W. Kusumah, jailed in 2005 for attempting to bribe an auditor of the Supreme Audit Agency to give the General Election Commission (KPU), for which he worked, a clean bill of financial health, appealed for a Judicial Review by the Constitutional Court.[8] One month later it was the turn of Professor Nazaruddin Sjamsuddin, the KPU's former chairman, who admitted in a fraud inquiry into the 2004 elections to having

2 *Kompas* (Indonesia), 21 May 2007.
3 *Kompas* (Indonesia), 3 May 2007.
4 *Reuters* (UK), 10 October 2006; *Jakarta Post* (Indonesia) 18 October 2007.
5 See english.peopledaily.com.cn/200506/23/eng20050623_191796.html.
6 Further information about Indonesia's anti-corruption institutions and their failure to coordinate is available at kemitraan.or.id/newsroom/staff-articles/fast-tracking-anti-corruption-management.
7 TI, Corruption Perceptions Index 2004, 2005 and 2006.
8 *Tempo* (Indonesia), 27 August 2007.

accepted US$2.1 million as a 'tactical fund' from various IT suppliers as 'thank you money',[9] and another US$1.5 million from an insurance agency as a kickback.[10] Altogether, six judicial petitions were taken out challenging the existence of the KPK, which recovered R200 billion (US$20.4 million) in 2005 and has promoted the concept of 'islands of integrity', subsequently adopted by seven provinces.[11]

On 19 December 2006 the Constitutional Court declared the Special Corruption Court unconstitutional on the grounds that it established the concept of 'duality' in the judiciary, by which a defendant tried in two different courts could expect different treatment. 'It shows there is a double standard in fighting corruption,' said Chief Jimly Asshiddiqie, 'which leads to the absence of legal certainty.'[12] The court gave the government three years to establish a separate legal foundation for the Special Corruption Court, causing one senior lawyer, Adnan Buyung Nasution, to wonder: 'How can an unconstitutional court be given time to exist for another three years?'[13]

This served to drive a coach through the existing legal framework, to the consternation of anti-corruption NGOs. Professor Sjamsuddin's judicial challenge to his corruption case, mentioned above, questioned the legality of the notion of 'material offence' (*delik materiil*), in which an action may not be regulated by law, but is still punishable because it is perceived as a disgraceful act that disrespects social norms.[14] The upshot is that KPK officers will henceforth need very explicit evidence to press on with prosecution – no easy matter in Indonesia.

A weakening of the Special Corruption Court has consequences for the fight against corruption. The main difference between the general and special courts in corruption cases is that the latter demands exemplary sentences with an obligation to pay significant fines and return stolen funds, while defendants in the former are under no obligation to make restitution. In general courts, corruptors tend to escape with minor penalties, and some elude the law altogether. Out of 126 corruption cases with 362 suspects brought by the general courts in 2006, for example, 117 corruption suspects were freed and thirty-seven were sentenced to less than two years with no additional fines.[15] In addition, general courts are not bound by time limitations, whereas the special court must rule on a case within ninety days at the first court level.

The achievements of the Special Corruption Court include the successful case against PT Industri Sandang Nusantara, a company that marked down its assets in Bandung, West Java, with estimated losses to the state of R60 billion (US$6.6 million). The same court sentenced a KPK investigator, known as Suparman, to eight years and fines of R200 million (US$22,000) for extorting money, phones, a car and '24 sets of prayer beads' from a witness in the case.[16]

The Supreme Court also has its own share of problems. In 2001 Judge Syafiuddin Kartasasmita was shot dead by assassins working for Hutomo 'Tommy' Mandala Putra Suharto, the son of the former president, Suharto. The killing was allegedly for failing to honour a

9 See dpr.papua.go.id/index.php?mod=news&page=3&id=36.
10 *BBC News* (UK), 14 December 2007.
11 *Jakarta Post* (Indonesia), 30 December 2005.
12 *Jakarta Post* (Indonesia), 21 December 2006.
13 Ibid.
14 Constitutional Court no. 003/PUU-IV/2006, available at www.mahkamahkonstitusi.go.id.
15 See www.antikorupsi.org/mod.php?mod=publisher&op=viewarticle&artid=9733.
16 *Tempo* (Indonesia), 6 June 2006.

promise after taking a bribe.[17] Kartasasmita had been handling the case. The Supreme Court was also rocked when Probosutedjo, Tommy's uncle and the elder Suharto's half-brother, admitted to providing his lawyer, Harini Wijoso, with R6 billion (US$644,000) to bribe the chair of the Supreme Court, Bagir Manan, and other officials to rule in favour of his appeal against a conviction for graft. Wijoso denied this, but was sentenced to four years in prison in June 2006.[18]

Money, politics and impunity

The TI Global Corruption Barometer 2005 and 2006 named parliament and political parties among the country's most corrupt institutions. These findings were confirmed by an investigation into the activities of Rokhmin Dahuri, a former minister of fisheries and marine affairs in the Megawati administration.

According to KPK prosecutors, since 2002 Dahuri had ordered the collection of a 'tactical fund' from every Fisheries Department office and project, based on a rate of 1 per cent of the total budget. Officially, the R30 billion (US$3.2 million) amassed would be used to finance projects for small fishermen.[19]

The former minister had taken little of the fund, which was distributed to MPs and candidates for the 2004 presidential election campaign. The special committee for fishery bills received R5 billion (US$550,000), while the campaign team for presidential candidate Amin Rais received R400 million (US$44,000), as did the election teams of other presidential and vice-presidential candidates.[20] Hundreds of millions of rupiahs also made their way to Dahuri's family, his alma mater and as aid to fishing communities.[21]

Contributions to MPs were a hot topic for anti-corruption activists. Led by Indonesia Corruption Watch, they insisted that the funds be investigated to establish whether corruption exists in the legislature. The idea was met with little enthusiasm in the Honourable Council and the head of the House turned down the suggestion, arguing that a donation to an MP could not be classified as corruption – even if the funds originated from a government department. After this avenue was exhausted the ICW filed a case, and the Honourable Council was forced to open an inquiry, which sanctioned three MPs.[22]

What turned a corruption case into a constitutional issue, however, was the admission by National Mandate Party presidential candidate Amin Rais that he had accepted a donation from Dahuri.[23] This was illegal on two counts: first, under article 43 of the Law on Presidential Elections, private financial contributions to presidential and vice-presidential hopefuls must not exceed R100 million (US$11,000); second, article 45 prohibits candidates from accepting donations from government-, state- or district-owned firms.

Dahuri told the court that such gifts are commonly exchanged between those with interests connected to parliament.[24] If true, it means that this example of the systematic distribution of public funds to MPs is merely the tip of the iceberg, and suggests that parliamentary decisions are routinely subject to 'purchase'.

17 *Tempointeraktif* (Indonesia), 28 October 2006.
18 *Tempo* (Indonesia), 31 December 2006.
19 *Tempo* (Indonesia), 22 April 2007.
20 Ibid.
21 Ibid.
22 *Kompas* (Indonesia), 13 July 2007.
23 *Tempointeractif* (Indonesia), 18 May 2007.
24 *Tempo* (Indonesia), 22 April 2007.

One reason for this trend is the fact that political parties are not allowed to solicit their own funds to the required amount. Ministers and other holders of political office are expected to solicit funds, usually corruptly, to feed their political party's demands. It was hardly surprising, therefore, when Jamaluddin Karim, chair of the Bintang Pelopor Demokrasi, in addition to informing on the inner workings of government, candidly declared his expectation that the minister of forestry, M. S. Kaban, would also be on the lookout for fresh party funding.[25]

Ministers and office-holders who do not provide party funding suffer the consequences. When former minister of state-owned enterprises Sugiharto faced dismissal in mid-2006 there was no defence from his party, because he had made no financial contribution to his party, despite supervising hundreds of state enterprises, including national oil company Pertamina, with combined assets worth billions of dollars.[26]

A similar fate befell Widjanarko Puspoyo, director of the extraordinarily wealthy Bulog state logistics agency. After his party, the Indonesia Democracy Party of Struggle, was defeated in the 2004 election, Widjanarko was abandoned to face the corruption charges that eventually sent him to prison.[27] Puspoyo was arrested on 20 March for alleged bribery in a cattle-import scheme from Australia in 2001, according to prosecutors losing the state US$1.2 million, as well as for allegedly undisclosed shipments of rice from Vietnam during 2001–3.[28]

Tommy's billions

In May 2007 the district court in the United Kingdom's offshore banking centre of Guernsey froze for a further six months the assets of Tommy Suharto, son of the former Indonesian strongman, and invited the government to inquire into the legality of the money.[29] The attorney general had prepared documents in March to recover the funds, estimated to amount to US$60 million, from BNP Paribas, but he was having difficulty pinning down a paper trail.[30]

In conjunction with the court process, the attorney general planned to reopen the investigation into Tommy Suharto, who was released on parole in October 2006 after completing a fraction of his fifteen-year sentence for killing Supreme Court Judge Kartasasmita in 2001. Suharto spent less than five years in prison, thanks to a pardon by two former ministers, Yusril Ihza Mahendra and Hamid Awaluddin, who were later dismissed from Cabinet for abetting his money-laundering.[31]

The investigation is likely to centre on the young Suharto's monopoly of the clove trade in the 1990s, which saw tobacco companies compelled to buy the spice for a local variety of cigarettes at marked-up prices. In 1998, shortly after his father's fall from power, Tommy's fortune was estimated at US$800 million.[32] He claimed to be too ill to attend when prosecutors called him in for formal questioning in July 2007.[33]

25 *Republika* (Indonesia), 16 May 2007.
26 *Fajar Online* (Indonesia), 3 June 2006.
27 *Tempo* (Indonesia), 8 April 2007.
28 *Tempointeraktif* (Indonesia), 27 March 2007.
29 *Detik.com* (Indonesia), 23 May 2007.
30 *Jakarta Post* (Indonesia), 21 May 2007.
31 *Washington Post* (US), 8 November 2006; *Tempo* (Indonesia), 11 November 2006; *Jakarta Post* (Indonesia) 24 April 2007; *Tempo* (Indonesia), 18 March 2007.
32 See edition.cnn.com/2002/WORLD/asiapcf/southeast/03/19/tommy.profile/index.html.
33 *Antara News* (Indonesia), 8 July 2007.

Local water boards and corruption

The issue of corruption in the local water boards (PDAMs) has exploded in many provinces of Indonesia, but it is hard to separate it from poor management, the water boards' deep indebtedness or the opportunities for graft that accompanied President Suharto's decision to privatise the sector in 1997. For example, Komparta, a water consumers' group militantly opposed to privatisation, reported inexplicable losses of R800 million (US$88,000) at PDAM DKI in Jakarta from 1997 to 2004, but admitted it could not decide whether this was due to real corruption or administrative incompetence.

This is not to deny the multitude of high-profile cases that have emerged in the period under review. On 10 December 2006 the former director of PDAM Indramayu, west Java, Deddy Sudrajat, was jailed for two years, fined R20 million (US$2,200) and ordered to refund R500 million.[34] In February 2007 an audit of PDAM Saumlaki in Maluku Tenggara Barat discovered that R300 million of an R700 million subsidy was missing from the local budget.[35] The list goes on. PDAM Ende NTT,[36] PDAM Kutai Kertanegara, east Kalimantan,[37] and PDAM south-east Aceh, among others, came under scrutiny for missing funds or, in the case of PDAM Semarang,[38] colluding with external contractors.

And yet the sums are comparatively small and should be seen in the context of the PDAMs' deep indebtedness, a process set in motion by the ninefold devaluation of the rupiah in 1998. The currency crisis made it impossible for water boards to import inputs and specialist machinery, led them to hike consumer prices to unacceptable levels and forced them to apply for international and domestic loans.[39]

As of 31 March 2000 63 per cent of the loans from the Ministry of Finance to PDAMs were still outstanding and only 48 per cent of the R1.37 trillion (US$151 million) owed had been repaid.[40] By 2003 the Ministry of Environment was looking to the private sector to rescue 90 per cent of Indonesia's 292 PDAMs, although the privatisation in Jakarta has been fraught by 'failing targets and rising prices', in the words of one analyst.[41]

The absence of firm leadership in the sector, the PDAMs' uncertain future as locally owned entities and the lack of supervision from central government have certainly helped debase the qualities of management and probity that would at least contribute to a reduction of corruption in the water provision sector. But crying 'corruption' at local employees is often a way of distracting attention from mismanagement higher in the hierarchy.

Anung Karyadi (TI Indonesia)

Further reading

ICW, 'Corruption Trends in 2006' (Jakarta: ICW, 2007).

34 See www.republika.co.id/koran_detail.asp?id=274418&kat_id=&kat_id1=&kat_id2=.
35 *Tempointeraktif* (Indonesia), 22 February 2007.
36 *Tempointeraktif* (Indonesia), 25 February 2007.
37 *Kaltim Post* (Indonesia), 29 March 2007.
38 *Suara Merdeka* (Indonesia), 15 July 2006.
39 The exchange rate of the rupiah fell from R2,317 to R16,950 to the US dollar in June 1998; see lnweb18.worldbank.org/eap/eap.nsf/Attachments/Water-PDAM/$File/Current+Efforts+to+Revitalize+PDAMs.pdf.
40 Ibid.
41 N. Ardhianie, 'Jakarta Water Privatisation: Seven Years of "Dirty" Water', in B. Balanyá and B. Brennan *et al.* (eds.), *Reclaiming Public Water* (Amsterdam: Transnational Institute and Corporate Europe Observatory, 2005).

T. Rinaldi *et al.*, 'Combating Corruption in Decentralized Indonesia' (Washington, DC: World Bank, 2007).

TI Indonesia, 'Corruption Perception Index of 32 Cities in Indonesia 2006' (Jakarta: TI Indonesia, 2007).

TI Indonesia: www.ti.or.id.

Japan

Corruption Perceptions Index 2007: 7.5 (17th out of 180 countries)

Conventions

ADB–OECD Anti-Corruption Action Plan for Asia-Pacific (endorsed November 2001)

OECD Convention on Combating Bribery of Foreign Public Officials (signed December 1997; ratified October 1998)

UN Convention against Corruption (signed December 2003; not yet ratified)

UN Convention against Transnational Organized Crime (signed December 2000; not yet ratified)

Legal and institutional changes

- The **Anti-Monopoly Law** was amended in April 2005 and came into force in January 2006. The amendments include raising surcharges for violations of the law, introducing a leniency programme for voluntary reporting of violations and giving the Fair Trade Commission (FTC) powers to investigate those violating the act. The surcharge for large-scale manufacturers was raised from 6 to 10 per cent, for retailers from 2 to 3 per cent and for wholesalers from 1 to 2 per cent. Under the leniency system, the first whistleblower before an official inspection will receive full exemption of the final penalty, while the second wins a 50 per cent reduction and the third a 30 per cent.[1]
- The **Bid Rigging Elimination and Prevention Act of 2002 was amended** in December 2006 and now empowers the FTC to demand that central and local governments take corrective measures when they discover officials or public entities engaging in bid-rigging. The amendment expands the scope of the law to include public corporations, while those liable may face a new charge of abetment. Officials can be sentenced to five years in prison and fined up to ¥2.5 million (US$20,000).
- The **Political Funds Control Act of 1948 was amended** in December 2006. Under the

1 See www.omm.com/webdata/content/publications/client_alert_antitrust_2005_04_26.htm.

previous law, companies with more than 50 per cent foreign ownership were prohibited from making financial contributions to Japanese politicians and political parties. Nippon Keidanren, one of Japan's largest business associations, lobbied for the government to deregulate the restriction amid growing external investment in Japan's stock market. The amendment makes it legal for foreign-owned companies that have been listed in Japan for more than five years to contribute to the campaign funds of parties and politicians.

Graft in public office

Japan's score in the Corruption Perceptions Index improved in 2006, but it made less progress in the field of money in politics. The Political Funds Control Act has been amended almost annually in the past fifteen years, with twenty amendments alone in the last three. Despite this frenzy of legislative activity, the act signally failed to establish accountability for politicians and their factions, because lawmakers intentionally leave loopholes in drafting so as not to restrict their activities.[2] Since 2006 many falsified items on the ordinary expenditure account have been criticised in the newspapers.[3]

The discovery of an act of abuse by one politician often leads to revelations about others, but the media's attention rarely goes beyond the immediate scandal, leaving other related instances of corruption and individuals unaccountable. Politicians who channelled illicit funds to their secretaries' accounts were in the media spotlight in 2004 (see *Global Corruption Report 2004*). The following year saw investigations into a wing of the ruling Liberal Democratic Party (LDP) for failing to report an illegal ¥100 billion donation (US$910 million) from a national dentists' association (see *Global Corruption Report 2005*). The faction's treasurer and a senior politician were subsequently indicted for violating the Political Funds Control Act, but top executives in the faction – all MPs – were not considered accountable.[4]

Throughout 2006 and 2007 the focus was on MPs' office expenses. In December 2006 a political support management company owned by Gen'ichiro Sata, minister in charge of administrative reforms, fell under suspicion of falsifying accounts after it claimed ¥78 million (US$700,000) in utility charges in 1999–2000, despite the office being non-existent.[5] Faced with growing criticism, he resigned after admitting misrepresentation. Two other ministers, for education and agriculture, and an LDP executive were found to have similarly misreported utility expenses.[6]

While some MPs and ministers admitted 'mistakes' and corrected their accounts, the agriculture minister, Toshikatsu Matsuoka, whose fund management group registered ¥28 million (US$161,500) in office expenses for 2001–5, refused to disclose in detail what was actually spent. Matsuoka insisted he had dealt with the funds appropriately and refused to resign, a position that the prime minister, Shinzo Abe, supported. The minister allegedly received donations from companies connected to a bid-rigging scandal involving the government's Japan Green Resources Agency. Matsuoka hanged himself in late May 2007.[7]

2 *Japan Times*, 7 January 2006.
3 *Tokyo Shimbun* (Japan), 24 January 2007; *Tokyo Shimbun* (Japan), 26 January 2006.
4 *Japan Policy and Politics*, 24 November 2004.
5 *Japan Times*, 1 January 2007.
6 *Japan Times*, 13 January 2007 and 17 March, 2007.
7 See www.cdnn.info/news/eco/e070528.html.

The reason why the 'office expense' category has been so widely used to camouflage expenses used for other purposes is clear. Under the Political Funds Control Act, politicians must attach receipts for expenses of more than ¥50,000 (US$400) with detailed explanations as to how the money was used. They are not required to submit receipts for office expenses, however, only to give the amount they claim to have spent.

In spring 2007, after this series of misrepresentations, the bigger parties started considering amendments that would oblige politicians to provide proper receipts for office expenses. MPs opposed such measures, saying it would further restrict their activities and make procedures even more troublesome. The government and parties may succeed in closing the immediate loophole in the near future, but they will not close others. What the government really needs is a comprehensive programme to establish the accountability of politicians. The first step would be to introduce a stricter system of punishments for violators of the existing Political Funds Control Act.

Public officials involved in bid-rigging

Japanese society has a long tradition of *amakudari*, or 'golden parachuting', whereby senior officials solicit semi-government or private corporations for post-retirement jobs at higher remuneration. A retired official may go through two or three post-retirement jobs, each time receiving a higher lump sum retirement allowance.

A clear picture of this engrained practice has never emerged, and those with the information are unlikely to disclose one of the hidden perks of state employment. It is commonly believed, however, that the practice does more harm than good, because *amakudari* is a major cause of *kansei-dango*, or bid-rigging, by public officials. As recent revelations show, public officials play important roles in awarding contracts in exchange for post-retirement jobs in the companies they favour. In other cases retired officials who have been promised jobs in private companies leak the bidding prices to likely bidders, who arrange between themselves which of their companies will win, based on the insider information.[8]

Late 2006 witnessed a flurry of cases in which corrupt prefecture governors decided the winners of bids for public works projects. In October and November 2006 three were arrested for their involvement in rigging bids in Fukushima, Wakayama and Miyazaki prefectures.[9] Some governors are on the lookout for political contributions because they do not enjoy the same level of funding as state politicians. The revelations triggered a public debate about capping the terms served by governors, who currently face no official time limit. This has led to the passing of some by-laws on the length of terms of some officials, but no widespread acceptance that limits should be enforced.[10] More considerate governors step down after serving two or three terms in order to avoid the appearance of cronyism.

The central government has belatedly started to tackle the practices of *kansei-dango* and *amakudari* in earnest. Amending the legislation against bid-rigging is one approach, and incorporating the newly introduced leniency system for whistleblowers in the Anti-Monopoly Law is another. As a further anti-*amakudari* measure, the Abe administration submitted a bill to initiate reforms including measures to secure more effective control over retiring and retired public

8 *Japan Times*, 6 January 2007.
9 *Japan Times*, 19 November 2006 and 6 December 2006.
10 *Japan Times*, 6 December 2006.

officials. The bill passed in the July session of the 2007 Diet, but is unlikely to be implemented until 2008.

Bankruptcy of local government

In June 2006 Yubari City, Hokkaido, a formerly prosperous mining town, applied for designation as a Fiscal Rehabilitation Body (FRB), meaning that it had effectively declared bankruptcy. With help from the central government under the Fiscal Rehabilitation Law, Yubari has since embarked on a drastic rescue package that includes substantial cuts in public services.

The news thundered across the nation as reports emerged about the austerity the city will endure for years to come, and the imminent reality of curtailed services to residents and reduced salaries for civil servants. Yubari is no exception, however. Many local governments in Japan are on the brink of bankruptcy.[11]

Many factors contributed to Yubari's bankruptcy, including its ageing demographic: nearly 40 per cent of the population is over sixty-five, according to the 2005 census. But another reason was the lack of transparent fiscal management and, possibly, corruption. The city's administration kept its rising deficit secret by taking out bridging loans or floating debt at the end of the financial year, while in reality the total debt had snowballed to ¥63 billion (US$509 million), fourteen times its annual budget. The balance sheet appeared positive because the mayor resorted to using the fiscal tool known as 'temporary borrowing' each year. The real state of the deficit was shielded from local parliament members and voters until 2006.[12]

Corrupt management is rampant throughout Japanese municipalities. Examples include unnecessary foreign travel by members of local assemblies; public investment in unwanted roads and bridges; off-the-books money for wining and dining officials; and a lack of discipline with regard to guarantees for 'third-sector companies' (public–private joint venture businesses), most of which suffer excessive losses.[13] All of this adds up to the prospect of bankruptcy for many other municipalities.

Local government deficits increased meteorically after 1991, when the bubble economy burst. Today it stands at about ¥200 trillion (US$1.8 trillion), or equivalent to 40 per cent of Japan's GDP as of the end of 2006. Coupled with central government borrowings of about ¥600 trillion (US$5.5 trillion), total public sector debt now stands at around 150 per cent of GDP, making Japan the world's most indebted country in terms of debt-to-GDP ratio. The deficits were mainly financed by government bonds, central and local, mostly held by Japanese citizens.

The rise in local debt accelerated Tokyo's decision to pump about ¥40 trillion (US$364 billion) every year into public works to stimulate – or at least to keep afloat – an economy badly hit by the collapse of the land market. Since local governments shoulder a large percentage (around 60 per cent) of the administrative responsibility for the public budget, they, and not central government, implement a major part of public spending. At Tokyo's insistence, cities financed their public works budgets by issuing local bonds, which the central government guaranteed through the Local Allocation Tax (LAT). With redemption guaranteed, municipalities launched work projects one after another in the 1990s.

In 2001 the administration of Junichiro Koizumi altered the way of drawing up and implementing economic policy, giving greater power to the prime minister's office. This more centrist style of management led to some successful reforms,

11 *Japan Times*, 30 June 2006.

12 Ibid.

13 I. Shirakawa, *Jichitai Hasan: Saiseino Kagiha Nanika* [Bankruptcy of Municipal Governments: What is Key to their Rehabilitation?] (Tokyo: NHK, 2007).

including the LAT. Since the LAT is a negative incentive for municipalities to strengthen their fiscal competence, the government introduced a package of further decentralisation measures that included decreasing the LAT. Many local governments were left with obligations to pay back public debts amid decreasing fiscal revenues owing to the reduced LAT.[14] Thus, hundreds of municipal bodies now face the possibility of bankruptcy. While corrupt management may not necessarily be the main reason for the bankruptcy of local governments, it is a crucial part of the background behind the already fragile fiscal standing of cities such as Yubari.

Toru Umeda, Keiichi Yamazahi and Minoru O'uchi (TI Japan)

Further reading

Ministry of Internal Affairs and Communications, Financial Management Division, Local Public Affairs, 'White Paper on Local Public Finance 2006 Illustrated' (Tokyo: Ministry of Internal Affairs and Communications, 2004).

T. Furukawa, *Nihon-no-Uragane,* vol. I, *Kensatsu-Keisatsu-hen* (Tokyo: Daisanshokan, 2007).

Nihon-no-Uragane, vol. 2, *Shushokantei-gaimusho-hen* (Tokyo: Daisanshokan, 2007).

T. Hobo *et al., Yubari Hatan to Saisei – Zaisei Kikikara Chiikiwo Saiken Surutameni* (Tokyo: Jichitai Kenkyusha, 2007).

TI Japan (ed.), 'National Integrity System TI Country Study Report of Japan 2006' (Tokyo: TI Japan, 2006).

I. Shirakawa, *Jichitai Hasan: Saiseino Kagiha Nanika* (Tokyo: NHK, 2007).

TI Japan: www.ti-j.org.

14 *Japan Times*, 22 April 2006.

Malaysia

Corruption Perceptions Index 2007: 5.1 (43rd out of 180 countries)

Conventions

ADB–OECD Anti-Corruption Action Plan for Asia-Pacific (endorsed November 2001)

UN Convention against Corruption (signed December 2003; not yet ratified)

UN Convention against Transnational Organized Crime (signed September 2002; ratified September 2004)

Legal and institutional changes

- In October 2006 parliament approved the **Electronic Commerce Act**, which supplements the Consumer Protection Act of 1999 by enabling consumers who make purchases via the internet to bring cases to the tribunal without going through the courts. The act stipulates that 'any information shall not be denied legal effect, validity or enforceability on the ground that it is wholly or partly in an electronic form'. Before the amendment no laws governed internet transactions, and the public was not entitled to bring cases before a tribunal when fraud did occur.

- The prime minister, Abdullah Badawi, launched the five-year **National Integrity Plan** (NIP) in April 2004. It identifies five key objectives: to reduce corruption and abuse of power; to increase the efficiency of public service delivery; to enhance corporate governance; to strengthen the family; and to improve citizens' quality of life. The first, collectively known as Target 2008, was tied to improving Malaysia's ranking in TI's Corruption Perceptions Index from 37th position and a 5.2 score in 2003 to 30th and 6.5 in 2008. Unfortunately, Malaysia has headed in the opposite direction in the past three years, falling to 39th and 5.0 in 2004, and 44th and 5.0 in 2006.[1]
- Mohamed Nazri Aziz, minister for parliamentary affairs in the Prime Minister's Department, said on 23 April 2007 the government was **considering introducing laws to protect whistleblowers**, although he had previously said that there were no plans to include a whistleblowers' provision in the Securities Industry Act 1983. Malaysia Airlines adopted a whistleblowers' policy for employees to report corruption, security infringements and other malpractices in January 2006, but there is little evidence so far that the policy has been implemented.[2]
- According to a recent report by Article 19 and the Centre for Independent Journalism (CIJ), Malaysia is bucking global trends towards improved access to information by **'the tendency of the government to revert to secrecy** whenever it faces challenges'.[3] It is assisted in this by the draconian conditions of the colonial-era Official Secrets Act (see below). The report, which focuses on access to environmental information, notes that the Air Pollutant Index remained a state secret from the 'haze crisis' of 1997–8 to the one of August 2005.
- On 17 May **Bernard Dompok, a federal minister, resigned as chairman of the Parliamentary Select Committee on Integrity** (PSCI) after National Registration Department officials did not attend a hearing to explain the mass issuance of national identity cards to Indonesian migrants in Sabah, one of two Malaysian states in northern Borneo. By one account the number of non-Malaysians in Sabah amounted to 1.75 million, outnumbering the 1.5 million locals.[4] The PSCI has twelve members – ten from the ruling Barisan Nasional (BN) coalition, two from the opposition – and allows MPs, *inter alia*, to inquire into cases of corruption and abuse of power, although it has no authority to investigate. Dompok said he resigned because he 'would not be able to do justice to the tasks assigned to the committee by parliament'.[5] The PSCI had also summoned former Sabah Anti-Corruption Agency (ACA) chief Mohamed Ramli Manan and former ACA director general Datuk Seri Zulkipli Mat Noor to attend an inquiry after Ramli accused Zulkipli of corruption. The hearing was cancelled, however, reportedly due to political pressure, and Zulkipli's contract was not renewed.[6] These cases of 'no

1 While it is encouraging that Malaysia is recognising the importance of the CPI, using the CPI scale for setting quantitative targets is not encouraged by Transparency International.

2 *Malaysiakini*, 2 February 2006.

3 Article 19 and CIJ Malaysia, 'A Haze of Secrecy: Access to Environmental Information in Malaysia' (London and Kuala Lumpur: Article 19 and CIJ Malaysia, 2007).

4 *Malaysiakini*, 27 June 2006.

5 *The Star* (Malaysia), 18 May 2007.

6 *The Star* (Malaysia), 31 March 2007; *Malaysiakini*, 12 March 2007.

show' are disrespectful to parliament, show lack of accountability in state institutions and, if not reversed, may encourage other agencies and individuals to ignore calls to appear before parliamentary select committees.

Media muzzled by crony ownership

Malaysia's constitution guarantees freedom of expression for its citizens, but parliament also has the right to impose laws to restrict this freedom 'in the interest of security, friendly relations with other countries, public order or morality'. The regulatory structure of freedom of expression restriction, inherited from British colonists, is not only still intact but has been made more stringent by a further set of laws, such as the Sedition Act 1948, the Defamation Act 1959, the Internal Security Act 1960, the Official Secrets Act 1972 and the Printing Presses and Publications Act (PPPA) 1984.

The PPPA grants discretion to the minister of internal security to grant, refuse, suspend and revoke annual publication licences. A licensing requirement also applies to users of printing machines. In addition, the minister's power also extends to banning publications and prohibiting their importation.

In December 2006 the ministry confiscated issues of the bimonthly opposition tabloid *Harakah*, on the grounds that it was not allowed to have articles in Jawi, or traditional Malay, script. In April 2007, according to the editor-in-chief, police raided the office of the tabloid *Putra Post*, which had carried an attack on the government by the former prime minister, Mohamad Mahathir. Police alleged that the publication did not have a valid permit, and seized computers and printing plates, despite the editor producing his licence.[7]

In January 2007 police detained a photographer from Malaysiakini.com, an influential news website, who was covering a protest, and demanded he surrender the pictures he had taken. Two journalists covering nomination day for the by-election in Ijok (see below) in April 2007 were barred from taking pictures of a fracas between supporters of contesting parties.[8]

Along with such repressive micro-management, ownership patterns have shaped a media landscape in which the government's view of events is upheld in varying degrees by Malaysia's thirty-four mainstream newspapers. This is in comparison to six opposition party journals, whose circulation is restricted to a members-only readership, a paper run by supporters of Mohamad Mahathir, and two web-based dailies (Malaysiakini.com and MerdekaReview).

The fourteen parties that make up the Barisan Nasional ruling coalition account for several of Malaysia's most widely read newspapers, including *Kosmo*, *Mingguan Malaysia* and *Utusan Malaysia* (United Malays National Organisation, or UMNO); *The Star*, *Sunday Star*, a stake in *Nanyang* daily and *China Press* (Malaysian Chinese Association, or MCA); and *Tamil Nesan* (Malaysian Indian Association). The biggest player in the media industry is Media Prima, a government-affiliated conglomerate whose New Straits Times Press subsidiary publishes *New Straits Times*, *New Sunday Times*, *Malay Mail*, *Sunday Mail*, *Berita Harian*, *Berita Minggu*, *Harian Metro* and *Metro Ahad*. Media Prima also owns all the private free-to-air TV networks (Sistem Tevisyen Malaysia (TV3), Natseven TV (ntv7), Metropolitan Station (8TV) and Ch-9 Media (TV9), and has a stake in the newest pay TV channel, MiTV Corporation (MiTV).

Tiong Hiew King, the Malaysian-Chinese timber tycoon, owns four of the major Chinese language

7 See tunkuaisha.blogspot.com/2007/04/cops-raid-putra-post-tabloids-office.html.
8 *Malaysiakini*, 21 April 2007.

papers: *Sin Chew Daily*, *Guang Ming Daily*, *China Press* and *Nanyang Siang Pau*. Astro All Asia Networks monopolises the pay TV market and its subsidiary, Satellite TV Astro, and owns ten subscriber and eight free-to-air radio stations. Astro is reportedly owned by the apolitical billionaire, Ananda Krishnan.

NexNews, under the Berjaya Group of Companies, publishes *The Edge* weekly and *The Sun* daily. It also holds shares in MiTV. Berjaya Group's chairman is Vincent Tan, who also holds shares in Media Prima.

Lau Swee Nguong, chairman of the KTS Group of Companies, has stakes in *Oriental Daily News*, *Borneo Post* (Sabah), *Borneo Post* (Sarawak), *See Hua Daily News* (Sabah), *See Hua Daily News* (Sarawak) and *Utusan Borneo*.

The primary concern in Malaysia is that the concentration of media control in the hands of a small group of companies with close links to the ruling political elites will make it difficult for the press to play its role as a key platform for exposing corruption in the system. Alternative views are discouraged and the government issues gag orders on certain sensitive topics from time to time to prevent open and civilised discussion by civil society. Corruption easily creeps into closed systems and the general public is deprived of its democratic rights when permits to publish favour companies that toe the dominant political line.

Democracy takes a back seat to political drivers

The by-election in Ijok, Selangor, triggered by the death of Datuk Sivalingam, was held on 28 April 2007 and won by the Barisan Nasional candidate, K. R. Parthiban, with a narrow 1,850 majority on an 81.9 per cent turnout. The constituency was clearly significant to the BN's self-esteem because it marked the return to electoral politics of Anwar Ibrahim, Malaysia's former deputy leader, who was released from jail in September 2004 after being imprisoned for what were commonly considered trumped-up charges of corruption and sodomy.[9]

A government fund of RM36 million (US$10.3 million) was created to pave and widen roads, install street lights, construct drains and lay water pipes for a community of just 12,000 during the week set aside for campaigning. Election law limits the maximum expenditure per candidate in a state assembly election to RM100,000.[10] There were widespread allegations of land-for-votes and low-cost housing offers to influence voters, in addition to reports of intimidation of opposition leaders and their supporters, abuse of government machinery, phantom voting and bribery.[11]

The Election Commission chairman, Rashid Rahman, dismissed these abuses in procedure by declaring the allocation of development funds to a voting district as acceptable electoral practice, rather than the 'tsunami of money politics to buy votes' that it appeared to be to one blogger.[12] Other alleged irregularities included the following.

- Many Malay addresses had apparently been occupied by people with Chinese names who had lived there for decades, and vice versa. In one village, thirty-five such cases were identified.
- Voters could not be found at the listed addresses, and occupants had no knowledge of the persons who had been registered at those addresses.

9 BBC (UK), 2 September 2004.
10 *Bernama* (Malaysia), 18 April 2007.
11 See politics101malaysia.blogsome.com/2007/04/30/ijok-by-election-real-losers-and-winners.
12 See bersih.org/?p=96.

- The electoral roll was stuffed with improbable voters. There were thirty-one voters aged above 100, and more than 200 voters aged over ninety.
- Many voters were long deceased.[13]

On 23 April police moved in to break up opposition rallies,[14] while applying standards of a very different order to meetings hosted by parties in the ruling coalition. The clampdown resulted in many opposition rallies being halted. The only positive outcome was vigorous campaigning by opposition parties in the last phase of the election, which augurs well for cooperative action in the future. The return to the political scene of Anwar Ibrahim, Mohamad Mahathir's former heir, and his Keadilan (National Justice) party was widely expected to win him a parliamentary seat, but he was barred from contesting due to his conviction and imprisonment.

A general election is expected to be called soon, although it is not due until April 2009. It will be tough for the opposition, as the BN has won all five by-elections since the last election in March 2004. If the BN chooses to hold the election before April 2008, it would mean that Anwar Ibrahim would be ineligible to stand since his disqualification period does not end until that month.

The Election Commission, which is responsible for ensuring a free and fair vote, and the police, which is responsible for maintaining law and order, must ensure a level playing field in future elections if voters are not to feel threatened or bribed.

A tug of war in the water industry

Water policy continues to be a tug of war between conflicting special interests. The Water Services Industry Act of 2006 transferred the control, regulation and distribution of water and sewerage services throughout Malaysia from state authorities to the federal government, on the grounds that such a move would 'curb corruption' through improved supervision.

This and another water-related bill aim to create a water assets management company, Wamco, owned by the Finance Ministry, to buy up existing water infrastructure in Malaysia for lease to state-owned or private operators. The government set a budget of RM16 billion (US$4 billion) to improve water infrastructure over the following five years.[15]

The key concern of the National Water Services Commission, set up in March 2007, is to ensure the provision of clean water at a fair price. Suppliers that fail to meet specific benchmarks on price, procurement, quality and technical standards will be penalised. The commission's members are appointed by the minister of energy, water and communications, potentially compromising its independence.[16]

On 30 May the minister, Datuk Seri Dr Lim Keng Yaik, announced that the privatisation of water supply would be put on hold and no more concessions would be given out. The government is still open to the idea of joint ventures between state governments and foreign corporations, however, so privatisation is effectively a policy that has been devolved from the federal to the state level.[17]

All this follows what was a disastrous flirtation with the free market in water. In January 2005 the federal government agreed to pay R2.9 billion (US$829 million) to Syarikat Bekalan Air Selangor (Syabas) as a settlement of pre-existing

13 *Malaysiakini*, 18 July, 2007.
14 Bangkit.net, 25 April 2007.
15 See Inter Press Service News (Rome), 5 May 2006.
16 For more information, see www.dapmalaysia.org/english/2005/jun05/lks/lks3503.htm.
17 For one example of the contradictory status in which water finds itself, see *Daily Express* (Malaysia), 15 May 2007.

debts to three water treatment plants in the thirty-year Selangor water concession.[18] Syabas subsequently increased water tariffs by five to 37 per cent.

In 2007 it came to light that Syabas had contracted two relatively unknown companies, Laksana Wibawa and Musa & Rahman Plastics, to supply all its needs in mild steel and polyethylene ducting for its pipe replacement programme. According to *The Edge*,[19] Laksana Wibawa turned out to be a subsidiary of Puncak Niaga Holdings, a 70 per cent shareholder in Syabas, suggesting a potential conflict of interest.

A few months earlier, in January 2007, TI Malaysia had 'note[d] with concern' a statement by the energy, water and communications minister, Dr Lim Keng Yaik, that the terms of the government's thirty-year concession agreement with Syabas and its own audited accounts were classified under the Official Secrets Act and could not be made public without Cabinet approval.[20]

Richard Y. W. Yeoh and Natalie P. W. Ng
(TI Malaysia)

Further reading

Article 19 and CIJ Malaysia, 'A Haze of Secrecy: Access to Environmental Information in Malaysia' (London and Kuala Lumpur: Article 19 and CIJ Malaysia, 2007).

R. Behari *et al.*, 'Curbing Corruption in Public Procurement' (Kuala Lumpur: TI Malaysia, 2006).

V. Capulong *et al.*, *Corporate Governance and Finance in East Asia: A Study of Indonesia, Republic of Korea, Malaysia, Philippines and Thailand* (Manila: ADB, 2001).

S. Gill, 'National Integrity System TI Country Study Report of Malaysia 2006' (Kuala Lumpur: TI Malaysia, 2007).

TI Malaysia, 'Malaysia Transparency Perception Survey 2007' (Kuala Lumpur: TI Malaysia, 2007).

TI Malaysia: www.transparency.org.my.

18 See www.dapmalaysia.org/english/2006/marc06/lks/lks3825.htm.

19 *The Edge* (Malaysia), 31 May 2007.

20 See www.transparency.org.my/press30.htm.

Nepal

Corruption Perceptions Index 2007: 2.5 (131st out of 180 countries)

Conventions

ADB–OECD Anti-Corruption Action Plan for Asia-Pacific (endorsed November 2001)
UN Convention against Corruption (signed December 2003; not yet ratified)
UN Convention against Transnational Organized Crime (signed December 2002; not yet ratified)

Legal and institutional changes

- An **interim parliament** of 330 MPs, created on 15 January 2007, included 209 members from the previous assembly, eighty-three Maoists and forty-eight representatives from CSOs, demonstrating the shift of power to the people. It has abandoned the principle of constitutional monarchy, a notion inscribed in the 1990 constitution. The interim parliament focused on creating an anti-corruption framework, passing an amendment to the Special Court Bill in June 2007 increasing the number of judges in order to speed up hearings on corruption charges (see *Global Corruption Report 2007*). The Constituent Assembly Election Bill, which was passed in August 2007, debars members of the royal government, wilful bank defaulters and people indicted on corruption charges from contesting election. New proposals, in the form of the Good Governance and Operations Bill, were introduced to make the government more transparent and accountable. Under a bill dealing with the right to information, approved on 18 July 2007, a Nepali is entitled to receive information within fifteen days of applying to any government body or public enterprise.
- The interim parliament promulgated an **interim constitution** on 15 January 2007 that bestowed on citizens the rights to health and education, and legal entitlement to their enforcement.
- A **Commission to Investigate the Wrongdoings of the Royal Government** was formed on 5 May 2007 under a retired Supreme Court judge, K. J. Rayamajhi. It heard testimony from 200 ministers, senior civil servants, and army and police officers, including former vice-chairs and members of the royal Cabinet. Cases involving financial irregularities and corruption are being investigated by the statutorily mandated Commission for the Investigation of Abuse of Authority (CIAA). The commission's report charged the royal Council of Ministers with using excessive force in suppressing the popular uprising.[1]

1 *Nepal News.com*, 4 August 2007.

New hope, but impunity prevails

Nepal is passing through an uneasy transition as a result of a tripolar power struggle between the king, the political parties and Maoist rebels. Autocratic measures, such as the dissolution of parliament, the formation of a royal government and the exclusion of political parties from governance, resulted in a 'twelve-point' agreement between the seven official parties and the Maoists. Three weeks of popular protest, known as the April Movement, forced the king in 2006 to dismiss his government, reinstate parliament and invite the eight political groupings to form a new government.

Changes to the structure of power altered the institutional landscape. The adoption of an interim constitution set in motion radical changes, such as the suspension of the monarchy, the secularisation of the state and the election of a Constituent Assembly charged with framing a new constitution. These changes paved the way for the Maoists to join the government on 1 April 2007.

The interim constitution became effective on 15 January 2007 and will remain in force until the Constituent Assembly adopts a new constitution. Transparency and accountability are promised in a subsection dealing with 'directives of state policy'.[2] Judging from the past, however, it is difficult to conceive how the commitment to eliminate corruption will be realised. The 1990 constitution had similar provisions but was no more effective in improving political integrity.

At present the main transition actors – the political parties, the coalition government, parliament and civil society – have not accorded priority to corruption control. The strong current of transitional politics has made leaders myopic to such issues, and the interim government's failure to punish a single corrupt person among the many identified by the common people, Special Court and the 2007 Rayamajhi inquiry has convinced many observers that the guilty will continue in power. Impunity is both the legacy and the rule of the game in Nepal. Short-term commissions are short-term ploys to assuage public anger.[3]

Ex-ministers, such as Chrinjibi Wagle and Govinda Raj Joshi, who both have cases pending in the courts, have nonetheless been inducted into the legislature as parliamentarians. Serial bank defaulters who wilfully fail to pay back loans face no punitive action, although the Amatya Group was forced to pay back more than US$44 million.[4] Warrants to arrest embezzlers and fraudsters are not acted upon. Allegedly corrupt influential people are freed by the Special Court on flimsy pretexts. It is a situation in which the rule of law has been compromised and impunity prevails.

Disappointment with the Special Court

The Special Court traditionally deals with corruption, and this is where the corruption watchdog, the Commission for the Investigation of Abuse of Authority, files its cases. The court has cleared several ministers accused of graft in the past year. For example, the CIAA filed a case against the former information minister, Jaya Prakash Gupta, in March 2002, accusing him of accumulating around US$300,000. He was cleared of all charges on 11 June 2007.[5] Other senior politicians, similarly accused but later freed, include ex-ministers Khum Bahadur

2 See 'Part 4, Responsibilities, Directive Principles and Policies of the State' of the *Interim Constitution of Nepal 2007* (Kathmandu: Parliament Secretariat, 2007).

3 *NepalNews.com*, 4 August 2007.

4 *Tribune Online* (India), 5 May 2007.

5 *Himalayan Times* (Nepal), 13 June 2007.

Khadka, Govinda Raj Joshi and Rabindra Nath Sharma and former heads of police Motilal Bohara and Achyut Krishna Kharel.[6]

Though CIAA cases are certainly not perfect, its investigations are generally painstaking and founded on a decade of experience. The Special Court's recent acquittals have undercut the CIAA's efforts to punish corrupt officials, however, and transformed anti-corruption laws into a code of 'victor's justice'. The efficacy of these laws can be improved only if the Supreme Court of Nepal decides to intervene in Special Court rulings.

What is notable is that the Special Court has cleared accused ex-ministers who are all members of the political parties that participated in the April Movement in 2006 and who are now in government. With the exception of Rabindra Nath Sharma, all the aforementioned individuals are members of the interim parliament. Despite the Corruption Control Act 2002, which bars leaders involved in corruption from parliament, they have faced no restrictions so far.

Against this background of possible impunity, journalist Kiran Chapagain published a revealing article about Chief Justice Dilip Kumar Paudel in June 2007.[7] According to Chapagain, Chief Justice Paudel had met ex-ministers Rabindra Nath Sharma and Khum Bahadur Khadka, and other accused individuals at his home before the final hearing. Subash Nembang, the speaker of parliament, called this a violation of the national code of conduct and the Bangalore Principles, which govern judicial integrity.[8] Chief Justice Paudel denied the meetings and, in any case, violating the code of conduct is more a moral than a criminal issue. Nonetheless, people look at the Special Court with dismay, and their faith in the judiciary has been shaken.

Defaulters enjoy impunity

Two government-owned commercial banks, Nepal Bank and Rastriya Banijya Bank, came under scrutiny as the number of defaulters continued to grow and the prospects for loan recovery became bleaker. The banks hired foreign firms to manage their business, streamline their organisation and realise bad and non-performing loans. Losses for Rastriya Banijya Bank and Nepal Bank amount to a staggering US$315 million and US$154.8 million, respectively.[9] Most were due to the non-payment of loans and interest, mostly by big business houses.

Nepal Rastra Bank, the central bank of Nepal, issued a blacklist of fifty-three defaulters owing more than US$770,000 each. Efforts to recover the loans have been fruitless.[10] The budget for fiscal year 2006/7 also included a special timeline to encourage defaulters to clear their debts. No action has been taken to recover the bad loans, however, and the blacklisted companies organised a strike protesting even against the measures that Nepal Rastra Bank has taken.[11]

Business houses have undoubtedly taken advantage of the limited liability provisions for companies regarding loans. There is a strong Nepali tradition of inflating project costs and the valuations of securities. Loans of extraordinary

6 See nepallaw.blogspot.com/2007_06_01_archive.html; *The Rising Nepal*, 16 November 2006; and www.blog.com.np/united-we-blog/2006/11/26/corrupts-get-clean-chit-in-nepal-even-in-democracy. See also Conflict Study Centre (CS Centre), 'Situation Update XVII: Nepal's Culture of Impunity' (Kathmandu: CS Centre, 2006).

7 *Kathmandu Post* (Nepal), 21 June 2007.

8 Ibid.

9 TI Nepal, *Paradarshi* (newsletter), no. 6 (Kathmandu: TI Nepal, 2007).

10 Ibid.

11 Ibid.

size are taken out with such tweaking, but businessmen rarely face legal and financial repercussions. The complexity of the situation is further deepened by the links between defaulters and politicians.

On the supply side, bank owners and loan evaluators systematically siphon off public money for their own benefit. In 2006 the then chair of the Cottage and Small Industries Development Bank was accused of embezzling US$4.3 million from his own bank and extending loans on inadequate securities. To the police he was 'untraceable' for a year, although a number of leading politicians were spotted at his son's extravagant wedding. After his 'controversial' arrest and handover by the Young Communist League, an offshoot of the Nepal Maoists, charges were filed against some fifty people, including promoters of the bank, loan evaluators and officials in the district court.[12]

Bank defaults, financial frauds and other irregularities strangle economic growth, discouraging savings and entrepreneurialism. The small number of defaulters responsible for such large amounts of debt signals entrenched economic inequality.

Will Melamchi ever quench Kathmandu's thirst?

The 4 million people of the Kathmandu Valley have suffered from acute water shortage for almost two decades. The valley's daily water need is around 250 million litres, while supply is less than a half of that amount. The Melamchi Drinking Water Project was conceived more than seventeen years ago to address this shortfall. The government attempted to divert water from the river Melamchi and bring it to Kathmandu through a 27-kilometre tunnel. This would have augmented the existing supply by 170 million litres per day. The project became bogged down by vested interests, however.

In the past, different phases of the project were funded by the World Bank and the ADB. The sheer size of the project, the related environmental issues and the parties generated considerable controversy. In spite of the vast amount of money already spent, there has been no tangible progress.

Controversy flared during the direct rule of King Gyanendra, when the former prime minister, Sher Bahadur Deuba, and his Cabinet colleague, Prakash Man Singh, were imprisoned for corruption in awarding a contract. The case was terminated when the Supreme Court designated as unconstitutional the agency responsible for the verdict, the Royal Commission for Corruption Control (see *Global Corruption Report 2006*).

The Melamchi project again dominated the media when the incumbent Maoist minister for physical planning, Hislia Yami, cancelled the contract awarded to UK company Severn Trent, in August 2007. (Under a US$120 million loan, the outgoing government and the ADB had awarded the contract to Severn Trent.) Yami claimed that Severn Trent did not have a sufficiently strong international track record. After renewed negotiations the ADB agreed to re-advertise the project.[13]

Incoming governments appear to cancel previous commitments and award contracts to their supporters. It is still to be seen what the effects of the decision to cancel the contract will be, but it is likely to delay the Melamchi project further, by at least a year.

Ramesh Nath Dhungel (TI Nepal)

12 *Nepal News.com*, 21 June 2007.
13 *Indo Asian News Services* (India), 15 July 2006.

Further reading

Good Governance, vol. 5, no. 6 (2007).
Government of Nepal, *Interim Constitution of Nepal 2007* (Kathmandu: Parliament Secretariat, 2007).
N. Manandhar, 'Corruption and Anti-corruption: Further Reading' (Kathmandu: TI Nepal, 2006).
ReMac Nepal, *Susashan Barsa Pustak 2006* (Kathmandu: ReMac Nepal, 2006).
TI Nepal, 'Progress Report 1995–2000' (Kathmandu: TI Nepal, 2006).
Bhrastachar Birodi Talim Pustika (Kathmandu: TI Nepal, 2006).
Bhrastachar ka Ayam: Patrakar ko Khoj (Kathmandu: TI Nepal, 2007).
TI Nepal: www.tinepal.org.

Pakistan

Corruption Perceptions Index 2007: 2.4 (138th out of 180 countries)

Conventions

ADB–OECD Anti-Corruption Action Plan for Asia-Pacific (endorsed November 2001)
UN Convention against Corruption (signed December 2003; ratified August 2007)
United Nations Convention against Transnational Organized Crime (signed December 2000, not yet ratified)

Legal and institutional changes

- Sindh became the **first of Pakistan's four federal provinces to introduce regulations ensuring transparency in procurement** on 1 November 2006, when the chief secretary promulgated the Public Procurement Rules 2004. The rules provide bidders with pre-qualification and tender documents that include all the relevant information, including detailed evaluation criteria, the bid award method, the mandatory signature of an integrity pact, bidders' complaints rights, standard conditions of contract, and so on. The rules allow the widest possible competition and discourage the favouring of any single contractor or supplier, while ensuring that the contract is awarded to the lowest bidder.[1]
- On 9 March 2007 **President Parvez Musharraf filed a reference against Chief**

1 See www.transparency.org.pk/documents/PPRARules2004.pdf.

Justice Iftikhar Muhammad Chaudhry, head of the Supreme Court, on charges of misuse of authority and misconduct after failing to obtain his voluntary resignation.[2] Musharraf's subsequent suspension of the chief justice triggered a four-month constitutional crisis over the independence of the judiciary (see *Global Corruption Report 2007*), which saw lawyers and barristers demonstrating alongside opposition politicians, journalists, doctors, engineers, accountants and other middle-class professionals (see below).[3]

Military grabs land, companies

The military has held unaccountable power for most of Pakistan's sixty years of existence and has also been engaged in, or planned for, open or proxy conflicts with the Soviet Union, India and Afghanistan for much of that time. It is unsurprising, therefore, that weapons procurement – with the secrecy such deals attract – has provided a flourishing channel for corruption, as it also has in India and Sri Lanka. In August 2000, for example, allegations of accepting kickbacks from French and Ukrainian companies were made against two army chiefs, one naval chief and two air force chiefs during negotiations for US$2.7 billion worth of submarines, jets and tanks (see *Global Corruption Report 2001*). The National Accountability Board (NAB) finally made the former chief of naval staff, Mansur-ul-Haq, return his US$7.5 million share of the commissions in 2001.[4]

What is less well known is the scale of the inroads made by the military into 'civilian' sectors of Pakistan's economy, including land, construction, property, manufacture, fertilisers, agriculture, road-building, trucking, health, education, insurance and banking. The military's interests in these areas are held by a group of charitable foundations set up in the colonial era to look after retired members of the armed forces, particularly members of the officer class. The wealthiest is the Fauji Foundation (FF), one of the army's two investment funds.[5]

For purposes of comparison, the Fauji Foundation is Pakistan's largest corporation, with a turnover of US$500 million in 2001, as well as the country's largest landowner. Askari Bank, owned by the Army Welfare Trust, is the country's most successful bank, and the trust has other interests in farming, milling, insurance and retail. The Shaheen Foundation, run by the air force, specialises in aviation-related services, including owning the country's second largest airline, Shaheen Air, aviation maintenance and in-flight catering. It also has interests in commercial property, TV and radio, and computer technology services. Meanwhile, the National Logistics Cell, the army's transport fleet, enjoys a near-monopoly on all large haulage contracts.

The army also houses 'defence colonies', which provide officers with prime housing at peppercorn prices and acquires farmland for cultivation by ex-servicemen. Since President Parvez Musharraf came to power by coup in 1999, the military has extended its economic reach through the appointment of ninety-two senior officers to key posts in the public sector, diplomatic service and leading utilities, providing further scope for illegal enrichment.[6]

Most Pakistanis know of the military capture of the civilian economy, but just how large a share it controls was a carefully guarded secret until July 2007. Dr Ayesha Siddiqa, a civil servant who

2 See www.thenews.com.pk/top_story_detail.asp?Id=9128.
3 Subsequent events in Pakistan took place beyond the reporting period of this country report.
4 See *Global Corruption Report 2001*.
5 Much of what follows can be found in A. Siddiqa, *Military Inc.: Inside Pakistan's Military Economy* (London: Pluto Press, 2007).
6 *Sindh Today* (Pakistan), no. 44, 1–15 October 2003.

once worked on defence accounting, published *Military Inc.*, a book that sought not only to quantify the scale of the armed forces' plunder of the civilian economy, but also the wealth of leading generals.

According to Dr Siddiqa, full generals enjoy individual wealth in excess of US$8.3 million, while President Musharraf has converted US$690,000 of army-granted farmland in Islamabad into US$10.34 million of movable assets.

A reliable parliamentary source said the Senate was informed in 2006 that a military officer received one plot of land after fifteen years of service, another after twenty-five years, a third after twenty-eight years and a fourth after thirty-three years.[7] The Defence Housing Authority (DHA), which allocates the plots, routinely encroaches on provincial authority lands to replenish its stocks without either negotiation or payment.[8]

The DHA, the five main foundations and the thousands of companies, small and large, that they control benefit from hidden subsidies from a national budget controlled by their ultimate beneficiary – the military elite – and can plead national interest as a way of justifying their activities. In February 2006 the Senate was informed that the government had paid back Rs13 billion (US$222 million) in loans by the Fauji Fertiliser Company Jordan out of taxpayers' money.[9]

According to opposition parliamentarian Farhatullah Babar, Khoski Sugar Mills, owned by the FF, was sold for US$300 million in 2005 to a company that had not even participated in the bidding process, while the highest official bid had been US$387 million. 'When the senate defence committee asked the head of the FF, himself a former chairman of NAB, to appear before it, he refused,' wrote Babar. 'We were told to shut up, as FF was a private enterprise.'[10]

The ambiguous status that the foundations enjoy – part tax-free charity, part private company and part military cartel – assumed surreal dimensions during the privatisation programme. When bidding opened for the sale of National Refinery Limited and Pakistan State Oil, FF lined up against Kuwait Petroleum, Lukoil and Chevron Texaco.[11] FF lost in both cases, not because the international financial institutions protested at conflict of interest, but because the offers were pitched too low.

Dr Siddiqa estimates that the armed forces control one-third of all heavy manufacturing, up to 7 per cent of private assets and around 12 million acres (4.8 million hectares) of land. Their private business empire could be worth as much as US$20 billion. 'So much has been grabbed by the military,' wrote Lord Patten, the former EU commissioner for external relations, in 2006, 'that it will take years just to catalogue it.'[12]

Justice and the general

On 9 March 2007 Chief Justice Iftikhar Muhammad Chaudhry was summoned by President Musharraf, who read out a list of offences and ordered him to resign.[13] When the chief justice refused, he was suspended from office and confined for five days in his house. The president

7 *South Asian Pulse* (Pakistan), 2 July 2007.
8 Ibid.
9 Ibid.
10 Ibid.
11 See www.rediff.com/money/2005/jun/09guest2.htm.
12 *Wall Street Journal* (US), 10 May 2006.
13 For the chief justice's deposition, see www.ahrchk.net/statements/mainfile.php/2007statements/1040.

took out a 'reference', or charge, against him for abuse of authority and 'ignorance of merit' in some of his decisions.[14]

Since his appointment in June 2005 Chief Justice Chaudhry had worked to clear the backlog of cases before the Supreme Court, then standing at 25,808. He had demonstrated an independence of mind that had been missing from the Supreme Court since 1999, when General Musharraf, having taken power, demanded that judges issue a Provisional Constitutional Order barring the court from challenging his authority on constitutional grounds (see *Global Corruption Report 2007*). This they did.

Justice Chaudhry had reversed the privatisation of Pakistan Steel Mills to friends of highest government functionaries and actively pursued the case of the several hundred 'missing persons' who were believed either to have been subject to 'extraordinary rendition' by US secret services or actually murdered since the launch of the 'war on terror'.[15] Justice Chaudhry was expected to become an even greater thorn in the president's side in 2007. First, the latter faced a delicate, constitutional manoeuvre if he were to be permitted to stand for election to the presidency – without surrendering his other job as head of the armed forces. Second, two exiled former prime ministers, Nawaz Sharif and Benhazir Bhutto, would be seeking rulings on whether they would be able to come home and contest the elections.

There followed an extraordinary four-month period in which Justice Chaudrhry, his barrister and advisers criss-crossed the country to address bar associations and gauge popular reaction to his suspension and Musharraf's rule. From Islamabad to Lahore, Faizlabad and Multan, they were greeted by hundreds of thousands of supporters, throwing rose petals in a rare case of Pakistani 'people power'.

Eighteen hours before the Supreme Judicial Council, the judiciary's regulatory body, was due to adjudicate his case, Justice Chaudrhry's barrister lodged a petition in the Supreme Court on behalf of his client. On 20 July the chief justice was reinstated in a unanimous judgment by the thirteen-member bench, which declared that the president's suspension order and reference had been illegal.[16]

The judgment was acclaimed as a victory for an independent judiciary and a guarantee that the complexities surrounding the forthcoming election and the legal standing of the leading candidates would be dealt with in an impartial and even-handed manner. At the time of writing, however, there was no evidence that President Musharraf would abide by Chief Justice Chaudrhry's future rulings if they did not entirely suit his own political purposes.[17]

Survey results

TI Pakistan conducted its second National Corruption Perceptions Survey from April to July 2006, gathering the views of 4,000 respondents through a twenty-four-page questionnaire. As in the 2002 survey, police and power utilities topped the list of Pakistan's most corrupt sectors, but taxation, customs and health improved their ranking. The majority of respondents were of the view that corruption had increased since 2002, and attributed this trend to a lack of accountability, the low salaries of public officials and their wide discretionary powers. From the testimony provided, TI

14 *Pak Tribune* (Pakistan), 9 March 2007.
15 *Daily Times* (Pakistan), 11 March 2007.
16 *Pak Tribune* (Pakistan), 27 July 2007.
17 Subsequent events in Pakistan bearing out this analysis took place beyond the reporting period of this country report in late 2007.

Pakistan estimated the average bribe expenditure at Rs2,303 (US$38) for each of the country's 20 million households.[18]

Some success in implementing water integrity

The successful application of an integrity pact in the Karachi Water and Sewerage Board's modernisation programme in 2003 resulted in the mandatory inclusion of similar pacts in all government procurements with effect from June 2004. In the following year the Sindh Ministry of Irrigation and Power signed a memorandum of understanding with TI Pakistan to strengthen transparency in its procurement by acting as an observer in all tenders. This led to considerable savings in two high-value projects.

Under the terms of the agreement, TI Pakistan agreed, *inter alia*, to ensure: that bid-opening is done according to the schedule; that the bidding-opening venue is accessible to all bidders; that bid documents are made available to all interested parties and also available on the procurement website; that bids are received in sealed forms, opened publicly and kept in a safe place to avoid any tampering after opening; that bid evaluations are carried out confidentially; that complaints are processed through the established complaint redress system; and, finally, that anomalies in the procurement process should be fully reported.[19]

In the first of the two projects – a US$265 million project to reline the province's canal system, partially funded by the World Bank – a US$9.6 million bid to provide tendering services was pulled back after TI Pakistan discovered that one of the company's associates had previously been blacklisted, but had not disclosed this. After calls for a new round of bidding, the evaluation found not only that M/s Nespak of Lahore had provided a technically superior offer, but also that their quoted cost of US$6.2 million was the least expensive.

In the second procurement process, TI Pakistan was asked to monitor the purchase of US$10.3 million worth of heavy earth-moving machinery and workshop equipment, and to compare the 2007 prices with a similar procurement made earlier in 1994. Even after a 15 per cent devaluation of the rupee, TI Pakistan was able to ensure savings of 22.7–42 per cent on excavators and bulldozers, compared to their cost thirteen years earlier. This is partially explained by the fact that TI Pakistan ensured that the tender terms complied with the World Bank's international standards and were in line with the Public Procurement Rules of 2004.

Syed Adil Gilani (TI Pakistan)

Further reading

M. Khan, 'Bureaucratic and Political Corruption in Pakistan', report submitted for the Operations Evaluation Department (Washington, DC: World Bank, 2001).

A. Mahmood, 'Pakistan's National Anti Corruption Strategy: Background, Status, Opportunities and Challenges', background paper (Bergen, Norway: U4, Chr. Michelson Institute, 2007).

'National Anti-Corruption Strategy' (Islamabad: National Accountability Bureau, 2002); www.nab.gov.pk/Downloads/Doc/NACS.pdf.

A. Siddiqa, *Military Inc.: Inside Pakistan's Military Economy* (London: Pluto Press, 2007).

TI Pakistan, 'National Corruption Perceptions Survey 2006' (Karachi: TI Pakistan, 2006).

TI Pakistan: www.transparency.org.pk.

18 TI Pakistan press release, 11 August 2006; see www.transparency.org.pk.
19 See www.transparency.org.pk/documents/PPRARules2004.pdf.

Papua New Guinea

Corruption Perceptions Index 2007: 2.0 (162nd out of 180 countries)

Conventions

ADB–OECD Anti-Corruption Action Plan for Asia-Pacific (endorsed November 2001)
UN Convention against Corruption (signed December 2004; ratified July 2007)

Legal and institutional changes

- As part of the implementation of the Proceeds of Crime Act 2005, attempts are under way to establish a **Financial Intelligence Unit** (FIU) within the police's fraud squad. Technical assistance is being provided by the Australian Agency for International Development (AusAID). Procedures for activating the FIU are due by September 2007. It will operate as a specialised body tasked with analysing transactions by banks in order to detect illegal activities, such as money-laundering. The Proceeds of Crimes Act was one of three anti-money-laundering measures drawn up under the 1992 Honiara Declaration of the Pacific Islands.
- The **National Anti-Corruption Alliance (NACA)**, which coordinates the efforts of all government agencies focusing on grand corruption, appointed its first full-time director, Mathew Yuangu, in March 2007. Yuangu, who previously served as the Office of the Prime Minister and National Executive Council's (NEC's) principle adviser on governance and justice sector matters, assumed responsibility for the NACA secretariat. He works closely with the NACA Technical Working Group to manage the day-to-day administration of investigations. In June 2006 the NACA launched an investigation into alleged systematic fraud in the Southern Highlands provincial government, focusing on 'ghost employees', excessive overtime payments, misappropriation of funds and suspicious financial transactions. AusAID has indicated that it will continue to support NACA's activities, although budgetary support from the government has been less than enthusiastic.
- In September 2006 amendments were passed to the **National Capital District Commission Act 2001** (see below). The amendments were harshly criticised by anti-corruption groups and the opposition because they would have left decision-making vulnerable to abuse. A coordinated campaign, including a petition of 17,500 signatures, persuaded the government to announce in March 2007 that it would revoke the new legislation. During the next parliamentary session, however, it reneged on that commitment, arguing that, because it was a private member's bill, it was the MPs and not the government that should rescind the amendment.[1]
- Parliament amended **the Organic Law on the Duties and Responsibilities of Leadership**,

1 *Pacific Magazine* (Papua New Guinea), 19 May 2007.

which requires the Leadership Tribunal to conduct its inquiries in strict compliance with rules of evidence. The tribunal is a quasi-judicial mechanism in which breaches of the Leadership Code are prosecuted by the Ombudsman Commission. If the leader is found guilty of misconduct, the Leadership Tribunal has the discretion to dismiss him or her from office and to impose a fine. Previously, formal evidentiary requirements had been set lower than the strict judicial standard in order to reflect the moral – rather than legal – tone of the Leadership Code. The practical effect of the amendment is that an increased number of breaches of the code are expected to fail due to evidential technicalities. Conversely, it will also require leaders to comply with the rules of evidence.

- The Office of the Prime Minister and NEC circulated a **draft National Anti-Corruption Strategy** for initial comments by key stakeholders at the end of April 2007. The draft strategy is a five-year plan that details short-, medium- and long-term objectives with an emphasis on political, legislative and administrative reform. The Fraud and Corruption Activity Management Team (AMT), a subcommittee of the Law and Justice Sector Working Group, is responsible for workshopping the strategy. The AMT comprises government and non-government delegates, including the police, the public prosecutor, the public solicitor, the magisterial service, national courts, the Ombudsman Commission and Transparency PNG.
- In March 2007 the NEC approved a Justice Ministry White Paper calling for the **strengthening of the law enforcement agencies' powers to prosecute fraud and corruption**. The NEC is responsible for defining policy and approving bills for presentation to the legislature. It sought better coordination between the Ombudsman Commission and the police, urging the latter to take prompt action to prosecute rather than leaving it to the commission. It sets out procedures for settling claims against the state as determined by the NEC and Supreme Court in recent cases. Abuse of procedures has led to major fraud, as disclosed at the inquiry into the Finance Department (see *Global Corruption Report 2007*). On the eve of the 2007 election the government had still failed to prove its commitment to anti-corruption measures.

Parliament fails to rein in executive

In September 2006 parliament amended the National Capital District Commission Act of 2001. Whereas appointments had previously been made by the Cabinet and reviewed by independent bodies, the amendments give sole power of appointment to the district governor. This risked turning the Port Moresby government into an unsupervised 'cash cow' at the beck and call of local political interests.

The changes were neither subtle nor disguised, as evidenced by a swiftly mobilised popular campaign against them, as well as a declaration from the Ombudsman Commission that any decisions under the amended act would be challenged in court as unconstitutional. Indeed, the amendments were so blatant in their intent to minimise accountability that many wondered how they could have been approved in the first place.

Although the amendments were previously rejected by the NEC, the country's policy-making body, the leader of government business suspended standing orders after a lengthy parliamentary session to allow the member for the National Capital district (who is also its governor) to introduce the amendments as a private member's bill. There were accusations both that the bill was deliberately introduced when MPs were tired and wanting to go home, and that the decision to introduce it as a private member's bill was coordinated by the government to avoid parliamentary scrutiny and negate the need for NEC approval.

Adding to the impression that parliament is subordinate to the government is the role played by the leader of government business with regard to the parliamentary schedule. As illustrated above, it is the government's representative, rather than the speaker, who determines which legislation is introduced to the House. This is despite the fact that the leader of government business is an active member of government, and often a minister.

This paucity of parliamentary input with regard to legislation is repeated in the legislative process. The parliamentary committee system is largely inactive, whether due to indolence, inability or habit. The parliamentary select committee on the powers of the Ombudsman Commission is currently over twelve months old, but has yet to deliver a final report – a fairly common fate, which passed largely unnoticed.

The parliamentary function of oversight is also neglected. In early 2007 there were concerns over the handling by the prime minister, Michael Somare, of a defence board of inquiry, established to investigate the illegal removal of the fugitive Julian Moti from PNG on a defence force aircraft. After being implicated in the inquiry, the prime minister pressured the defence minister into dissolving the board of inquiry, sacked him and later appropriated his portfolio, thereby gaining control over the contents of the relevant report. Members of the governing coalition party, Pangu (of which the former defence minister had been a member), threatened to secede if the sacking went through, but failed to follow through on the threat.[2]

The floor of parliament should have been the toughest critic of the episode, but it was the media and the Trade Union Congress that kept the issue in the spotlight. Parliament was equally mute when the government failed to follow through on its commitment to revoke the National Capital District Commission Act amendments in the final parliamentary session. The sitting was postponed three times in five days due to the lack of a quorum.

Public Accounts Committee delivers damning verdict

In April 2007 the Public Accounts Committee (PAC) undertook the final sitting of its five-year term, delivering its recommendations for the future.[3] The PAC is a constitutional body that had largely been dormant until MP John Hickey revived it in 2002. Given responsibility for monitoring the spending of public money, it maintained a relatively clean image through difficult circumstances. In delivering its final report, the PAC swims against a tide of official indolence by continuing to function as an effective oversight mechanism in the midst of a dysfunctional committee system. This integrity is largely attributable to individual MPs on the committee, rather than any effective support. Indeed, the PAC operates in an environment of strong non-compliance, and outright hostility in some cases.

The PAC's findings were damning. Though not included in the final report, committee members stated at its launch that an estimated 25–50 per cent of all public money had been misappropriated or misapplied in the previous five years. The report concluded that the PAC was unable to find a single department that had even begun to comply with the Public Finance (Management) Act of 1995 or the regulations authorised by section 117 of the act. It found that instances of corruption were not isolated events in the public sector, but characteristic of the operating environment. 'This committee found evidence of

2 *Pacific Magazine* (Papua New Guinea), 27 February 2007.

3 Chairman's address, 'Final Sitting of the Permanent Parliamentary Public Accounts Committee of the 7th National Parliament of PNG', 2007; *Post Courier* (Papua New Guinea), 15 June 2007.

misapplication, fraud, negligence, dishonesty and disregard for the law, and for the welfare of the state and its citizens by public servants at every level in every inquiry that we have held – with only one exception.'

This exception was the Department of Labour and Industry, but even that was thought to require improvement. The PAC found that this uniformity of corrupt and negligent practice was worsened by evidence of sophisticated collusion: 'We detect a clear web of organised and systemic illegality reaching across and involving several departments, which is designed to access public money in an illegal manner.'

The report noted that the conduct of senior civil servants served as an example of acceptable practice to lower-ranking officers, concluding that corruption filtered from the top down. The tolerance of illegal conduct at all levels meant that wrongdoers effectively operated in a culture of immunity, openly admitting their violations without fear of prosecution.

In addressing how the situation had deteriorated to this stage, the committee highlighted a flawed appointment process and a general lack of morale. Those appointed to senior positions were ill suited to their jobs. The committee found that the worst-performing officers were those in an 'acting capacity', who lacked job security or the power to act independently. Accountability mechanisms, such as the Public Service Commission, did not function properly because they exhibited the same levels of incompetence and venality as the agencies and officials they were intended to police.

The committee's recommendations were predictable – and its warnings dire. Identifying the problems and recommendations for change was not novel, it said, urging the government to halt its tolerance of such behaviour and honest public servants to speak up. It said that the government has to take a central role in the reform process, given that the public service had shown no desire to improve itself. The report encouraged 'ruling nothing out', going as far as to suggest the privatisation of service delivery and recruitment of foreign expertise.

It underlined the need for adequate funding for the auditor general, the Ombudsman Commission and PAC, as well as recommending the creation of an Independent Commission against Corruption (ICAC). The report specifically referred to a failed ICAC bill in 1997 that envisioned a specialised, dedicated agency, incorporating educative, preventive, investigative and prosecutorial functions.

The committee's report ended in the same way as its final sitting – with an impassioned plea for change. Members predicted Papua New Guinea declining to the level of a failed state if it did not change its ways. 'Such failings cannot be allowed to continue. If they do continue, we hold grave doubt that a viable, cohesive state can exist. Indeed, the signs of civil unrest in deprived areas are already evident.'

TI Papua New Guinea (Inc.)

Further reading

S. Dinnen, 'Building Bridges: Law and Justice Reform in Papua New Guinea', SSGM Working Paper no. 01/3 (Canberra: Australian National University, 2001).

P. Hughes, 'Issues of Governance in Papua New Guinea: Building Roads and Bridges', SSGM Discussion Paper no. 00/4 (Canberra: Australian National University, 2000).

J. Ketan, 'The Use and Abuse of Electoral Development Funds and their Impact on Electoral Politics and Governance in Papua New Guinea', Centre for Democratic Institutions Policy Paper no. 2007/02 (Canberra: Australian National University, 2007).

P. Larmour, 'Transparency International and Policy Transfer in Papua New Guinea', *Pacific Economic Bulletin*, vol. 18, no. 1 (2003).

B. Mana, 'An Anti-corruption Strategy for Provincial Government in Papua New Guinea', Asia Pacific School of Economics and Management Working Paper no. Gov99-5 (Canberra: Asia-Pacific Press, 1999).
TI Papua New Guinea: www.transparencypng.org.pg.

Philippines

Corruption Perceptions Index 2007: 2.5 (131st out of 180 countries)

Conventions

ADB–OECD Anti-Corruption Action Plan for Asia-Pacific (endorsed November 2001)
United Nations Convention against Corruption (signed December 2003; ratified November 2006)
United Nations Convention against Transnational Organized Crime (signed December 2000; ratified May 2002)

Legal and institutional changes

- With funding support from the European Commission, the Office of the Ombudsman, the Department of Budget and Management and the Civil Service Commission, the Development Academy of the Philippines undertook the second round of **integrity development reviews** (IDRs) for eleven government agencies from January to July 2007.[1] Through the IDR, agencies are able to use two tools, the corruption resistance review and the corruption vulnerability assessment, in order to pinpoint corruption in their systems. The first cycle was completed in August 2006 with agencies identified either as high-procuring, revenue-generating or involved in infrastructure and so considered vulnerable to corruption. These agencies are now implementing integrity and transparency reforms.[2]
- The Office of the Ombudsman, implementing agreements made during the Convergence Summit in March 2006 to activate the **National Anti-Corruption Program of Action** (NACPA), has set up the NACPA secretariat and is currently organising the Multi-Sectoral Advisory Council. The NACPA secretariat aims to provide coordination, planning, communications, information, project

1 The IDR is the ombudsman's response to its legal mandate on corruption prevention under sec. 13 (7) article XI of the 1987 constitution in relation to sec. 15 (7) of RA 6770 (Ombudsman Act of 1989).
2 Office of the Ombudsman, press release, 27 July 2007.

management, human resource development and other support to government agencies, civil society organisations and the private sector participating in the implementation of the NACPA. The organisation of the secretariat involved training and planning with financial support from the European Union.

- On 19 June 2006 President Gloria Macapagal-Arroyo pledged a P1 billion (US$22.1 million) **Anti-Corruption Fund** as a counterweight to the US$20.6 million Millennium Challenge Account – Philippines Threshold Program (MCA-PTP). Under the programme, the government will accelerate its anti-corruption campaign, plug revenue leaks, increase tax collection and channel more resources into health care, education and social services.
- The president signed the **Anti-Red Tape Act** on 18 June 2007, confirming the government's resolve to stamp out corruption in bureaucracy. The law directs agencies, corporations and financial institutions with a government interest to develop citizens' charters to serve as conduct guidelines for employees. Each agency is also required to set up a public complaints desk. Employees who refuse to accept applications or requests, attend to clients' needs, give written notice of rejections of applications, or impose additional, non-official requirements will be penalised, climaxing in permanent disqualification from public office. Other penalties under the law include a maximum six-year prison sentence for grave offences and a maximum fine of P200,000 (US$4,420).

Over a 'pork' barrel: Arroyo's struggle for legitimacy

The most important trend in corruption in the Philippines over the past six years has been President Gloria Arroyo's struggle for legitimacy. The 2004 elections were meant to settle once and for all the debate around the Arroyo takeover from President Estrada in 2001 in the wake of 'People Power II', but this controversy has since been overtaken by questions surrounding the elections.

President Arroyo won the May 2004 presidential election by a margin of 1.1 million votes. The legitimacy of the election has repeatedly been questioned following the July 2005 'Hello Garci' scandal, however, which suggested that the president might have talked to the election commissioner Virgilio Garcillano, regarding vote-padding arrangements during the elections (see *Global Corruption Report 2007*). Instead of consolidating power, the president found her administration in deeper political trouble, as ten senior officials, in a vote of no confidence, resigned en masse.[3]

Allegations of corruption led to impeachment attempts from 2005 to 2006. In June 2006 eight identical complaints were filed by concerned citizens and organisations, and endorsed by opposition congressmen from 26 June to 27 July to impeach the president. Aside from the named plaintiffs, the eight complaints were identical in content, including allegations that Arroyo had stolen the 2004 vote, stifled dissent, abetted political killings and distributed an agricultural fund as patronage to congressional and local allies, so as to ensure victory in the election.[4]

In August 2006 the House of Representatives dismissed the case without hearing the evidence, and so the president avoided a full trial in the Senate for a second time. The opposition alleged that the administration had bought congressional support by releasing pork-barrel funds

3 In February 2006 elements of the military led by General Danilo Lim publicly withdrew their support from President Arroyo, citing the persistence of large-scale corruption in the administration.

4 *Philippine Daily Inquirer*, 9 August 2006.

and straight bribes.[5] The impeachment moves in 2005 and 2006 were defeated by roughly 90 per cent of the House.[6]

After the first impeachment attempt in 2005 the administration attempted to change the constitution in order to move the current form of government to a parliamentary system. A Consultative Commission recommended stalling the 2007 elections until 2010, enabling the two Houses of Congress to convene as a parliament.[7] This was seen as an attempt to keep Arroyo in power by garnering support from politicians up for re-election in 2007.[8] Amending the constitution in this way would have taken two years, however, so the administration supported the People's Initiative for Reform, Modernisation and Action, which sought to push through the required changes by a petition of at least 12 per cent of all registered voters and 3 per cent of voters in each legislative district.[9] In October 2006 the Supreme Court ruled that the campaign was 'void and unconstitutional'.[10]

The administration attempted the final avenue available for reforming the constitution: the creation of a Constituent Assembly, comprised of the House of Representatives and the Senate. The majority of senators were opposed, but their objections were overruled by the House of Representatives, which interpreted the rules to mean that the Senate was required only to provide the bicameral dimension of the Constituent Assembly.

A howl of protest was heard as the mass media broadcast the calls of political, economic and religious leaders, who spoke out at street demonstrations. Finally, the Arroyo administration was left to prepare for the 2007 elections, attempting to obtain as many of the seats in the legislature as possible while forestalling a third impeachment attempt in 2007.[11]

Incompetence or corruption in the 2007 elections

Comelec is the constitutional body in charge of conducting elections. In 2004 the Supreme Court nullified Comelec's contract with Mega-Pacific eSolutions, Inc., for the purchase of nearly 2,000 ballot-counting machines costing P1.2 billion (US$26.5 million), citing irregularities in the contract. It also directed the ombudsman, Merceditas Gutierrez, to determine criminal liability and recover the money already paid for 1,991 machines.[12] On 30 June 2006 the ombudsman absolved Comelec's chair, Benjamin Abalos, and other officials of wrongdoing, saying that the case showed 'no evidence of malice, bad faith and partiality to warrant an indictment'.[13]

In May 2007 some 75 per cent of the 45 million registered voters turned out to vote. The senatorial election went to the opposition, which was interpreted as a vote of no confidence in President Arroyo, but the administration won a strong majority in the House of Representatives. One way of reconciling these inconsistent results would be to argue that voters prioritised parochial issues in local elections and national interests in the senatorial contests. But an alternative perspective may suggest that the

5 *Philippine Daily Inquirer*, 24 August 2006.
6 Ibid.
7 *Asian Analysis* (Singapore), 6 February 2006.
8 Ibid.
9 See www.opinionasia.org/PeoplesInitiativePetition.
10 Ibid.
11 *Philippine Daily Inquirer*, 25 July 2007.
12 *Manila Times* (Philippines), 1 July 2006.
13 Senate of the Philippines, press release, 3 October 2006.

administration used the resources of the state to benefit its own candidates.

The military, which was severely criticised for its role in the May 2004 elections, signed an agreement with Comelec on 12 October 2006 limiting the role of its units during elections to manning checkpoints, enforcing the weapons ban and providing security to areas that Comelec and defence officials jointly identified as having 'serious armed threats'. In an apparent attempt to handicap the progressive party-list parties, however, the armed forces openly linked some to the New People's Army, which is considered a terrorist organisation.[14]

Not only did these parties claim that the armed forces campaigned against them, the government arrested a number of their leaders. Crispin Beltran and Satur Ocampo, both veteran labour leaders and MPs, were arrested on charges of rebellion and multiple murder, respectively, dating back to the mid-1980s.[15] The Supreme Court ordered Ocampo's release after he posted a P100,000 (US$2,000) bond on 3 April 2007 and the Makati Regional Court judge, Elmo Alameda, released Beltran after the Supreme Court dismissed the charges against him.[16]

There are also allegations that the administration supported some party-list organisations, propping them up with public funds. An Akbayan party-list representative asked Comelec to disqualify groups that were accused of being supported by the Arroyo government. Comelec refused. Akbayan then asked Comelec to release a list of the nominees of all party-list groups, as this would be necessary for the 'filing of formal disqualification charges'. Comelec again refused. After consistent pressure from Akbayan and its senatorial support, however, Comelec agreed to release the information, but only after the elections. Akbayan took its petition to the Supreme Court, which ruled in its favour on 3 May.[17]

Comelec's lack of preparedness for a complex electoral exercise was obvious. Voters' lists were released only the day before the election in some areas. Many voters who were unwittingly disenfranchised discovered it at the last minute, with no remedy available to them. Members of the Board of Election Inspectors were appointed only days before the election. The complaints, protests and petitions received by Comelec were not acted upon until the last moment. Notable among them was senatorial candidate Alan Cayetano's petition for the disqualification of Joselito Cayetano as a nuisance candidate. Despite previously acknowledging that the latter sought only to exploit the confusion in names, Comelec failed to remove his name from the candidate list.[18]

The election result ultimately went in the president's favour. While the opposition won seven out of twelve Senate seats, with two independent candidates winning, Arroyo increased her hold on the House of Representatives with 195 out of 220 seats, making further impeachment proceedings highly unlikely.

Water: a mild rapping of knuckles in Rapu-Rapu

A major scandal involved the Australian mining firm Lafayette NL, which started its Rapu-Rapu Polymetallic Mining Project in mid-2005. The project approval allegedly contained some irregularities. In particular, it was alleged that the Lafayette country manager, Roderick Watt, threatened to withhold US$55 million in capital

14 *Philippine Daily Inquirer*, 13 March 2007; *Malaya* (Philippines), 24 March 2007.
15 *Inquirer.net* (Philippines), 5 June 2007 and 4 April 2007.
16 See www.ipu.org/hr-e/181/phi01.htm.
17 See www.supremecourt.gov.ph/jurisprudence/2007/may2007/177271.htm.
18 See www.gmanews.tv/story/56727/Cayetano-missing-Alan-Peter-spearheads-search.

investments from Lafayette Ltd and the LG Group of Korea if the company were not granted special economic zone status. President Arroyo duly capitulated and issued Proclamation 625, declaring the mining area a special economic zone. This effectively deprived the local community of royalties from the mining revenues and sanctioned a six- to eight-year income tax holiday on top of a value added tax exemption.[19]

Although environmental compliance certification had been sought, the local community complained that they were either not consulted, or had not been told of the project's adverse effects. In October 2005 the residents learnt of two cyanide spillages from the mining site, which contaminated nearby creeks and poisoned the water source. Local farmers noted a marked decrease in water supply, resulting in a miniature water war in Barangay Poblacion. Water supply for drinking, washing and other domestic needs was also scarce.

Lafayette was found grossly negligent by a government-convened fact-finding commission on Rapu-Rapu, which called for a halt to all mining activities. It recommended that the Bureau of Internal Revenue investigate the company for underreporting ore production and violating tax laws. It also suggested that all financial and economic incentives should be rescinded and that the company should be required to pay back the previously waived taxes.[20]

The government rejected these recommendations and in February 2006 lifted the cease-and-desist order earlier issued against the company in the wake of the spills. In doing so, it went against the express interests of local residents and its own commission. As things stand, Lafayette will continue to operate until 2013.[21]

Segundo Romero, Dolores Español and Aileen Laus (TI Philippines)

Further reading

Asian Institute of Strategic Studies, Inc. (AISSI), 'Grappling with Graft and Corruption: The Philippine Experience' (Quezon City: AISSI, 2003).

E. Co *et al.*, *Minimizing Corruption* (Manila: Ateneo de Manila University Press, 2007).

Concerned Artists of the Philippines, 'Culture of Corruption: The Corruption of Culture: A Critique of GMA's Anti-corruption Campaign', presentation at the National Study Conference on Corruption, Quezon City, January 2005.

Office of the Ombudsman, 'Compilation of Laws on Graft and Corruption' (Manila: Office of the Ombudsman, 2004).

G. Quimson, 'National Integrity Systems TI Country Study Report of the Philippines 2006' (Berlin: TI, 2006).

Social Weather Stations (SWS), '2005 SWS Survey of Enterprises on Corruption' (Makati City: SWS, 2005).

TI Philippines: www.transparencyintl.org.

19 *Philippine Daily Inquirer*, 25 July 2007.

20 Rapu-Rapu Fact-Finding Commission, 'Findings and Recommendations of the Fact-Finding Commission in Rapu-Rapu Island', 19 May 2006; available at pcij.org/blog/wp-docs/Rapu-Rapu-Commission-Report.pdf.

21 See www.corpwatch.org/article.php?id=14488.

7.4 Europe and Central Asia

Armenia

Corruption Perceptions Index 2007: 3.0 (99th out of 180 countries)

Conventions
Council of Europe Civil Law Convention on Corruption (signed February 2004; ratified January 2005)
Council of Europe Criminal Law Convention on Corruption (signed May 2003; ratified January 2006)
UN Convention against Corruption (signed May 2005; ratified March 2007)
UN Convention against Transnational Organized Crime (signed November 2001; ratified July 2003)

Legal and institutional changes

- Armenia's anti-corruption policy is based on the **Anti-Corruption Strategy Programme and Action Plan**, adopted by the government in November 2003. Improvements to the establishing documents are critical to the fight against corruption, since their primary focus was the passage of legislation rather than the introduction of efficient enforcement measures. The main anti-corruption institutions are an Anti-Corruption Council – headed by the prime minister – and the Anti-Corruption Strategy Monitoring Commission, established in June 2004 to strengthen the implementation of anti-corruption policy.[1] These institutions scarcely functioned in 2006–7, although they were supposed to meet twice-quarterly and monthly, respectively. The post of head of the monitoring commission lay vacant for three months after Bagrat Yesayan, an adviser to President Robert Kocharyan, was removed in June 2006 to become deputy minister of education and science. Amalia Kostanyan, chair of the Center for Regional Development (CRD), TI's Armenia chapter, resigned from the monitoring commission in February 2007,[2] arguing that the anti-corruption programme had failed and corruption had become 'more politicised and large-scale'.[3] These concerns were also reflected in the World Bank report 'Anticorruption in Transition 3: Who is Succeeding . . . and

1 RA President Decree NH-100, 1 June 2004.
2 CRD/TI Armenia, 'CRD/TI Armenia Informs', letter of resignation, 20 February 2007; available at www.transparency.am/news_storage.php?month=2+2007.
3 *ArmeniaLiberty.org* (Armenia), 17 October 2006.

Why?', which pointed to increasing bribery in certain areas, including tax and customs.[4]

- In December 2006 the government **submitted a report to the OECD's Anti-Corruption Network** (ACN) for Eastern Europe and Central Asia. This report was on the implementation of twenty-four expert recommendations by the OECD to improve Armenia's anti-corruption policies and institutions.[5] The sixth OECD Monitoring Meeting in December 2006 highlighted a number of positive aspects, including strengthening money-laundering controls and improved integrity in public service. It also noted that the number of convictions for corruption was low, however, especially of senior officials; the declaration of assets by public officials required greater transparency; and cooperation between law enforcement and financial control institutions needed improvement.[6]
- An **Anti-Corruption Public Reception opened in Yerevan in April 2007** with the assistance of the Organization for Security and Co-operation in Europe's (OSCE's) office in Yerevan. Citizens are given free legal, procedural and practical advice concerning corruption by a coalition of fourteen civil society groups involved in anti-corruption work. Similar receptions also operate in Lori and Gegharkunik regions.

Political party finance and the May 2007 election

Political activity in Armenia has focused squarely on the legislative election of 12 May 2007. Since Armenia committed to the European Neighbourhood Policy in 2004, free, fair and transparent elections have been considered critical to building common values with neighbouring states. Stronger democratic processes are also a condition for continuation of the Millennium Challenge Account, a five-year assistance programme, aimed at reducing rural poverty, worth US$236 million.

In December 2006 and February 2007 the National Assembly revised the Electoral Code based on the recommendations of the Council of Europe's Venice Commission and the OSCE Office for Democratic Institutions and Human Rights (OSCE/ODIHR).[7] The changes included increasing the number of seats determined by political party lists, compared to those by majority vote; strengthening the procedures for nominating members of the Central Electoral Commission; improving the distribution of tasks within election commissions; abolishing the recall of election commission members; reducing bureaucratic procedures for election observation missions; strengthening the role of proxies; regulating video recording; improving voting and counting procedures; and cancelling the right to vote of Armenians living outside the national borders.

OSCE/ODIHR and the Venice Commission evaluated the changes as positive steps towards an adequate legal framework, but reported that loopholes remained in the legislation that allowed corruption to flourish. In the opinion of certain local and international experts, the Electoral Code does not sufficiently regulate political party finances prior to the start of campaigns. In particular, there is no clear distinction between 'pre-election campaign' and 'political advertis-

4 J. Anderson and C. Gray, 'Anti-corruption in Transition 3: Who Is Succeeding . . . and Why?' (Washington, DC: World Bank, 2006).

5 OECD ACN, 'Istanbul Anti-corruption Action Plan for Armenia, Azerbaijan, Georgia, the Kyrgyz Republic, the Russian Federation, Tajikistan and Ukraine, Review of Legal and Institutional Framework for Fighting Corruption – Armenia: Summary Assessment and Recommendations' (Paris: OECD ACN, 2004).

6 OECD ACN, 'Istanbul Anti-Corruption Action Plan for Armenia, Azerbaijan, Georgia, the Kyrgyz Republic, the Russian Federation, Tajikistan and Ukraine – Armenia: Monitoring Report', (Paris: OECD ACN, 2006).

7 See www.venice.coe.int/site/dynamics/N_Country_ef.asp?C=42&L=E.

ing', providing significant opportunities for infringements, which prove difficult to identify.[8]

Legal reforms were accompanied by other activities aimed at ensuring free and fair elections. Voters' lists were verified with the help of the International Foundation for Electoral Systems. Training for members of electoral commissions was organised by the International Institute for Democracy and Electoral Assistance. The Council of Europe, OSCE, UNDP and the American Bar Association's Europe and Eurasia Program contributed to projects aimed at building capacity, including publishing a manual on election legislation and raising citizens' awareness of their rights and the voting process through televised public service announcements. The election process was monitored by more than 500 long- and short-term international observers and 13,000 local monitors.[9]

Despite these multiple efforts aimed at ensuring a more democratic process, the 2007 election was accompanied by corrupt practices that were more sophisticated than in previous elections. Monitoring work throughout the year revealed violations of freedom of expression, freedom of assembly and association, and the right to privacy.[10] The institutions responsible for ensuring fair elections, such as electoral commissions, the police and prosecutor's offices, failed to perform their duties effectively. Despite improvement to voters' lists, numerous instances of multiple voting were still identified.[11] There was evidence that employees of state institutions had been pressured to vote for certain political parties.[12]

Control over state institutions, the mass media and administrative resources created unfair conditions for opposition parties. Civil servants campaigned in working hours, both before and during the official campaign period, in violation of article 22 of the Electoral Code.[13] Representatives of local and regional government actively intervened in the electoral processes. It was reported that the murder or attempted murder of officials, the beating of demonstrators and journalists, the bugging of offices and raids on the homes of opposition leaders generated an atmosphere of fear among voters.[14]

The elections demonstrated that vote-buying has become institutionalised in Armenia. Bribery occurred before and during the election campaign through the free distribution of agricultural products or television sets by so-called charities loyal to leading politicians and parties, and also by gifts of money to citizens in exchange for their vote.[15]

---

8 CRD/TI Armenia, press release, 31 May 2007. See also OSCE/ODIHR International Election Observation Mission, 'Statement of Preliminary Findings and Conclusions', 13 May 2007.

9 See www.elections.am/?go=aj2007&lan=e.

10 Yerevan Press Club, 'Monitoring of Democratic Reforms in Armenia' (Yerevan: Yerevan Press Club and Open Society Institute, 2006).

11 *A1+* (Armenia), 13 May 2007.

12 *Armenia Now*, 18 May 2007. See also, Helsinki Citizens' Assembly Vanadzor Office (HCAV), 'Report on the Results of Observation Mission of Parliamentary Elections on May 12 2007' (Vanadzor, Armenia: HCAV, 2007).

13 Yerevan Press Club, 'Monitoring the Armenian Media Coverage of Parliamentary Elections 2007', (Yerevan: Yerevan Press Club and Open Society Institute, 2007). See also OSCE/ODIHR International Election Observation Mission, 2007.

14 Partnership for Open Society Statement, 14 May 2007, available at www.partnership.am/docs.php?Lang=E&ID=76.

15 '"It's Your Choice" NGO's Observation Mission of the May 12, 2007 National Assembly Elections in Armenia' (Yerevan: It's Your Choice, 2007). See also HCAV, 2007.

In spite of generally positive reviews of the training of electoral commissions, voter education, the computerised registration of voters and the extensive media coverage, preliminary and interim reports of the International Election Observation Mission highlighted major concerns during the election.[16] These included gaps in the regulatory framework; the domination of electoral commissions by members of the ruling Armenian Republican Party, the Armenian Revolutionary Federation 'Dashnaktsutyun' and presidential appointees; the manipulation of vote-counting procedures; discrepancies between protocols and tabulated results; and the late announcement of results. The same discrepancies were highlighted in reports by local observers.[17]

According to the preliminary results of CRD/TI Armenia, monitoring of political party campaign finances revealed that the two political parties that won the majority of the proportional seats – the Armenian Republican Party and the Prosperous Armenia Party (forty-one and eighteen seats out of ninety, respectively) – exceeded the D60 million (US$178,000) limit set by the Electoral Code. The two parties spent D79.1 million and D129.6 million on campaign materials and events, respectively.[18] In June 2007 these data were submitted to the Constitutional Court by two opposition parties that questioned the fairness of the elections. The court dismissed it as evidence of violations, but admitted that deficiencies remained in the legislation and called for further improvements.[19]

Broadcast media under strict control

The majority of Armenians receive their information from television, while the print media have a less significant role.[20] The existence of independent and pluralistic broadcast media is therefore of critical importance in ensuring democratic development.

Though Armenia has a number of laws guaranteeing access to information, including the Law on Freedom of Information 2003 and the Law on Mass Media 2003, certain trends, mainly associated with the Law on Television and Radio 2000, restrict those liberties and lead to corruption.[21] For example, the Council of the Public Television and Radio Company (PTRC) is appointed by the president and thus naturally reflects the government's political agenda. Until 2007 the president also appointed all the members of the regulatory body of the private broadcast media, the National Commission on Television and Radio (NCTR), which was established to ensure fair competition in the broadcast media.[22]

The tender of broadcast frequencies is not transparent, as it is subject to the NCTR's discretion and guided by political interests rather than the requirements of law. Although the legislation allows the engagement of NGO experts in the assessment of applications,[23] the NCTR has rejected the applications of several media organisations seeking to engage in the process. Furthermore, although the law sets out criteria for awarding frequencies, they are unclear, and

16 OSCE/ODIHR, 'International Election Observation Mission, Interim Report no. 2, 29 March–17 April 2007'; 'Interim Report no. 3, 18 April–2 May 2007'; 'Statement of Preliminary Findings and Conclusions, 13 May 2007'; and 'Post Election Interim Report no. 1, 13–22 May 2007'.

17 '"It's Your Choice Final Report"' (Yerevan: It's Your Choice, 2007); HCAV, 2007; and Yerevan Press Club, 2007.

18 *Armenia Now*, 25 May 2007.

19 Decision of the Constitutional Court of RA no. SDV-703 from 10 June 2007.

20 CRD/TI Armenia and UNDP, '2006 Corruption Perceptions in Armenia' (Yerevan: CRD/TI Armenia and UNDP).

21 Yerevan Press Club, 'Monitoring of Democratic Reforms in Armenia' (Yerevan: Yerevan Press Club and Open Society Institute, 2005).

22 This provision was modified in February 2007.

23 Law on Procedures of the National Commission on Television and Radio, article 26.

award decisions appear subjective and not justified under the law. The most famous example is the independent and successful television company A1+, which was deprived of airtime in 2002 and since then has been rejected various other frequencies as many as ten times, often losing to unknown companies.[24]

There is also inadequate oversight of compliance with laws and licence terms, and the punishment for violations is discretionary. Such practices affect free competition in the broadcast media and restrict freedom of speech. Constitutional amendments in November 2005 were partially designed to ensure more freedom for the NCTR to promote more pluralism in the broadcast media.[25] It was expected that the government would also amend the Law on Television and Radio in accordance with the amended constitution. This was not the case, however. In September 2006 the government produced a hasty draft amendment to the law, which was discussed neither with stakeholder organisations nor with the relevant parliamentary committee. Five media NGOs – including the Yerevan Press Club, Internews of Armenia, the Association of Journalists of Armenia, the Committee for the Protection of Free Speech and the Asparez Journalists' Club – expressed serious concerns over the implications for freedom of expression. In particular, the amendment provided for the PTRC Council's continued dependence on the president. The draft was discussed at an extraordinary session of the National Assembly on 27 September 2006 and voted on at a regular session on 3 October, but it did not pass. Only after it was rejected did discussions with relevant stakeholders finally begin.

The law was finally changed in February 2007 in accordance with the amended constitution. It provides that a half of the NCTR members are to be appointed by the president and the others by the National Assembly. The president is still expected to retain control for a few more years, however, as it could take six years to achieve this fifty-fifty ratio. The current members of the commission will continue to serve in their positions until their terms expire or their powers are terminated before the end of their term.[26] Terms for three of the members expired in March 2007, and one month later the president reappointed two members. Grigor Amalyan, the former chairman of NCTR, whose name was associated with numerous rejections of A1+, was reappointed and re-elected chairman, thus ensuring continuity of the current policy.[27] Earlier, in February 2007, the president had reappointed to the PTRC Council Alexan Harutyunyan, who the other members then re-elected chairman.[28]

Control of the broadcast media had a significant impact on the pre-election campaign of several parties and candidates. Some opposition parties were almost entirely deprived of airtime before the official campaign. Though public TV and other state-controlled broadcast outlets offered more balanced coverage during the actual campaign, it did not compensate for the previous damage.[29]

Survey results

According to a nationwide perception survey by the CRD/TI Armenia in 2006, 74.3 per cent of respondents considered the government's anti-corruption policy ineffective. Only 15.6 per cent were aware of the existence of the

24 Yerevan Press Club, 2006.
25 Constitution of the Republic of Armenia, article 83.2.
26 Yerevan Press Club, 2006.
27 Yerevan Press Club, newsletter, 6–12 April 2007.
28 Yerevan Press Club, newsletter, 2–8 February 2007.
29 Yerevan Press Club, 2007 ('Monitoring the Armenian Media Coverage').

Anti-Corruption Council and 8.6 per cent of the Anti-Corruption Strategy Monitoring Commission. The majority of interviewees thought the main causes of corruption were poor law enforcement (94 per cent), public tolerance of corruption (87.8 per cent) and inadequate control and punishment mechanisms (87.7 per cent). Nearly all respondents (93.9 per cent) thought that strengthening law enforcement was key to reducing corruption: 91.9 per cent highlighted punishment of those involved in corruption, and 91.3 per cent suggested promoting public awareness on citizens' rights and obligations.[30]

The rise of Lake Sevan

Lake Sevan is the largest freshwater reservoir in the Caucasus, and a crucial habitat for aquatic and migratory bird species. The water level started falling in the 1930s, however, due to overuse through power generation, irrigation and the drainage of surrounding wetlands. The level decreased by over 19 metres, resulting in further deterioration of water quality and loss of biodiversity. A variety of interventions have been launched to raise the level of the lake, including the construction of tunnels to divert rivers,[31] the adoption of a special Law on Lake Sevan in 2001 and a corresponding rehabilitation plan. The latter set a target of raising the water level by 6 metres within thirty years, in order to re-establish the ecological balance and prevent environmental catastrophe.[32]

The ongoing measures to save the lake have been accompanied by increased pressure from vested interests, seeking to use it for recreational purposes. Despite restrictions on lakeside development, the shores are cluttered with illegal buildings. Among the major developers are senior government officials and politicians.[33] In the early 2000s, when the amount of rainfall unexpectedly raised the water level in Lake Sevan by 2.44 metres in six years, the government began to reconsider the 6-metre target and introduced the notion of paying compensation to the illegal developers for damage to their property.[34]

Sona Ayvazyan (Center for Regional Development/TI Armenia)

Further reading

J. Anderson and C. Gray, 'Anti-corruption in Transition 3: Who Is Succeeding . . . and Why?' (Washington, DC: World Bank, 2006).

CRD/TI Armenia and UNDP, '2006 Corruption Perceptions in Armenia' (Yerevan: CRD/TI Armenia and UNDP, 2006).

European Bank for Reconstruction and Development (EBRD) and World Bank, 'Business Environment and Enterprise Performance Survey' (London: EBRD and World Bank, 1999, 2002 and 2005).

A. Kostanyan and V. Hoktanyan, 'National Integrity System TI Country Study Report of Armenia 2004' (Yerevan: CRD/TI Armenia, 2004).

'Anti-corruption Policy in Armenia' (Yerevan: CRD/TI Armenia, 2006).

TI Armenia: www.transparency.am.

30 CRD/TI Armenia and UNDP, 2006.

31 UN Economic Commission for Europe (UNECE) and Armenian Ministry of Nature Protection, 'National Report on the State of the Environment in Armenia 2002' (Yerevan: UNECE, 2003).

32 *Armenia Now*, 15 December 2006.

33 For further details on compensation for illegal developers, see *Armenia Now*, 29 June 2007.

34 For details of increases in water level, see www.armeniapedia.org/index.php?title=Lake_Sevan_water_level.

Austria

Corruption Perceptions Index 2007: 8.1 (15th out of 180 countries)

Conventions

Council of Europe Civil Law Convention on Corruption (signed October 2000; ratified August 2006)
Council of Europe Criminal Law Convention on Corruption (signed October 2000; not yet ratified)
OECD Convention on Combating Bribery of Foreign Public Officials (signed December 1997; ratified May 1999)
UN Convention against Corruption (signed December 2003; ratified January 2006)
UN Convention against Transnational Organized Crime (signed December 2000; ratified September 2004)

Legal and institutional changes

- As a result of the ratification of the Council of Europe Civil Law Convention on Corruption, Austria joined the **Group of States against Corruption (GRECO)** in December 2006.
- On 24 July 2007 the Justice Ministry introduced for public discussion a **proposal to amend anti-corruption legislation**. Institutions could submit legal comments and requests until 10 September 2007.[1] A revised version of the bill is scheduled to pass through parliament in late 2007 and become effective in 2008. The proposal aims to implement the provisions of the UN Convention against Corruption and the Council Framework Decision 2003/568 on combating corruption in the private sector, as well as facilitate ratification of the Council of Europe Criminal Law Convention on Corruption.
- The Justice Ministry plans to **create a Public Prosecutor's Office** in Vienna that will focus exclusively on corruption and exercise jurisdiction for the entire country. In addition, the ministry put up for discussion a proposed leniency programme in the field of corruption. Provisions for leniency programmes can be found in the Austrian legal system, particularly in the fight against organised crime. The proposed programme is expected to grant immunity from prosecution to those who collaborate with law enforcement agencies, and not just mitigate penal sanctions.[2]
- Despite Austria's reputation for only 'moderate' corruption, the issue has grown increasingly important in recent years. **Allegations of illegal party funding** tainted the federal election campaign in 2006, triggering two parliamentary committees of inquiry in October 2006. One investigated the procurement of eighteen Eurofighters in 2002/3 (see below) and the other scrutinised the external control mechanisms of the banking sector in the wake of the so-called BAWAG affair (see below). The

1 See www.bmj.gv.at/gesetzesentwuerfe/index.php?nav=13&id=95.
2 *DiePresse.com* (Germany), 24 July 2007.

work of both was impaired by controversies between the two coalition parties, the SPÖ and the ÖVP,[3] and the committees terminated their work in July 2007 without producing final reports.[4]

The Eurofighter procurement

The first parliamentary committee had to investigate the decision in 2002 to acquire twenty-four Eurofighter Typhoons to replace the Austrian air force's outdated SAAB Draken interceptors.[5] Austria is neutral and, at only 0.8 per cent of GDP, has one of the lowest defence budgets in the world. In addition, it was an open secret that the military preferred the Gripen, also made by SAAB, a company that enjoys close relations with the Austrian army and its political parties. SAAB's prices were not much lower than Eurofighter's, but it was clear that the latter – a multi-role aircraft more in keeping with out-of-area missions than Austria's defensive posture – would be far more costly to operate. Nonetheless, in 2002 the government argued that it was time for Austria to give up its neutrality and contribute to EU and North Atlantic Treaty Organization (NATO) operations. The government defended its decision by presenting the Eurofighter as a genuinely 'European' product and emphasising the offset agreements with Austrian companies offered by the European Aeronautic Defence and Space Company (EADS) and other members of the Eurofighter consortium, which would amount to €4 billion (US$5.4 billion) by 2018.[6]

These were convincing arguments and, under later scrutiny, the court of auditors could find no serious reasons to challenge them. Nevertheless, the circumstances of the procurement were surprising. Up until the final decision in July 2002, the minister of finance, Karl-Heinz Grasser (then a member of the FPÖ), had apparently opposed the purchase of new aeroplanes, arguing instead for used Lockheed Martin F-16s[7] on grounds of cost. Financial considerations were also uppermost with military decision-makers (including the minister of defence, Herbert Scheibner, also of the FPÖ) who seemed to prefer the Gripen instead. The upshot, however, was the decision by the ÖVP–FPÖ Cabinet to purchase the Eurofighter, which was not only the most expensive option, but would not become available until 2007, making it necessary to lease Northrop F-5E interceptors for two years.[8] To limit the price tag to €2 billion, Austria reduced the number of

3 In 2000–6 the (Christian-Democrat) Austrian People's Party (Österreichische Volkspartei, or ÖVP) and the (right-wing) Freedom Party (Freiheitliche Partei Österreichs, or FPÖ) formed the government. When the latter split in April 2005, most FPÖ members of government joined a newly formed Confederation for the Austrian Future (Bündnis Zukunft Österreich, or BZÖ). The Social Democratic Party (Sozialdemokratische Partei Österreichs, or SPÖ) and the Greens were in opposition until the end of 2006. In January 2007 a new 'grand coalition' government was formed between the SPÖ and the ÖVP.

4 The Eurofighter committee produced a brief written report (192 d.B, XXIII. GP), but the parties drew diverging conclusions only in their own minority reports (see Parlamentskorrespondenz/01/05.07.2007/no. 561). The committee on the financial sector was terminated without a written report; the chair of the committee gave only an oral report in the plenary debate of the national parliament (see Parlamentskorrespondenz/01/06.07.2007/no. 568).

5 For diverging interpretations of the Eurofighter purchase, see M. Rosenkranz, 'Österreich kauft Abfangjäger', at www.airpower.at/flugzeuge/beschaffungsstory.htm; and P. Pilz (chairman of the Eurofighter committee of inquiry), 'Mein Luftraum', at www.peterpilz.at/luftraum (this site also publishes the protocols of the parliamentary committee).

6 See www.wirtschaftsblatt.at/home/specials/eurofighter/247333/index.do?_vl_backlink=/home/specials/eurofighter/index.do.

7 See news.orf.at/?href=http%3A%2F%2Fnews.orf.at%2Fticker%2F242375.html.

8 See www.airpower.at/flugzeuge/beschaffungsstory.htm.

aircraft to eighteen, and did not order the options necessary for international missions.[9]

There were calls for a formal committee of inquiry, and, after the ÖVP-led government was defeated in the 2006 election, that demand was met. The inquiry attracted broad public attention, because the SPÖ, of which the minister of defence is a member, and the Greens, which chaired the investigating committee, explicitly searched for reasons to terminate the contract.[10]

Intervention by the Canadian car parts manufacturer Magna, which had a commercial interest in offset agreements with DaimlerChrysler (parent of DASA, a major shareholder in the Eurofighter consortium), led to suspicions that Grasser, a former Magna employee, had been acting throughout in the company's interests. The committee revealed several payments that gave grounds for suspecting corruption. Most prominent were a €6.6 million (US$8.9 million) contract between EADS (which handled the lobbying for the Eurofighter in Austria) and a former FPÖ party manager, Gernot Rumpold; a consultancy fee (or, alternatively, an interest-free loan) of €87,600 (US$118,260) from a Eurofighter lobbyist to the wife of the Austrian air chief after the purchase had been decided; additional payments of some €10,000 (US$13,500) for minor contracts with former FPÖ party secretaries; and the payment of €1 million (US$1.35 million) per year since 2003 to the Viennese soccer club Rapid, which employed several prominent SPÖ politicians as officials.[11]

A convincing reason to step out of the procurement was not found, however. According to the purchasing contract, only bribery payments by Eurofighter GmbH itself or its direct representatives (but not parent enterprises such as EADS, and their representatives) would provide legal grounds to terminate the procurement. Though most of the investigation committee would have preferred to run the risk of legal action, the new minister of defence presented it with a fait accompli by settling with Eurofighter, reducing the order to fifteen aircraft and creating a price reduction of €370 million (US$540 million).[12] In military terms alone, this was a substantial paradox. In 2002 Austria had decided to buy the most sophisticated type of Eurofighter available, but in the final settlement downgraded the equipment to mere interceptor aircraft.

The BAWAG affair

The Bank for Labour and Business (BAWAG) is one of the biggest Austrian banks and was owned until 2007 by the Austrian Trade Union Federation (ÖGB), with a 45 per cent minority holding by the Bayerische Landesbank from 1995 to 2004. In March 2006 BAWAG admitted to having lost more than €1 billion (US$1.35 billion) after speculative transactions in the late 1990s through investment firms owned by Wolfgang Flöttl, the son of a former chief executive officer (CEO). It was also made public that BAWAG had been rescued from bankruptcy in 2000 only by a declaration of liability by the ÖGB.[13]

In a bid to disguise its losses, BAWAG became involved with the US broker Refco. Since 2000 BAWAG and Refco had allegedly helped one another with balance sheet manipulations, involving other foundations owned by the ÖGB: BAWAG and, indirectly, the ÖGB technically

9 See www.peterpilz.at/luftraum/33UnterzweiM.htm.

10 See www.airpower.at/flugzeuge/beschaffungsstory.htm.

11 *Profil* (Austria), 26 March 2007; *Die Zeit* (Austria), no. 15 (2007); *Profil* (Austria), 16 April 2007; *Die Zeit* (Austria), no. 17 (2007); *Profil* (Austria), 11 June 2007; ORF, 14 May 2007, available at oe1.orf.at/inforadio/76223.html.

12 Eurofighter Typhoon press release, 26 June 2007.

13 *Times* (UK), 1 April 2007.

owned up to 50 per cent of Refco.[14] In May 2006 it was revealed that the ÖGB had secretly assumed liability for some €1.5 billion (US$2 billion) of BAWAG debts in 2005, and that BAWAG had had to settle with Refco's creditors to prevent a class-action lawsuit. Settling the Refco affair cost BAWAG (and therefore, indirectly, the ÖGB) over US$1.3 billion. By mid-2007 the union had spent its financial reserves (including a legendary strike fund) and had been forced to implement savage retrenchments.[15]

In a mid-2006 report, later published by a weekly magazine,[16] the Austrian National Bank questioned why all the speculative transactions since 1995 had ended in disaster. Auditors claimed that they were unable to trace the whereabouts of hundreds of millions of euros. A criminal lawsuit, involving former members of BAWAG's board, Flöttl and the former president and leading secretary of the ÖGB, started in July 2007.[17] The indictment contains allegations of embezzlement and accounting fraud.

Although the BAWAG affair is primarily an economic crime, it was facilitated by the lack of transparency enjoyed by Austria's trade unions.[18] Nominally a non-partisan association, the ÖGB is actually dominated by political interest groups, mainly the Fraktion Sozialdemokratische Gewerkschafter (FSG), which is closely associated to the SPÖ, and the Fraktion Christlicher Gewerkschafter (FCG), which is aligned with the Österreichischer Arbeiter- und Angestelltenbund (ÖAAB), a vehicle of the ÖVP. The FSG is the dominant faction in the ÖGB.

The ÖGB's president and financial secretary did everything possible to hush up the malaise at BAWAG, with the result that the rest of the board (including leading lights in the FSG and FCG) knew nothing about the ÖGB's liability for the losses. Indeed, so successful were they that BAWAG was even able to hide its losses from the Finance Ministry, and thus win approval for its takeover of the publicly owned Postsparkasse bank for €1.3 billion (money that BAWAG clearly did not have). Postsparkasse was ruthlessly stripped of its assets in the years that followed.[19]

During the election campaign in August and September 2006, rumours circulated of secret payments by BAWAG and the ÖGB to the SPÖ, although nothing ever came to light. The ÖGB already gives 3 per cent of its membership fees to parties, according to their popularity within the workforce (though mainly to the SPÖ).[20] While this is legal party financing under Austrian law, it is questionable whether union members fully support the practice.

Party finance: more loopholes than rules

It remains an open question whether illicit payments to parties or politicians figured in the BAWAG and Eurofighter scandals, though both cases added to the debate about the transparency of party funding. Nor were these the only examples of possible illicit funding in the past half-decade.

The discovery of the transfer of €283,000 in 2001–3 from the Federation of Austrian Industry

14 *Financial News* (US), 15 December 2006.

15 *Der Standard* (Austria), 10 June 2006 and 12 June 2006; *News* (Austria) 34/2006; *Der Standard* (Austria), 13 December 2006; *Profil* (Austria), 2 April 2007.

16 See 'Das BAWAG-Dossier', at www.networld.at/prod/510/bawag/bawag_dossier.pdf.

17 See wien.orf.at/stories/207510.

18 For a discussion of the political reasons and consequences of the BAWAG affair, see F. Karlhofer, 'BAWAG und die Folgen', in A. Khol *et al.* (eds.), *Österreichisches Jahrbuch für Politik 2006* (Vienna: Oldenbourg, 2007).

19 'BAWAG Skandal: "Alles in Trümmern" ', *Austria Presse Agentur* (Austria), 12 June 2006.

20 *Austria Presse Agentur* (Austria), 20 September 2006.

to the then finance minister, Karl-Heinz Grasser (known as the Homepage affair, because the alleged purpose of the money was to set up his web page), demonstrated the absence of transparency in political donations.

A split in one of the ruling parties, FPÖ, in April 2005 revealed the lack of internal financial controls, the generous lump sum 'allowances' for leaders and the high debts after the party's election defeats. It also demonstrated the fragmentary nature of Austria's party finance regulations.

Austria has an extraordinarily generous method of public party funding. Federal subsidies amounted to €40 million (US$54 million) in 2007, with a further €110 million (US$148.5 million) from the nine Austrian states and over €20 million (US$27 million) from municipal sources, or a total equivalent of €28–29 per voter per year.[21] For this reason, Austrian parties are far less dependent on private donations than those in other European countries and, as a further result, there are fewer regulations on the size of party donations.

Political parties that receive federal funding under the Political Parties Act are required only to fulfil some trivial disclosure criteria: two public accountants must approve the party's budget, and a simple income–expenditure balance sheet must be published in the official gazette, the *Wiener Zeitung*.[22] The party must also submit a list of donations exceeding €7,260 from individuals, private associations and corporations (though not business associations, chambers or trade unions) to the president of the Court of Auditors, but this list need not be published. The published balance sheets contain only summarised information on donations and nothing about the amount of single donations or the identity of the donors.

These disclosure requirements do not include parliamentary groups, ancillary organisations or party-owned companies. In practice, the 'rules' hardly qualify as an appropriate tool for public control of party finance. The Political Parties Act does not stipulate the kind of sanctions that can be applied if a party does not meet its obligations, only the consequences should a party not deliver its report by the deadline of the following September. In that event, the federal subsidy is withheld until the party delivers its report. The act does not lay down what should happen when a party delivers an incorrect report (an incomplete donation list, for example). The same is true for nearly all other legislation governing party and parliamentary activities.

By no measure does Austria's current system of regulating party finance meet the fundamental requirements of transparency, as laid down in the Council of Europe's Recommendations on Common Rules against Corruption in the Funding of Political Parties and Electoral Campaigns of April 2003.[23] Austria's media are becoming increasingly critical of the predicament.

Hubert Sickinger (TI Austria)

Further reading

F. Karlhofer, 'BAWAG und die Folgen', in A. Khol *et al.* (eds.), *Österreichisches Jahrbuch für Politik 2006* (Vienna: Oldenbourg, 2007).

M. Kreutner (ed.), *The Corruption Monster: Ethik, Politik und Korruption* (Vienna: Czernin, 2006).

H. Sickinger, 'Überlegungen zur Reform der Österreichischen Parteienfinanzierung', *Österreichische Zeitschrift für Politikwissenschaft*, vol. 31, no. 1 (2002).

'Austria', in T. Grant (ed.), *Lobbying, Government Relations, and Campaign Finance Worldwide:*

21 Updated calculation for 2007 by the author.
22 Political Parties Act 1974, article 4.
23 Available at www.coe.int/t/dg1/greco/general/Rec(2003)4_EN.pdf.

Navigating the Laws, Regulations and Practices of National Regimes (Dobbs Ferry, NY: Oceana Publications, 2005).
'Politische Korruption', in H. Dachs *et al.* (eds.), *Politik in Österreich: Das Handbuch* (Vienna: Manz, 2006).
'Korruption in Österreich. Verbreitung, Ausgewählte Problembereiche, Reformbedarf', in Bundesministerium für Justiz (ed.), *35. Ottensteiner Fortbildungsseminar aus Strafrecht und Kriminologie, 19. bis 23. Februar 2007* (Vienna: Neuer Wissenschaftlicher Verlag, 2007).
TI Austria: www.ti-austria.at.

Georgia

Corruption Perceptions Index 2007: 3.4 (79th out of 180 countries)

Conventions
Council of Europe Civil Law Convention on Corruption (signed November 1999; ratified May 2003)
Council of Europe Criminal Law Convention on Corruption (signed January 1999; not yet ratified)
UN Convention against Transnational Organized Crime (signed December 2000; ratified September 2006)

Legal and institutional changes

- On 25 July 2006 parliament adopted a **revised Customs Code that came into effect on 1 January 2007**. The new code defines the duties of customs and other government agencies, sets zero tariff rates for most imports (and 5 and 12 per cent rates for others) and establishes a one-stop window that allows importers to obtain required documentation more easily. The code regulates procedures for appealing decisions to the Customs Department of the Ministry of Finance or the courts. The customs agency has also opened a hotline where clients can get information or report violations. In 2005 the customs revenues had steadily risen to L989 million (US$560.4 million), a 54 per cent increase on the previous year.[1] Much of the difference would previously have been siphoned off through corruption and organised smuggling.
- On 29 December 2006 **parliament adopted a law that merged the Tax Department, customs service and financial police** into a single structure, called the revenue service (see below). Businesses generally supported the initiative, though there was lingering concern

1 See www.customs.gov.ge; www.customs.gov.ge/reforma.htm#Semosavlebis_mobilizacia.

over the future role of the financial police. The new agency has powers to investigate, search facilities and make arrests in cases where financial crimes are suspected. In the past these responsibilities were assumed by the financial police, which allegedly abused its powers.[2]

- Parliament passed a number of amendments over the past year that **simplified licences and permits**, helping to improve the business environment. The total number of licences and permits now required in all fields of activity has fallen from over 900 to fewer than 110.[3] Apart from streamlining economic activity, the purpose of the changes was to reduce opportunities for corruption due to excessive red tape, thereby making Georgia more attractive to foreign and domestic investors.
- In November 2006 and April 2007 the government published its first (six-month and one-year, respectively) progress reports on the **action plan for the implementation of the National Anti-Corruption Strategy**, adopted by presidential decree in March 2006. Given the action plan's original failure to specify concrete benchmarks by which to measure progress, its lack of detail on implementation methodology and lack of dialogue with the public, the reports were not overly informative.[4] In April 2007 the government presented a revised draft action plan that identified responsible agencies, partner agencies, implementation timeframes and, importantly, implementation indicators – the elements missing from the previous version. The new action plan eliminated important provisions that had been in the original document, however, including establishing a clear distribution of functions within the civil service and ensuring the transparency of the courts. Overall, the April version emphasised the transparency of public finances, improved revenue administration, the development of law enforcement and the harmonisation of anti-corruption legislation with international conventions. The prime minister and president have yet to approve the new plan, intended to cover 2007–9.
- At the **Sixth Monitoring Meeting of the Istanbul Anti-Corruption Action Plan**, on 13 December 2006, the OECD's Anti-Corruption Network for Eastern Europe and Central Asia presented a report that reviewed Georgia's progress in implementing the recommendations made in June 2004.[5] Of the twenty-one recommendations, Georgia was fully compliant with two, largely compliant with nine, partially compliant with six and non-compliant with four. The four non-compliant recommendations were: the responsibility of legal persons for criminal offences under the Georgian Criminal Code; aligning the Georgian Criminal Code's treatment of active and passive bribery with international standards; criminalising the bribery of foreign or international public officials; and adopting measures to protect whistleblowers.

Cautious welcome for one-stop revenue service

One of the most significant changes in the fiscal field since 2003 has been the expansion of the

2 Civil.ge (Georgia), 16 August 2005.

3 The law concerning licences and permits, passed on 24 June 2005, was subsequently amended on 25 May 2006, 30 June 2006, 24 July 2006, 29 December 2006 and 28 March 2007.

4 TI Georgia collected individual action plans from the ministries, translated them into English and Russian, and distributed them to domestic and international NGOs, the media and the public, along with an analysis of their content. In addition, TI Georgia evaluated and publicised information about the government's anti-corruption initiatives and their consistency with the National Anti-Corruption Strategy. The reports can be accessed at www.ti.itdc.ge/index.php?lang_id=ENG&sec_id=190&info_id=223.

5 See www.oecd.org/dataoecd/10/11/38009260.pdf.

national budget, demonstrating the state's increased revenue-collecting capacity. This has been coupled with significantly lower taxes since the adoption of a new Tax Code in 2004 (see *Global Corruption Report 2004*). State revenues, including grants, have more than tripled, from L933 million (US$528.6 million) in 2003 to L2.95 billion (US$1.7 billion) in 2006.[6]

In July 2006 parliament adopted a new law consolidating the Tax Department, Customs Department and financial police into a single, unified 'revenue service', effective from 1 January 2007. The main functions of the new agency are to:

- inform taxpayers about their rights and obligations;
- inform taxpayers in a timely manner about amendments and modifications in revenue and customs legislation;
- administer taxes (including customs tax) and those local or state payments within its competency;
- participate in the preparation of draft bills and interstate projects on tax and customs tax rating issues;
- supervise and control goods imported to and exported from the territory of Georgia;
- uphold and fulfil tax legislation and customs law;
- ensure the safety of personnel while they fulfil their duties as employees of the Ministry of Finance;
- reveal and prevent administrative law violations within its competency; and
- prevent or reveal crimes, execute preliminary investigations, and organise and conduct expert analysis.[7]

The consolidation led to staff restructuring[8] as well as the introduction of new practices within the service designed to make its function more efficient and user-friendly.

The revenue service is not a new phenomenon. A similar service existed until its abolition in 2003, when separate entities were established to regulate tax and customs revenues. The new law introduced two tools to address the most corruption-prone areas, however: the one-window principle and a reduction of person-to-person interactions. Citizens had previously been required to submit documents to various departments in different buildings at different sites, which greatly lengthened the process of paying. Giorgi Isakadze, president of the Federation of Georgian Businessmen (FGB), says that the new service will have more unified decision-making capabilities and better coordination and information-sharing.[9]

Furthermore, unlike the previous system, the new service requires that, when submitting forms, the initial contact is not with the official responsible for the final review, but for simply verifying that the right documentation has been presented. Documents are actually reviewed by a third party who has had no contact with the client or citizen.

In March 2007 Robert Christiansen, the International Monetary Fund's representative in Georgia, summarised the IMF's recommendations for the need to improve taxpayer services,

6 The data used are based on reports by the budgetary department of the Ministry of Finance, available at www.mof.ge.

7 See www.parliament.ge.

8 According to Koba Abuladze, deputy head of the revenue service, around 45 per cent of positions were eliminated, but retaining qualified staff was necessary for the expansion of regional branches in Gori, Telavi and, eventually, Batumi. Camrin Christensen interview with K. Abuladze, 18 May 2007.

9 C. Christensen interview with Giorgi Isakadze, president of the FGB, 17 May 2007.

create a dispute resolution process and clarify taxpayers' rights.[10] 'We think that the new revenue service,' he said, 'has a potential to realise a significant progress in tax administration.'[11] Giorgi Isakadze of the FGB said that the revenue service needs to do a better job informing the public of the innovations introduced by the new revenue service law and devote more time consulting stakeholders regarding the reforms. There are also significant staffing concerns. Although opportunities to make extra money ('salary supplements') through corruption have been sharply reduced, actual salaries have not been increased. The revenue service is rapidly losing qualified staff to the private sector and personnel turnover is high.[12]

One concern that the opposition raised when the new law was being debated was that the financial police would inevitably overshadow other entities in the new revenue service. These concerns stemmed from the financial police's prior reputation and evidence of some disproportionately violent raids in 2005.[13] 'The financial police is a tool in the hands of authorities to carry out racketeering,' argued Kakha Kukava, an MP of the opposition Conservative Party. 'This is a reality and we must not turn a blind eye on this reality . . . President Saakashvili himself said that the decriminalisation of relations between state and business is a priority. But that is not reflected in this draft law.'[14]

Who is knocking down Tbilisi's booths – and why?

Article 21 of the Georgian constitution addresses the protection of property rights and asserts that any abrogation of the right to own property is impermissible.[15] This article's effectiveness has been called into question after many incidents involving the seizure and demolition of private property in late 2006 and early 2007.[16]

The confiscations in 2006 mainly affected restaurant owners in Tbilisi and Mtskheta, just outside the capital. Without formally pressing charges for possible infractions, the owners were simply 'reminded' that they had obtained their business licences through allegedly corrupt deals with officials in the era of the former president, Eduard Shevardnadze. To 'correct' these errors owners were 'invited' 'voluntarily' to hand over their properties to the state.[17] The transfers were officially registered as 'gifts' rather than expropriations, and the former owners were not reimbursed. The 'gift agreement' was signed between the owner and the Ministry of Economic Development, and notarised.

According to the ombudsman, Sozar Subari, the authorities throughout the country used various tactics to intimidate the former proprietors.[18] In many cases the threat of possible criminal proceedings was made solely for the purpose of intimidation. Cooperation with the authorities and the voluntary surrender of property were presented to victims as the least distressing option.

10 *Georgia Online*, 2 March 2007.
11 Ibid.
12 C. Christensen interview with K. Abuladze, 18 May 2007.
13 On 16 June 2006 dozens of officers from the special unit of the financial police, masked and armed with Kalashnikov assault rifles, raided two fast food restaurants in Tbilisi owned by suspected tax evaders.
14 *Georgia Online*, 24 November 2006.
15 See www.legislationline.org/legislation.php?tid=1&lid=6099.
16 Much of what is reported below is taken from a report by TI Georgia, 'Property Rights in Post Revolutionary Georgia' (Tbilisi: TI Georgia, 2007).
17 *Georgia Online*, 21 December 2006.
18 *Georgia Online*, 25 May 2007.

Later confiscations targeted the small shops, booths and stalls that mushroom around metro stations. Tbilisi City Hall's supervision agency tore down the structures with only a few days' verbal notice, and with no court order or written notice. The operators had no chance to appeal the decision in court. City Hall claimed that the buildings did not have proper permits and/or were not registered with the public registry; that demolition was needed to free up space for 'public use'; or that the structures 'did not correspond with the city's image'. City Hall staff maintained that the demolitions in no way constituted a breach of the law and suggested that anyone who believed his or her rights had been violated should file a complaint in court.[19]

The ombudsman and opposition MPs, on the other hand, argued that most contested buildings were legal, that land ownership was certified and the relevant city agencies had agreed to architectural plans. According to Subari, there may have been some infractions pertaining to the implementation of architectural projects or lease conditions, but these in no way gave grounds for the ensuing demolition blitz.

Georgian legislation defines very specific conditions under which private property may be confiscated, ranging from natural disaster, epidemics or other circumstances that endanger human health to conventional public needs, such as building roads, railways, pipelines, sewage systems or structures required for national defence purposes. The legal process is lengthy and complicated. Confiscating property without adequate reimbursement is an explicit breach of Georgian legislation. The opposition has pressed for the establishment of a parliamentary inquiry to investigate the confiscations, but the ruling United National Movement blocked the proposal.[20]

But questions remain. Will the government restore the rights of owners whose property has already been demolished, or will it offer them compensation? Will the officials and agencies responsible for the demolitions be held to account? In light of the impunity apparently granted to those who trampled upon article 21 of the constitution, it remains uncertain whether new legislation would be any better at securing property rights when vested powers have a clear interest.

Misuse of administrative resources in the 2006 local government elections

Misuse of administrative resources[21] has been a problem in every Georgian election since 2 November 2003. Indeed, that election's shortcomings were the main source of popular discontent that triggered the Rose Revolution – the resignation of the former president, Edward Shevardnadze, and the election of President Mikheil Saakashvili in January 2004.

While the January 2004 election was an overall improvement in quality, a number of problems that had cast a shadow over the previous election were still evident, including the lack of separation between government and party, and the resulting potential for misuse of state resources.

The municipal elections of 5 October 2006 were considered an improvement on previous electoral outings, although the OSCE and the Council of Europe's Congress of Local and Regional Authorities again highlighted that the blurred

19 City Hall maintains that no one has sought to take legal action against it.

20 *Georgia Online*, 21 December 2006.

21 Administrative resources range from 'hard' resources, such as the coercive powers of law enforcement agencies, to 'soft' resources, such as the financial and material resources of the state, etc. The misuse of soft resources is not strictly illegal.

distinction between government and governing party reinforced the incumbents' advantage.[22]

In June 2006 parliament passed further amendments to the Election Code that influence campaign behaviour. Most significantly, the heads of executive agencies and local government bodies were freed up to participate legally in election campaigning. As a result, enthusiastic support for the ruling party and its candidates by President Saakashvili and senior public officials dominated campaigning in 2006. Distinguishing party from state resources became extremely difficult, as government officials campaigned for their protégés during working hours.[23]

There were also cases of public funds being used to influence the election outcome. One of the most visible examples took place in September, when Tbilisi City Hall distributed L100 (US$57) vouchers to teachers and lecturers, ostensibly to pay for scarce gas. The voucher was signed by the mayor, Gigi Ugulava, and displayed his picture.[24] Shortly before the municipal elections, Tbilisi City Hall ran a television public information item, *Tbilisuri Ambeb* ('Tbilisi Stories'), promoting government successes. City Hall paid L600,000 (US$340,000) to produce the series. The adverts showed Ugulava and President Sakashvili summarising the main activities carried out by the government since it had come to power.[25]

By law, election candidates are required to submit financial reports on their donations and campaign expenditures, which are then scrutinised by a special commission set up by the Central Election Commission (CEC). The CEC's role in preventing or detecting the misuse of administrative resources in the pre-election period has been minimal to date, however.

A technical working group is currently developing Election Code amendments that focus on political party financing, the CEC's budget and the publication of election results, among other subjects. The amended code is expected to be ready by spring 2008, ahead of the presidential and parliamentary elections at the end of that year. Misuse of administrative resources, however, does not always violate legal regulations. Governments committed to democracy and fair competition limit their powers not only with laws, but also with high ethical and democratic standards, drawing a clear line between the affairs of state and electoral competition.

Camrin Christensen and Tamuna Karosanidze (TI Georgia)

Further reading

American Bar Association's Central European and Eurasian Law Initiative, 'Judicial Reform Index for Georgia' (Washington, DC: ABA/CEELI, 2005).

J. Anderson and C. Gray, 'Anticorruption in Transition 3: Who Is Succeeding . . . and Why?' (Washington, DC: World Bank, 2006).

T. Karosanidze, 'National Anti-Corruption Strategy and Action Plan: Elaboration and Implementation' (Tbilisi: U4 Anti-Corruption Resource Centre, 2007).

G. Nodia, 'Georgia', in J. Goehring (ed.), *Freedom House Nations in Transition 2006* (Budapest: Aquincum, 2006).

OECD ACN, 'Georgia: Update on National Implementation Measures' (Paris: OECD/ACN, 2006).

TI Georgia: www.transparency.ge.

22 See OSCE/ODIHR Limited Election Observation Mission, Final Report, 'Georgia: Municipal Elections 5 October 2006' (Warsaw: OSCE/ODIHR, 2006).

23 See www.transparency.ge/index.php?lang_id=ENG&sec_id=190&info_id=222.

24 Ibid.

25 Ibid.

Germany

Corruption Perceptions Index 2007: 7.8 (16th out of 180 countries)

Conventions

Council of Europe Civil Law Convention on Corruption (signed November 1999; not yet ratified)
Council of Europe Criminal Law Convention on Corruption (signed January 1999; not yet ratified)
OECD Convention on Combating Bribery of Foreign Public Officials (signed December 1997; ratified November 1998)
UN Convention against Corruption (signed December 2003; not yet ratified)
UN Convention against Transnational Organized Crime (signed December 2000; ratified June 2006)

Legal and institutional changes

- In September 2005 Germany ended its isolation regarding **freedom of information** by passing a law that allows access to information at a federal level, thereby adding momentum to existing initiatives at state level. Vigorously supported by TI Germany, four additional states made documents accessible between July and September 2006: Mecklenburg-Western Pomerania, Bremen, Hamburg and Saarland. Eight of the sixteen states have now passed access-to-information laws.[1]
- In September 2006 the Federal Ministry of Justice issued a first draft of the **Second Anti-Corruption Act**.[2] Although it suggests that important improvements can be made to existing anti-corruption law, critics point out that it lacks provisions that deal with corruption involving MPs. These would need to be extended in order to ratify the UN Convention against Corruption, as well as the Council of Europe's Criminal Law Convention on Corruption (see below). The federal legislature has reserved the right to issue its own draft on the subject. Even if parliament adopts the draft, therefore, the ratification and implementation of conventions signed more than seven years ago may be further delayed.
- In mid-2006 the **German Association for Freedom of Information** was founded in Berlin. The new association was initiated by members of the Green and Social Democratic Parties, and is supported by NGOs and the Bertelsmann Foundation. The association aims to promote the enforcement of existing access-to-information provisions and to cooperate with organisations engaged in the issue.
- In May 2007 the federal government approved drafts for submission to the parliament concerning a **'witness-of-the-crown' provision and the use of wire-tapping** of suspects in severe cases of corruption. Both are of major importance to broaden the investigative powers of prosecutors.

1 See www.aitel.hist.no/~walterk/wkeim/IFG.htm#Deutschland.
2 The first Anti-Corruption Act was established in 1997.

Siemens languishes in a sea of scandal

In mid-November 2006 more than 270 police and tax investigators raided the offices and homes of Siemens' staff as part of the biggest corruption scandal involving a German company in decades. The widening effects of corruption allegations against Siemens, the European Union's largest engineering company, generated headlines across the world and fostered a debate as to how German companies truly do business.

Siemens is at the centre of three separate investigations. The most wide-ranging concerns allegations that hundreds of millions of euros in bribes were paid to win the company contracts for telecommunications equipment. The case went public in autumn 2006, as Munich prosecutors began to investigate the disappearance of €200 million (US$270 million) from Siemens' accounts. In December 2006,[3] Siemens announced that it was looking into more than €420 million (US$560 million) of what it called 'dubious payments' to consultants over the previous seven years.[4] Thomas Ganswindt, head of the telecommunication division from 2004 to 2006, was arrested. Siemens cut its reported 2005/6 net profit by €73 million (US$97 million) as a result of the affair and hired outsiders[5] to examine its compliance systems. In April 2007 it announced that it expected a 'significant increase' in the number of possible bribes identified in an internal investigation after it had expanded the search for similar payments to other divisions.[6]

The second scandal unfolded in March 2007, after prosecutors raided Siemens' offices in Munich, Erlangen and Nuremberg following fresh allegations of suspicious payments. Johannes Feldmayer, a member of the board, was accused of diverting €50 million (US$67 million) of company money to the independent labour union AUB, which was perceived as friendly to management and sometimes acted as a counterweight to IG Metall, Germany's most powerful union.[7] As a result, Heinrich von Pierer, the head of Siemens' supervisory board, resigned in April 2007, and its chief executive, Klaus Kleinfeld, also resigned, in June. Both men denied any wrongdoing.

In May 2007 two former Siemens officials were convicted of bribery and abetting bribery in a multimillion-dollar deal with the Italian energy utility ENEL. Both received suspended jail sentences, but were ordered to pay €400,000 (US$533,000) to charity. The state court of Darmstadt ordered Siemens to forfeit €38 million (US$51 million) from its ENEL deals. During the trial the two men admitted paying kickbacks worth €6 million (US$8 million) to the utility's managers for contracts valued at €450 million (US$600 million) for Siemens gas turbines between 1999 and 2002. The defendants revealed that slush funds had existed at Siemens' power generation division for many years.[8] This was the first verdict in an ever-widening series of investigations that have shaken the company since 2006. Siemens declared that it would appeal the Darmstadt ruling on the grounds that the legal situation in Germany had not been clear when the bribery was committed. The bribing of foreign employees did not become illegal under German law until 2002.

In addition to these three cases, Siemens is among fifty-seven German companies listed in

3 Due to the widening corruption scandal, in December 2006 TI Germany asked Siemens to leave the organisation. Siemens had been a member of TI Germany since 1999.

4 *Reuters* (UK), 7 February 2007.

5 Among these outside advisers are the law firm Debevoise and Plimpton, and Michael Hershman, one of TI's founders.

6 *New York Times* (US), 27 April 2007.

7 *Süddeutsche Zeitung* (Germany), 20 April 2007.

8 *hr online* (Germany), 14 May 2007.

the Volcker Report[9] for alleged abuse of the UN's oil-for-food programme in Iraq. It is also being investigated for offering bribes to secure a construction project contract in Serbia.[10] These scandals are some of the most far-reaching in German corporate history. Questions are being raised as to whether conducting business in certain countries would be possible without the payment of bribes. Indeed, Siemens announced in February 2007 that it would decide which of the 190 countries in which it operates it would have to leave.[11]

It was only in 2005 that the media woke up to the fact that corruption was widespread in the private sector. Since then a number of prominent companies, such as VW, DaimlerChrysler, Infineon and BMW, have been accused of entanglement in private-to-private corruption.[12] As the *Global Corruption Report 2005* pointed out, there is reluctance in Germany to initiate investigations concerning allegations of foreign bribery due to staff resources and problems with international legal assistance. Should the accusations against Siemens prove true, it would highlight the failure of Germany's corporate governance structure, both by the company's supervisory board and its auditor, KPMG.[13]

The first sanctions over the ENEL case are just the beginning. More threatening are possible sanctions by the US Securities and Exchange Commission.[14] Due to the widening scandals, the SEC upgraded its informal inquiry to a formal investigation in April 2007.[15] This gives regulators powers to issue subpoenas. In past investigations the SEC has imposed serious sanctions, including compensatory damages and exclusion from public procurement contracts. The external specialists Siemens has commissioned to investigate violations of anti-corruption laws will also cost the company a good deal. Siemens has so far paid €63 million (US$85 million) to outside consultants.[16] Siemens has also paid €4.5 million of a €5 million bail bond for the release of board member Johannes Feldmayer, who was taken into custody in March 2007.[17] Siemens' reputation is also certain to suffer; the independent credit ratings agency Standard & Poor's has put the company on watch for possible downgrading.

The Siemens scandals illustrate a shift in the ethical climate of corporate Germany. The first phase of fighting transnational corruption involved the establishment of the main anti-corruption laws, and it was not until 2002 that the bribery of foreign companies was made illegal. The second phase involved a growing number of companies attempting to adapt to the new legal situation. In 2000, for instance, the Deutsche Bahn AG introduced extensive anti-corruption measures, including the appointment of two independent lawyers as ombudspersons, who can be approached by whistleblowers.[18] The company has also implemented a blacklist of business partners that have previously offered bribes and publishes a detailed corruption report each year. The discovery of the Siemens corruption affair is especially significant, however. It

9 Independent Inquiry Committee into the United Nations Oil-for-Food Programme, 'Report on Programme Manipulation of the Oil-for-Food Programme by the Iraqi Regime', 27 October 2005.
10 *Deutsche Welle* (Germany), 22 December 2006; *Wirtschaftsblatt* (Austria), 13 August 2007.
11 *Reuters* (UK), 7 February 2007.
12 *Handelsblatt* (Germany), 27 July 2005; See also www.goethe.de/ges/eur/prj/ejp/thm/kor/en965954.htm.
13 *Wall Street Journal* (US), 4 May 2007.
14 The SEC has responsibility for enforcing the federal securities laws and regulating the stock market. It also has powers to bring civil enforcement actions against companies found to have committed accounting fraud, provided false information, engaged in insider trading or other violations of the securities law. Since Siemens is listed on the US stock market, US securities laws apply to the company.
15 *Tagesschau.de* (Germany), 26 April 2007.
16 *Süddeutsche Zeitung* (Germany), 26 April 2007.
17 *New York Times* (US), 27 April 2007.
18 See www.db.de/site/bahn/de/unternehmen/presse/mediathek/audiodatenbank/korruptionsbericht_2005.html.

could have negative implications for the reputation of the German private sector, which according to the 2006 TI Bribe Payers Index was good.[19] The scandal also indicates that the sector still has a long way to go before complying fully with German anti-corruption laws.

MPs unwilling to surrender second jobs

German attitudes towards politicians and parliaments show a continuing decline in confidence. While no major scandal of party financing or bribery of MPs has occurred in some years, there were reports in 2005 of MPs entangled in conflicts of interest about outside employment. According to polls in 2005, only 17 per cent of Germans still had confidence in MPs,[20] and conflicts of interest were seen as a primary cause of disillusion. Some MPs received second salaries that they could not justify, while others continued to be paid by previous employers long after they had been elected. Such conduct was not actually illegal, but the public considered it a means of inappropriately influencing the political agenda and questioned the independence of parliamentarians. Some MPs resigned from office as a result of the public outrage.

TI Germany has urged lawmakers to ban such practices and introduce provisions that allow public access to information about their MPs' sources of income. As long ago as 1975 the Federal Constitutional Court said that the existing laws were inadequate to prevent parliamentary conflicts of interest.[21]

In June 2005 – thirty years later – the Social Democrat/Green Party coalition enacted a new 'parliamentarians' law' and a code of conduct that experts recognise as a substantial improvement in terms of disclosure obligations. MPs are allowed to hold additional jobs, but the new provisions oblige MPs to report all additional income at the beginning of their four-year term. The speaker is obliged to monitor the disclosures and display them on the parliamentary website. Information about MPs' additional income does not have to show precise figures, but is published in three tranches: between €1,000 and €3,500 per month; between €3,500 and €7,000 per month; and exceeding €7,000 per month. Shortly after the coalition of Social Democrats and Christian Democrats came to power in September 2005, nine MPs filed a lawsuit against the disclosure requirements, citing right-to-privacy issues and asserting that they would deter qualified candidates from running for office. Norbert Lammert, the new speaker and a Christian Democrat, decided not to enforce the new law, arguing that the Federal Constitutional Court would have to take a decision on the matter in summary proceedings. Experts argued that Lammert's refusal was illegal.[22]

In July 2007 – more than one and a half years after these events – the Federal Constitutional Court came to a decision. Eight judges dismissed the case against the 'parliamentarians' law' and related code of conduct, arguing that MPs' activities should be circumscribed by the conditions of their mandate.[23] Any additional activities were to be considered as secondary employment, which must be disclosed and made public. The fact that only 109 of the 614 federal MPs had

19 According to the 2006 TI Bribe Payers Index, German companies are currently perceived to be some of the least likely to pay bribes when doing business abroad. Bribes Payers Index 2006 score: 7.34 (7th out of 30 countries).

20 B. Weßels, 'Wie Vertrauen verloren geht', *WZB-Mitteilungen*, no. 107 (2005).

21 BVerfGE 40, 296 (Diäten-Urteil); Zur Verfassungsmäßigkeit der Entschädigungsregelung für Abgeordnete, 1975.

22 See interview with Professor Hans Herbert von Arnim at www.campact.de/nebenekft/home, and *Frontal 21* (Germany), 4 April 2006.

23 *Süddeutsche Zeitung* (Germany), 4 July 2007.

published their outside earnings prior to that decision reveals the extent of their reluctance to re-engage voter confidence.[24]

MPs have shown a similar resistance to preventing corruption elsewhere in politics. Germany signed the UN Convention against Corruption in 2003, but no real efforts have been made to enforce it.[25] A precondition to enforcement would be altering the definition of the act of bribery involving MPs. The existing provision is limited to buying or selling votes in the plenary or committees, which does not meet the requirements under the UNCAC. In fact, the provision is much weaker than those concerning members of foreign public assemblies of international organisations. While the UNCAC regards MPs as public officials, this is not the case in Germany. The July 2004 directive concerning the prevention of corruption in the federal administration that governs public officials, therefore, does not apply to MPs. Only one case of bribery involving an MP at local, state or national level had resulted in a conviction since 1994.[26] Reform is badly overdue.

Dagmar Schröder-Huse (TI Germany)

Further reading

A. A. van Aaken, 'Genügt das deutsche Recht den Anforderungen der VN-Konvention gegen Korruption', *Zeitschrift für ausländisches öffentliches Recht und Völkerrecht*, vol. 65, no. 2 (2005).

Control Risks and Simmons & Simmons, 'International Business Attitudes to Corruption Survey – 2006' (London: Control Risks and Simmons & Simmons, 2006).

Netzwerk Recherche (NR), *nr-Werkstatt: Dunkelfeld Korruption* (Wiesbaden: NR, 2006).

O. Pragal, *Die Korruption innerhalb des privaten Sektors und ihre strafrechtliche Kontrolle durch § 299 StGB* (Berlin: Carl Heymanns Verlag, 2006).

TI Germany: www.transparency.de.

24 *Focus.de* (Germany), 19 May 2007.
25 *Der Spiegel* (Germany), 21 May 2007.
26 *Ruppiner Anzeiger* (Germany), 4 April 2007; *Der Spiegel* (Germany), 21 May 2007.

Israel

Corruption Perceptions Index 2007: 6.1 (30th out of 180 countries)

Conventions

UN Convention against Corruption (signed November 2005; not yet ratified)

UN Convention against Transnational Organized Crime (signed December 2000; ratified December 2006)

Legal and institutional changes

- In July 2006 the Knesset (parliament) approved an amendment to the Courts Law 1984 that authorises the chief justice to determine a **code of conduct for judges**; to give the code an obligatory status; and to enhance public trust in the judiciary (see *Global Corruption Report 2007*).[1] First published in 2007, the new code is based on the work of a committee headed by Mishael Cheshin, a former judge.
- A court ruling in July 2006 set a precedent in interpreting the law on **protecting whistleblowers**. In an appeal on 25 July the court ordered the state to pay Assaf Grety, a former immigration official in northern Israel dismissed for exposing misdeeds, compensation equivalent to the income he would have received had he not been fired.[2] In addition, the court emphasised the state's obligation to return Grety to an equivalent unit and job (Grety was subsequently reappointed). The verdict was important for three reasons. First, the court emphasised that Grety should not suffer for his disclosures; second, by accepting Grety's claims the court signalled that it would protect other whistleblowers in future; and, third, by giving a whistleblower compensation the court encouraged others to come forward.
- On 6 February 2007 the Knesset legislated **amendments to the Parties Law** (intra-parties elections) (temporary measure) of 1992,[3] one of which requires the state comptroller to audit intra-party elections rather than the party itself. Another amendment determines that the law will apply to all parties that hold intra-party elections. The legislation, which aims to establish the monitoring of expenditure in party elections, is a partial implementation of the recommendations of a committee headed by the late judge Dov Levin that recommended

1 Code of Ethics for Judges, 5767-2007.

2 *Assaf Grety v. State of Israel*, CA 502/05, 25 July 2006 (unpublished).

3 A Hebrew explanation of the legislation is published at www.knesset.gov.il/Laws/Data/BillGoverment/280/280.pdf. The fact that the law is temporary means the Knesset is aware that the amendments do not meet the requirements of monitoring party elections.

changing the public financing of political parties after several corruption scandals during the 1990s.[4]

- On 16 April 2007 the government passed **amendments to the Tenders Law** of 1992. Each ministry is permitted to exempt itself from issuing tenders for contracts with a value of up to S4 million (US$937,000). The exemption will be considered by a ministerial committee headed by the general director of each ministry.[5] These tenders amount to around 20 per cent of total government spending. Around 80 per cent of all tenders previously handled by the accountant general in the Finance Ministry will therefore now be managed by the ministry concerned. It is not entirely clear what the result of the amendment will be. The accountant general believes that it will lead to corruption and harm competition, but the attorney general supported the move, as did many officials seeking to cut red tape in the civil service.[6]

Allegations of corruption in the tax office

Throughout 2006 the police conducted a secret investigation into what may be the biggest corruption case in Israel's modern history. If only a fraction of the allegations published so far turn out to be accurate, it will indicate that the scale of corruption in Israeli society has been woefully underestimated.

According to press reports, businessmen and political figures bribed senior officials of the revenue office to give them tax breaks and promote their associates to positions of power in the tax authority. In January 2007 Shula Zaken, personal secretary to the prime minister, Ehud Olmert, was placed under house arrest on suspicion that she and powerful figures in the Likud party had orchestrated the appointment of Jacky Matza to the post of director general of the Israel Tax Authority. Matza allegedly helped Zaken's brother, Yoram Karshi, a Jerusalem City councillor and member of the Likud central committee, and two other confidants – businessmen Simu Tubol and Koby Ben-Gur – to obtain tax breaks.[7] A senior official in the Finance Ministry, Yossi Bachar, said: 'We awoke to a nightmare.'[8]

The investigation was made public on 1 January 2007. Matza, previously known for his professional integrity, was remanded for six days, and Karshi and Ben-Gur were remanded for nine days. Shmuel Borbov, deputy director general of the Israel Tax Authority, was placed under house arrest for three weeks. On 10 February the civil service commissioner, Shmuel Hollander, suspended Zaken, six members of the tax administration and an official in the prime minister's office.[9] Matza resigned in February, protesting his innocence. The police later questioned Olmert as part of the investigation into Matza's appointment – which he had approved – and also Zaken, but were emphatic that he was not a suspect in the case.[10]

The scale of alleged corruption appears to date back to the extensive patronage system that emerged in the late 1990s and from 2001 to 2006, when the Likud dominated government. In November 1997 the party abolished the primaries system for selecting its list of candidates

4 The committee was appointed by the minister of justice and delivered its report in October 2000. A copy can be found in the library of the justice minister.
5 See *Yediot Aharonot* (Israel), 18 July 2007, and *Ha'aretz* (Israel), 15 April, 2007.
6 See *Ha'aretz* (Israel), 15 April 2007, and www.knesset.gov.il/protocols/data/rtf/kalkala/2007-04-18.rtf.
7 *Yediot Aharonot* (Israel), 3 January 2007; *Ha'aretz* (Israel), 3 January 2007.
8 *Yediot Aharonot* (Israel), 9 January 2007.
9 *Ha'aretz* (Israel), 11 February 2007; *ynetnews* (Israel), 10 February 2007.
10 *Ha'aretz* (Israel), 11 April 2007; *Jerusalem Post* (Israel), 10 April 2007.

to the Knesset,[11] conferring the power instead on the 2,500 members of the party central committee and greatly increasing its political leverage. One result was that it became much more difficult to win civil service promotion without the help of Likud activists; it also enhanced the influence of 'vote harvesters' and people such as Karshi, Ben-Gur, Tubol and Zaken.

Prime minister under suspicion

Various parties have accused Ehud Olmert of misuse of power for private gain based on four separate incidents during his time in various governmental roles. First, the state comptroller and accountant general accused him of intervening as minister of finance on behalf of his friend, the Australian retail tycoon Frank Lowy, in a tender to privatise the state's share in Bank Leumi in 2005.[12]

A second affair concerned Olmert's purchase of a house on Jerusalem's Cremieux Street at an alleged discount of several hundred thousand dollars. The house had previously been subject to a conservation order precluding development, but the permit had been revoked so the building can now be demolished, rebuilt or enlarged. Olmert served as mayor of Jerusalem from 1993 to 2003, and is alleged to have used his influence to have the permits altered. The state comptroller recommended in April 2007 that the attorney general open a criminal investigation into the matter.[13] The attorney general's decision has not yet been published.

The third involves allegations that Olmert illegally created posts for members of his former party, the Likud, in the Small Business Authority (MMB) at the Industry, Trade and Employment Ministry at a cost of hundreds of thousands of shekels.[14] According to a report by the state comptroller published in August 2006,[15] the MMB was exploited for 'appointments made with political considerations through an improper process' and without publishing public tenders. In early 2004 Olmert and Raanan Dinur, general director of his office, reportedly established a new post whose role was 'to implement projects'. The job went to Lilach Nehemia, partner of the then tourism minister, Abraham Hirchson, who had served as Olmert's finance minister until he temporarily stood down in April 2007 after the police had investigated him for corruption. He subsequently resigned in July 2007.[16] Three other Likud central committee members were appointed 'to implement projects' without formal tenders being issued. The state comptroller advised the attorney general to order the police to open a criminal investigation, but a decision has not yet been reached.[17]

The fourth, another case of potential conflict of interest, involved a factory that filed a request in 2001 to have its status upgraded so as to qualify for state benefits and grants.[18] Olmert is

11 The party chairman is still elected by primary, however.

12 *Ha'aretz* (Israel), 17 January 2007.

13 *Ha'aretz* (Israel), 30 April 2007. The difference between the amount Olmert reportedly paid for his house (US$1.2 million) and its true market value (US$1.6–1.8 million) is currently being investigated.

14 *Ha'aretz* (Israel), 4 September 2006.

15 See www.mevaker.gov.il/serve/contentTree.asp?bookid=470&id=186&contentid=&parentcid=undefined&sw=1280&hw=954.

16 Lilach Nehemia had been turned down for a post in the State Tourism Authority due to concerns it would appear to be a political appointment. See *Ha'aretz* (Israel), 1 July 2007 and 22 July 2007.

17 *Ha'aretz* (Israel), 30 April 2007.

18 Improper activities by senior civil servants are an offence of breach of trust under clause 284 of the Israeli Criminal Code, according to a high court judgment in November 2004. See *Israel v. Sheves*, P.D. 59 (4), 385, and the Israel country report in the *Global Corruption Report 2006*.

suspected of exerting pressure on the investment centre to have the factory's request approved shortly after it secured the services of attorney Ori Masar, his friend and former business partner. According to the state comptroller's report,[19] Olmert, as trade minister and head of the body responsible for the factory's status, reportedly discussed the matter 'actively and intensively. . .taking decisions and instructing the professional team, which expressed many reservations about this factory'. A criminal investigation is likely to be opened against Olmert in the coming months.[20] The prime minister attacked the state comptroller's report, saying that it bore 'no resemblance to reality'.[21]

There are a number of possible explanations for this series of allegations. Some allege that, because of Likud's massive patronage base, the MMB effectively was not being supervised by the civil service commissioner or any other gatekeeper, other than the state comptroller. Observers have commented that Olmert's style has been to emphasise 'results' at the expense of means – and even appearances.[22] There are other possible explanations, however. All the cases were initiated in the office of the state comptroller, Micha Lindenstrauss, who has adopted a severe policy on corruption since his inauguration in June 2005.

Several journalists and jurists criticised Lindenstrauss soon after his inauguration. Avraham Tal, a *Ha'aretz* reporter who worked in the state comptroller's office from 1958 to 1972, lambasted Lindenstrauss for publishing the names of audited individuals and ascribing personal liability instead of remaining detached and impartial.[23] He compared investigations of individuals to those of the police,[24] while other reporters denounced Lindenstrauss's 'lust for publicity'.[25] This all came to a head as Lindenstrauss sought to present the Knesset with his report on the 2006 war in Lebanon: Olmert had refused to testify in person.[26] Published on 18 July 2007, the report criticised the government with regard to the state of the home front before and during the Second Lebanon War.[27] The upshot was a request by Olmert's lawyers on 14 May for the attorney general to open a criminal investigation against Lindenstrauss for 'neglecting the basic rules of fairness and integrity that apply to every civil servant'.[28] He refused and asked Olmert to present his criticism outside the legal arena.[29]

Throughout the Olmert administration, beginning in March 2006, many gatekeepers have had their personal lives investigated by the government. In November 2006 the police launched an investigation of Jacob Borovsky, the state comptroller's consultant on corruption, on the grounds that he had offered in 2004 to go easy on an inquiry into the affairs of the former prime minister, Ariel Sharon, if he was appointed

19 *Ha'aretz* (Israel), 26 April 2007.
20 Ibid.
21 *Ha'aretz* (Israel), 25 July 2007.
22 The prime minister's explanation for this neglect is that people will be suspicious no matter what he does; see *Ha'aretz* (Israel), 30 April 2003.
23 *Ha'aretz* (Israel), 12 May 2006.
24 *Ha'aretz* (Israel), 28 September 2006.
25 *Ha'aretz* (Israel), 19 July 2007; see www.Ha'aretz.com/hasen/spages/786216.html; and *Yediot Acharonot* (Israel), 17 March 2006.
26 *Ha'aretz* (Israel), 6 March 2007.
27 See www.Ha'aretz.com/hasen/spages/883857.html. The full report is published in Hebrew at www.mevaker.gov.il/serve/contentTree.asp?bookid=493&id=188&contentid=&parentcid=undefined&sw=1280&hw=954.
28 *Ha'aretz* (Israel), 15 May 2007; *Yediot Acharonot* (Israel), 15 May 2007.
29 For an analysis of the respective behaviours of Olmert and the attorney general, see *Ma'ariv* (Israel), 18 May 2007.

police commissioner.[30] There have allegedly been threats against the accountant general, Yaron Zelicha, a prominent figure in the fight against corruption,[31] and concerted efforts to present him as an 'egomaniac'.[32] Nava Ben-Or, a deputy state prosecutor in the Justice Ministry, was prevented from becoming a judge in a Jerusalem district court in May 2007 because she was said to be surrounded by 'a dark cloud of corruption'[33] – although she is noted for her integrity.

The Supreme Court president, Dorit Beinish, summed up these trends that same month in a speech at the Knesset to a delegation of foreign judges. 'We tell people from the outside that our democracy is so stable that we need not fear for the independence of the courts, and that it is hard to harm us,' she said. 'But let there be no misunderstanding: there is a danger.'[34]

Doron Navot (Hebrew University and the Israel Democracy Institute)

Further reading

D. Barak-Erez, 'Judicial Review of Politics: The Case of Israel', *Journal of Law and Society*, vol. 29, no. 4 (2002).

M. Hofnung, 'Fat Parties – Lean Candidates: Funding Israeli Internal Party Contests', in A. Arian and M. Shamir (eds.), *The Elections in Israel 2003* (London: Transaction, 2005).

G. Rahat, 'Candidate Selection in a Sea of Changes: Unsuccessfully Trying to Adapt?', in A. Asher and M. Shamir (eds.), *The Elections in Israel 1999* (Albany, NY, and Jerusalem: SUNY Press and the Israel Democracy Institute, 2002).

TI Israel: www.ti-israel.org.

30 See *Ha'aretz* (Israel), 17 January 2007. The attorney general closed the case against Borovsky after finding no evidence to support the allegations against him, although he was reprimanded for meeting with Likud politicians when he was a senior officer running for the post of police commissioner.

31 *Ha'aretz* (Israel), 16 January 2007.

32 *Ha'aretz* (Israel), 3 October 2005; *Yediot Acharonot* (Israel), 11 December 2006; *Ha'aretz* (Israel), 2 March 2007.

33 *Ha'aretz* (Israel), 27 May 2007.

34 *Ha'aretz* (Israel), 29 May 2007.

Latvia

Corruption Perceptions Index 2007: 4.8 (51st out of 180 countries)

Conventions

Council of Europe Civil Law Convention on Corruption (signed February 2004: ratified April 2005)
Council of Europe Criminal Law Convention on Corruption (signed January 1999: ratified February 2001)
UN Convention against Corruption (signed May 2005; ratified January 2006)
UN Convention against Transnational Organized Crime (signed December 2000; ratified December 2001)

Legal and institutional changes

- On 12 September 2006 the Cabinet adopted a **procedure of public procurement** on contracts priced between L1,000 (US$1,980) and L10,000. The new regulation is designed to increase accountability, because it eradicates the tendency to organise low-price contracts that do not require open tender.
- On 15 November 2006 the **parliamentary Corruption, Enforcement and Organised Crime Prevention and Combating Supervision Committee** (often referred to as the Anti-corruption Commission) **was abolished**. Corruption issues will now be handled by a committee that works with the Defence and Interior Ministries. The Anti-corruption Commission was established on 5 November 2002 when the ruling party was New Era, one of whose platforms was anti-corruption policy. The commission actively engaged in anti-corruption legislation and prioritised it as an issue on the public agenda. The abolition is possibly related to the election of October 2006, when New Era was forced into opposition. Before its abolition the commission was investigating two corruption scandals, notably cases of violating party financing rules and vote-buying in the Jurmala mayoral elections.[1]
- On 1 February 2007 the leading political parties, led by the People's Party (TP), agreed to **eliminate all ceilings on party spending during election campaigns** by amending the law on financing political organisations. TI Latvia criticised the move, saying the proposal would ensure the hegemony of certain wealthy parties and deepen their dependency on key finance providers. According to the Centre for Public Policy, PROVIDUS, the TP violated the existing limit of L500,000 (US$968,260) in the 2006 election.[2] Because of a Supreme Court ruling on 3 November 2006, however, it is likely that the Corruption Prevention and Combating Bureau (KNAB) will rule that the TP must return half a million

1 *Bloomberg News* (US), 25 September 2007.
2 TI Latvia, press release, 8 February 2007; see also www.providus.lv.

lats to the national budget. That would be impossible if the amendments were to take effect before the fine is ordered. The TP's proposal to eliminate party spending, therefore, should be seen as an attempt to legitimise the party's violations of campaign finance regulations during the 2006 election campaign. Moves to amend party financing rules have stopped and work has begun to examine the possibility of state funding (see below).

- In May 2007 parliament conceptually approved the **Law on Declaring Incomes and Property** in the first reading. Discussion on the need for such a law has been ongoing for several years, but the responsible institutions could find neither the most effective solution nor political support. Only civil servants are currently obliged to declare assets, but the new law will include all Latvians. The new law will facilitate anti-corruption work and also improve party financing transparency, because the declarations of income will allow more effective control of the source of political donations, based on the giver's average income over three years, as party financing law dictates.
- In May 2007 a working group led by KNAB developed a **draft proposal on the regulation of lobbying.** The draft offers three approaches: to develop a special law; to include basic principles of lobbying and its regulation in existing legislation; or to include the basic principles of transparency, equity and integrity of lobbying in state and municipal institutions' codes of ethics and normative acts. The draft was reviewed in the State Secretary Council and public hearings have started. There are still several problems to resolve with the new regulations, such as unequal access to public officials and public information, restricted information about lobbyists' activities, a lack of clarity of the legal requirements, and the distinction between the criminal offence of trading in influence and lobbying.
- On 1 January 2007 a new law **established the post of ombudsman with the goal of ensuring observance of human rights and good governance**. Parliament elected a former constitutional court judge, Romāns Apsītis, as Latvia's first ombudsman, in preference to the government candidate, Ringolds Balodis. Apsītis's first decision was to call for the retention of spending limits for political parties during election campaigns, arguing that the amendments proposed by the TP would weaken the country's parliamentary system.[3]

Changing party financing rules

In 2005 it was argued that amendments to the law regulating party financing had triggered a change of government.[4] The amendments restricted party financing to no more than L0.20 (US$0.39) per voter – or a total of around US$546,000 for a single party. The amendments were subsequently tested in the 2005 municipal elections and the 2006 parliamentary election. The first conclusions appear to show that the amendments effectively halted the 'arms race' in which political parties vied to outstrip each other's spending in election campaigns,[5] but new loopholes have appeared.

The experience of the last two elections revealed the problem of so-called third-party campaigning, which at that time was still unregulated. By introducing limits for party spending in

3 See also www.vcb.lv/index.php?open=viedoklis&this=270307.282.

4 See TI, *Global Corruption Report 2005* (London: Pluto Press, 2005).

5 According to PROVIDUS, campaigns by Latvian political parties are very expensive by EU standards. In the 2002 parliamentary election, parties spent around US$7.60 per resident voter, compared to US$3.90 in Austria, and US$2.00 in France and Sweden. See Soros Foundation and TI Latvia, 'Analyses of Expenditure and Revenue of Political Parties before the Elections of 8th Saeima' (Riga: Soros Foundation and TI Latvia, 2002).

elections, clear rules should also have been defined for third parties – organisations, other than political parties, that campaign for or against a party, alliance or candidate. A few third parties tested the controller's reactions before the 2005 municipal elections by launching short campaigns, but they elicited no response from KNAB.[6] Since it is politicians who determine financing regulations, few of them wished to tie their own hands by regulating spending. As a result, two 'front' NGOs that defined themselves as third-party groups were formed shortly before the 2006 parliamentary election. It was clear that they had direct relationships with their respective political parties.

The Society for Freedom of Speech, for example, was established by Jurģis Liepnieks, head of the Prime Minister's Office and creative leader of the TP campaign. The organisation launched a costly campaign, in which well-known people were invited to express their opinions about a specific minister or social issue. Liepnieks used the freedom of expression angle to answer criticism in the media of his links to the TP and violation of campaign spending limits. The organisation's largest donor was Andris Šķēle, former prime minister and founder of the TP, who gave L300,000 (US$581,000), dramatically exceeding the L10,000 limit on political donations in a single year. (The donation was noted in the Supreme Court decision.)[7]

The same was the case with the second NGO, *Pa Saulei* (Towards the Sun), founded by Ēriks Stendzenieks, creative leader of the First Party of Latvia's election campaign. According to the state register of enterprises, Stendzenieks is a post-holder in this NGO, but the campaign for the First Party was actually organised by the advertising agency Zoom, where he is also creative director. Other parties offered only token resistance, since, according to PROVIDUS, most spent within the legal limits and could not afford to buy media advertisements in bulk, as the two leading parties had.[8] The 2006 election has since been dubbed the 'cynical election'. The Supreme Court's response to an appeal by four parties, the New Democrats, Our Land, the Party for Social Justice and Fatherland's Union, concerning the legitimacy of the election results seemed to confirm that nickname.

The Supreme Court determined that the principle of free elections had not been violated and the election results were therefore legitimate.[9] Due to freedom of information legislation and the fact that violations of spending limits had been publicly discussed, voters had a chance to make up their own minds during the campaign, and therefore the principle of free elections had not been broken. The court decided that the two third-party organisations had to be considered as extensions of their associated parties, however, and their expenditure therefore had to be considered party expenditure. In response, the TP proposed the amendment to the law on political financing that would have removed spending limits in elections.

TI Latvia organised a campaign in which people were invited to write to their MPs and ask their opinion on the attempt to change the law. PROVIDUS mobilised public discussions on the issue. A survey discovered that 68.3 per cent of voters thought that election expenditures should be limited, while 8.9 per cent disagreed.[10] As a clear voice of Latvian society, the newly

6 *Diena* (Latvia), 25 July 2006.
7 The Republic of Latvia's Supreme Court Department of Administrative Cases judgment on 3 November 2006, no. SA-5/2006.
8 Ibid.
9 Ibid.
10 These figures come from a survey of Latvian public opinion towards campaign spending limits. The survey was undertaken in March 2007 by SKDS, a Latvian marketing and public opinion research centre.

elected ombudsman, Romāns Apsītis, called upon MPs to retain expenditure limits for election campaigns.[11]

In April 2007 a parliamentary working group was established to work on the Law on Political Financing. Discussion so far focuses on the level to which expenditure limits should be raised, with mention also of state funding and a ban on political advertising on television and radio. The deadline for proposals has been extended twice, and parliamentarians were to return to the matter in autumn 2007. Politicians have called for the introduction of the new rules as soon as possible. That means that Latvian parties will probably work with the new financing system in time for the municipal and European Parliament elections in 2009.

The sale of the Social Democrats

On 14 March 2007 a prime ministerial candidate for the Greens and Farmer's Association, Mayor Aivars Lembergs of the port city of Ventspils, was arrested and imprisoned for a period of investigation. Lembergs is one of Latvia's most powerful and influential businessmen, and his arrest was widely viewed as the first step in a war against the so-called 'oligarchs'.

In July 2006 the prosecutor's office charged Lembergs with bribery, money-laundering and abuse of entrusted power. The allegations focus on bribery between 1993 and 1995, when Lembergs allegedly accepted shares in the Swiss-registered firm Multi Nord AG. It was also alleged that Lembergs held a stake in Kalija Parks, which handles potassium exports, and used his political influence to make important decisions concerning the company. Kalija Parks' headquarters is in Ventspils.[12]

More unofficial information came to light during the investigation. Lembergs is believed to have sponsored up to three dozen politicians at a cost of 'hundreds of thousands of lats', according to the prime minister, Aigars Kalvitis.[13] Around sixteen MPs of various affiliations have been questioned so far, although the prosecutor's office has remained silent on the charges.[14] Punctuating this state of affairs was an explicit statement by the former president, Vaira Vike-Freiberga, that 'there is important evidence of the illegal financing of political parties'.[15]

In April 2007 the television programme *Kas notiek Latvijā* broadcast details of a document that indicated that Lembergs had made an agreement with the Latvian Social Democratic Workers Party (LSDSP) in 1999 that 'regulated' its options on such crucial issues as privatisation and the composition of the Cabinet. In return, Lembergs agreed to support the party both financially and in the pages of the daily he owned, *Neatkariga Riga Avize*.[16] According to Juris Bojars, former head of the LSDSP, Lembergs' funds were spent on advertising, including a campaign against the privatisation of the power utility Latvenergo.[17]

To justify his position, Bojars emphasised that 'the agreement actually featured 90 per cent of the party's programme including, for instance,

11 *Valsts Cilvēktiesību Birojs* (Latvian National Human Rights Office), 27 March 2007; see www.vcb.lv/default.php?open=jaunumi&this=270307.283.
12 *Baltic Times* (Latvia), 9 August 2006.
13 *Delfi* (Latvia), 11 July 2007.
14 *Diena* (Latvia), 3 July 2007.
15 *LETA* (Latvia), 5 July 2007.
16 *Baltic Times* (Latvia), 18 April 2007.
17 Bojars admitted this in a television interview on 12 April 2007, after the agreement between the LSDSP and officials from Ventspils municipality was published.

such points as delaying the privatisation of large, state-owned companies'.[18] According to the agreement, the LSDSP also promised not to participate in government with the People's Party. The amount of money the LSDSP received from Lembergs has not yet been disclosed, though the list of MPs he sponsored is expected to be publicised. Depending on who is on the list and their current positions in government, those revelations – if they come out – could trigger a profound constitutional crisis.

Līga Stafecka and Zanda Garanca (TI Latvia)

18 *LETA* (Latvia), 13 April 2007.

Further reading

V. Kalninš, 'Parliamentary Lobbying between Civil Rights and Corruption: An Insight into Lobbying Practice in Latvia and Recommendations for the Saeima' (Riga: Centre for Public Policy PROVIDUS, 2005).

(ed.), 'Corruption °C: Report on Corruption and Anti-corruption Policy in Latvia' (Riga: Centre for Public Policy PROVIDUS, 2005 and 2006).

R. Karklins, *The System Made Me Do It: Corruption in Post-Soviet Societies* (Armonk, NY: M. E. Sharpe, 2005).

J. Rozenvalds, 'How Democratic Is Latvia? Audit of Democracy' (Riga: University of Latvia, 2005).

TI Latvia: www.delna.lv.

Montenegro

Corruption Perceptions Index 2007: 3.3 (84th out of 180 countries)

Conventions

Council of Europe Civil Law Convention on Corruption (signed April 2005; not yet ratified)
Council of Europe Criminal Law Convention on Corruption (accession December 2002)
UN Convention against Corruption (succession October 2006)
UN Convention against Transnational Organized Crime (succession October 2006)

Legal and institutional changes

- **Changes to the Criminal Code and the Code on Criminal Procedures that increase penalties for corruption and the misuse of power were adopted** on 25 July 2006, but parliament turned down a request that the police be allowed to use undercover surveillance in cases of suspected corruption. It justified the latter decision by referring to the lack of public trust in the police force, while the opposition warned that such powers could be misused for political purposes. The Prosecutor's Office argued that the existing legislation allowed the police to use secret surveillance in organised crime cases, but that corruption crimes, notably bribery, were hard to prove without similar powers.

- On 31 July 2006 **MPs refused to adopt the Conflict of Interest Bill**, which was developed with the support of the Council of Europe. The majority of MPs, government and opposition alike, voted against the measure, in an extraordinary example of unanimity in Montenegrin politics. In its 2006 progress report,[1] the European Commission noted the 'influence of organised crime in certain spheres of economic and social life' and 'the lack of an appropriate legal framework to deal with the conflict of interests of officials'. The existing law does not provide an adequate definition of conflict of interest, nor lay down criteria on how to recognise it, allowing public officers, for example, to continue to serve as directors of major companies.
- On 24 August 2006 **the government adopted the Action Plan for the Fight against Corruption and Organised Crime**, establishing a commission five months later to monitor implementation. Membership is weighted in favour of the government, with four Cabinet members, four senior officials from the police and justice sector, two MPs and a single representative of civil society. As eight of its eleven members are directly or indirectly responsible for implementing the plan, some may question the commission's impartiality. The commission meets four times a year, raising concerns about its own ability to perform, and has limited resources at its disposal. This is far from answering the Council of Europe's original recommendation that the government create a specialised, independent, anti-corruption body to monitor implementation of the action plan.[2]
- In January 2007 parliament adopted the **Law on Responsibilities of Legal Entities for Criminal Acts**, which introduces criminal responsibility for legal entities, such as companies, and defines criminal acts, penalties and procedures for their enforcement. Until the adoption of the law, legal entities that had perpetrated criminal acts suffered no consequences for their actions, nor did procedures exist for the restitution of property obtained through criminal acts. The new law prescribes the circumstances in which legal entities can be sanctioned, ranging from financial penalties to the closure of a company. If enforced, it could become a powerful weapon in the fight against financial crime and bogus companies.
- Despite **surveys showing high levels of perception of corruption in the judiciary, not a single citizen reported an incident** from September 2006 to May 2007, according to official sources.[3] This could indicate higher than expected trust in the judiciary, but is more likely to reflect public fears of reporting corruption to the authorities. In July 2007 the supreme state prosecutor[4] identified the inefficiency of the judiciary as a key obstacle to the timely processing of crimes, while the ombudsman said the judiciary frequently violated the constitutional right to a trial within a reasonable time frame.[5]

1 European Commission, 'Montenegro 2006 Progress Report', Commission Staff Working Document, SEC (2006) 1388 (Brussels: European Commission, 2006).

2 Ibid.; for TI's assessment of the progress of the action plan by July 2007, see TI Montenegro, 'Action Plan to Fight Corruption and Organised Crime in Montenegro', press release, 19 July 2007.

3 Government of Montenegro, 'First Report on the Realisation of Measures from the Action Plan for Implementation of the Programme for the Fight against Corruption and Organised Crime' (Podgorica: Government of Montenegro, 2007).

4 'Annual Report of the Supreme State Prosecutor', adopted by Parliament of Montenegro in July 2007.

5 Ombudsman Office, 'Annual Report 2006' (Podgorica: Government of Montenegro, 2007); Ombudsman Office, 'Special Report on Violation of Right to Trial within Reasonable Timeframe by Montenegrin Judiciary' (Podgorica: Government of Montenegro, 2006).

Scramble for the Adriatic

Montenegro, the newest member of the UN, became independent after a referendum in mid-2006 in which 55.5 per cent of the 86.4 per cent of registered voters who voted supported independence from Serbia. There has been no change of government since the end of communism in 1991 and politics has become Montenegro's most profitable business.

A number of EU countries have accused Montenegro's elite of links with organised crime.[6] Italian and German investigators have established links between cigarette-smuggling and money-laundering networks and prominent Montenegrins, including the former prime minister, Milo Đjukanovic.[7] The news agency ANSA recently reported that prosecutors from Puglia, Italy, had discovered €500 million (US$675 million) in a Cypriot bank account, which was subsequently invested in Montenegro.[8]

Privatisation and construction have attracted large investments, particularly from Russia and offshore companies. Russian oligarchs have bought homes along the Adriatic and state inspectors have been prevented from entering their properties.[9] President Vladimir Putin recently valued Russian investments in the state at US$2 billion,[10] although less than US$200 million was invested through official channels.[11]

Construction has expanded vastly in the past few years, although buildings are often constructed without licences or inspection certificates. In a country whose future lies in tourism, and probably tourism alone, unsightly construction will dramatically affect tourist inflows. Meanwhile, the United States reports that proceeds from narcotics trafficking have been laundered through the real estate boom.[12]

Exact data are not officially published, but the Ministry of Environmental Protection and Urban Planning estimates that around 80 per cent of new buildings have been erected illegally over the past decade.[13] The showcase of illegal construction is the Hotel Splendid in Becici, which imposes a huge strain on an infrastructure already buckling from the plethora of new villas and apartments. It was built without a permit or inspections approval, yet was enthusiastically promoted by the government.[14]

The Hotel Splendid was initially sold to the Russian–Montenegrin joint venture Montenegro Stars Hotels Group, for €2.4 million (US$3.24

6 European Commission, 'Report on the Preparedness of Serbia and Montenegro to Negotiate a Stabilisation and Association Agreement with the European Union' (Brussels: European Commission, 2005).

7 Italian prosecutors also accused the former deputy prime minister, Miroslav Ivanisevic, the former head of Montenegro's trade mission in Italy, Dusanka Pesic-Jeknic, and the Montenegrin tycoons Veselin Barovic, Branko Vujosevic, Branislav Micunovic and Stanko Cane Subotic, as well as several Italian citizens who have already been arrested. See *Balkan Investigative Reporting Network* (Bosnia), 6 June 2007.

8 *Daily Vijesti* (Montenegro), 26 June 2007.

9 Building inspectors claim they were stopped from entering the property of a company reportedly connected to Yuri Luzhkov, the mayor of Moscow, and twice from the Hotel Splendid site. Further information is available from www.pravodaznam.info/publikacija/EN/6-EN.pdf.

10 *New York Times* (US), 24 December 2006.

11 *Daily Vijesti* (Montenegro), 29 August 2006. See also Government of Montenegro, 'Strategy for Promotion of Foreign Investments' (Podgorica: Government of Montenegro, 2005).

12 Bureau of International Narcotics and Law Enforcement Affairs, 'International Narcotics Control Strategy Report 2007' (Washington, DC: US Department of State, 2007).

13 Republic of Montenegro, Ministry of Environmental Protection and Urban Planning, 'Ministry perspective', presentation at ministerial conference on informal settlement in south-eastern Europe, Podgorica, September 2004.

14 See www.ens-newswire.com/ens/jun2006/2006-06-28-05.asp.

million) in 2004. According to media reports, the company's majority owner is Viktor Ivanjenko, former director of the Russian oil giant Yukos, who entered the hospitality business through Rašan Investors, a company registered in Switzerland.[15]

On the other side of Europe, the Montenegro Stars Hotels Group came under police investigation for allegedly laundering Yukos's money through projects in Spain,[16] where the police established a direct link between organised crime and the construction boom.

During 2005 three bombs were detonated at the Hotel Splendid site within several days, and the inspector in charge of the case was later murdered in front of his house. According to some analysts, the bombings were an extension of a war either between rival construction lobbies or money-laundering rackets.[17]

Transparency in privatisation

Officially, most public enterprises have been privatised through transparent tender processes, with the government seeking strategic partners with related experience. In practice, the obligations defined in the contracts are rarely fulfilled and the strategic partner often turns out to be an offshore company.

Ever since the law on access to information came into effect in late 2005 the authorities have shown a complete unwillingness to inform the public about the terms of privatisations. While 85 per cent of state-owned property has been privatised, only one contract – on Aluminium Plant Podgorica (KAP) – has ever been published, and that was only in response to a strong media campaign.[18]

The KAP privatisation showed how non-transparent processes and a lack of public accountability create conditions for strengthening the informal centres of power. Prior to privatisation, KAP was the country's largest economic entity, accounting for 10 per cent of employment, around 20 per cent of GDP and 40 per cent of exports. It is also the country's largest polluter of water, air and soil. KAP is located in the vicinity of Skadar Lake National Park, one of Europe's most precious wading bird habitats.[19]

Despite announcing a competitive public tender for its share of the plant, the government sold to a British offshore company, Eagle Capital Group, on 27 July 2005. Eagle's proprietor is a Cypriot offshore company, Salomon Enterprises, which allegedly owns Russia's Russal, the world's third largest aluminium producer. After signing, Eagle changed its name to En Plus Group and moved from the British Virgin Islands to Jersey. The new owners reportedly paid €48.5 million (US$65.5 million) for the state's share in KAP, a further €55 million (US$74.25 million) in assorted projects and €20 million (US$27 million) on environmental interventions. The conditions of the agreement remained secret until twenty months later, when, following a number of trials, they were finally published.[20]

KAP is Montenegro's largest consumer of electricity, so the price of energy was a key issue

15 *Monitor* (Montenegro), 29 July 2005.

16 *BBC News* (UK), 13 March 2007.

17 See *International Herald Tribune* (US), 29 November 2006, and *Daily Vijesti* (Montenegro), 5 October 2005 and 31 August 2005.

18 See iwpr.net/?p=bcr&s=f&o=242153&apc_state=henibcr2005.

19 T. Salathé, 'Mission Report, Skadarsko Jezero, Serbia and Montenegro', Ramsar Advisory Mission no. 56 (Gland, Switzerland: Ramsar Convention Secretariat, 2005).

20 For further information, see V. Calovic *et al.*, 'Free Access to Privatisation Information in Montenegro: Behind the Closed Doors, Case Study Aluminium Plant Podgorica' (Podgorica: MANS, 2007).

during negotiations.[21] The current minister of economy, Branimir Gvozdenovic, was vice prime minister for economy, president of the board of the National Power Supply Company (EPCG) and president of the tender committee at the time of the privatisation. Both EPCG and KAP hired consultancy firms, which provided wildly varying estimates of future electricity prices. In his capacity as president of the tender commission, Gvozdenovic selected the lowest available estimate; proposed it to himself – and accepted it – as president of the board of the EPCG; proposed it to himself again as vice prime minister; and finally signed the privatisation contract.[22]

Subsidised electricity for KAP's new owners, together with increased demand and high oil costs, forced up citizens' bills by over 70 per cent in January 2007 and 40 per cent in February. This created huge difficulties in a country where an eighth of the population is extremely poor and a third is economically vulnerable.[23] The government is now trying to use this 'electricity shock' to win support for the construction of twelve hydro and four thermo plants on the river Tara under the National Spatial Plan. Three investors have so far expressed interest in building new energy sources, all of whom are under investigation. Acussations range from involvement in organised crime to serious fraud and corruption, as follows.

The Russian oligarch and KAP's new owner, Oleg Deripaska, expressed interest in privatising existing plants and building new ones. Deripaska had his visa to travel to the United States cancelled last year because of Federal Bureau of Investigation suspicions of his links to Russian organised crime.[24] The Energy Financing Team (EFT), which supplies 70 per cent of Montenegro's imported energy, is being investigated by the United Kingdom's Serious Fraud Office, USAID, the Bosnian Special Department for Organised Crime, Economic Crimes and Corruption, and the UN High Representative in Bosnia. One of EFT's founding members is a former adviser to President Filip Vujanovic.[25] Montenegrin tycoon Vesko Barovic, who owns shares and is on the board of the National Electricity Company, has also shown interest. Like the former prime minister, Milo Đjukanovic, Barovic is wanted by the Italian prosecutor for his connections to organised crime and cigarette-smuggling.[26]

Vanja Calovic (the Network for Affirmation of the NGO Sector – MANS)

Further reading

V. Calovic, 'Corruption in the Construction Industry' (Podgorica: MANS, 2004).
'Squeezing the Poor: EBRD's role in Privatisation in Montenegro', *BankWatch Mail*, no. 32 (2007).
V. Calovic and M. Deletic, 'Right to Know in Montenegro: Experiences in the Application of the Law on Free Access to Information' (Podgorica: MANS, 2006).
'Free Access to Privatisation Information in Montenegro: Behind the Closed Doors, Case Study Aluminium Plant Podgorica' (Podgorica: MANS, 2007).

21 Two Russian companies, Russal and Sual, were involved in the final phase. Russal owned about 30 per cent of Sual's shares and subsequently the two companies merged. Sual later withdrew because the electricity price was too high, leaving Russal as the only bidder.

22 *Daily Vijesti* (Montenegro), *Pobjeda* (Montenegro) and *Dan* (Montenegro), 13 April 2005; *Daily Vijesti* and *Pobjeda* (Montenegro), 14 April 2005; *Daily Republika*, 25 April 2005.

23 See www2.undp.org.yu/montenegro/home/poverty.html and www2.undp.org.yu/montenegro/home/poverty/Data%20sources.pdf.

24 *New York Times* (US), 20 August 2006; *Financial Times* (UK), 16 December 2006 and 11 May 2007; *Wall Street Journal* (US), 11 May 2007.

25 *Guardian* (UK), 26 February 2005 and 23 July 2005.

26 Italian News Agency, 22 June 2007; *Daily Vijesti*, 23 June 2007.

D. Milovac, 'Eyes Wide Shut, or How the State Sanctions Illegal Construction' (Podgorica: MANS, 2007).
Z. Radulovic, 'Corruption in Montenegro', Global Integrity Index, 2007; available at www.globalintegrity.org.
World Bank, 'Republic of Montenegro: Public Expenditure and Institutional Review', Report no. 36533 (Washington, DC: World Bank, 2006).
MANS (the Network for Affirmation of the NGO Sector): www.mans.cg.yu.

Romania

Corruption Perceptions Index 2007: 3.7 (69th out of 180 countries)

Conventions

Council of Europe Civil Law Convention on Corruption (signed November 1999; ratified April 2002)
Council of Europe Criminal Law Convention on Corruption (signed January 1999; ratified July 2002)
UN Convention against Corruption (signed December 2003; ratified November 2004)
UN Convention against Transnational Organized Crime (signed December 2000; ratified December 2002)

Legal and institutional changes

- In 2006 there were two important **modifications to the Law on Free Access to Information of Public Interest.**[1] The law's application has been extended beyond the authorities that administer public finances to include companies under government ownership, or in which the government holds a majority stake. Other modifications exempt information about commercial and financial activities that could damage fair competition or endanger intellectual property.[2] Changes to the law also make access to information more explicit,[3] stating that contracting agents must provide public procurement contracts to interested parties, rolling back the practice of adding confidentiality clauses to such contracts.
- In July 2006 **parliament modified the Penal Code so as to criminalise conflicts of interest** (see below).[4] This raises serious problems of application, especially in providing evidence of intent (as formulated, the legal text requires

1 Law no. 544/2001.
2 Law no. 371/2006, modifies article 12, paragraph (1), sect. c. of Law no. 544/2001.
3 Law no. 380/2006.
4 Law no. 278/2006.

prosecutors to prove that the person under investigation *knew* he or she was in a conflict of interest and actively participated in making a decision that brought him or her benefits). The text also differs from administrative law,[5] which defines conflicts of interest more narrowly. The new penal law does not automatically annul decisions performed in a conflict of interest. Confusingly, both administrative and penal regulations are simultaneously applicable, so a civil servant could be sanctioned for a conflict of interest under criminal law, but the victim of an administrative decision issued under conflict of interest would be forced to file a civil suit to obtain relief.

- The Penal Code modifications[6] also **introduced provisions allowing for the penal responsibility of legal persons**. With the exception of the state, public authorities and public institutions in domains outside private sector activity, the new law provides that all legal entities, such as companies, trade unions and foundations, are considered criminally responsible. This provision applies only to the corruption infractions of bribe-giving and trafficking in influence; all other offences require a physical person to play an active role in criminal transactions. The adoption of the measure falls within a series of steps taken towards transposing into internal legislation the Council of Europe's Criminal Law Convention on Corruption and was particularly requested in the Group of States against Corruption Second Evaluation Report on Romania, issued in October 2005.[7]
- The new Law on Political Party and Campaign Financing, adopted in July 2006,[8] **enlarges political parties' obligations to declare income and expenses**, stipulating that contributions from members and other sources must be published in the official gazette. The format for reporting expenses will also be more strict, owing to a new definition of 'propaganda materials' that includes the cost of written, video or audio materials. Another positive development is a clearer system for donations, inheritances and campaign contributions. Limits are unchanged, but it is more difficult to exceed them and price deductions on goods or services are now considered as donations. More troubling was the government's decision to delay the law's application. In January 2007 the government postponed several of the provisions until July 2007.[9]
- The work of **the General Anti-Corruption Department** (DGA), an investigation unit in the Ministry of Administration and Internal Affairs, **was undermined in March 2007 when its director resigned in response to an unlawful performance review by a ministerial body**. Legislation governing the DGA requires an independent performance review at the minister's request. The DGA director is a magistrate, however, meaning that reviews belong in the jurisdiction of the Superior Council of the Magistracy (CSM). The ministry instead subjected the director to its own standing review committee, including some representatives actually under investigation by the DGA, constituting a clear conflict of interest.
- Despite the Anti-Corruption Department's (DNA's) intensive activity, **the justice system has not yet produced convictions in cases of high-level corruption**. More worryingly, recent practice has been to grant a large number of suspended sentences (i.e. with no prison time) in grand corruption trials, diluting the sanction to a simple mention on the individual's criminal record. In the absence of

5 Law no. 161/2003.
6 Law no. 278/2006.
7 Greco, 'Second Evaluation Round: Evaluation Report on Romania', adopted at the twenty-fifth Plenary Meeting, Strasbourg, October 2005.
8 Law no. 334/2006.
9 Emergency Ordinance, 1/2007.

decisive action by the judiciary, grand corruption cases are dealt with in the press rather than the court of law.[10]

The fight over the National Integrity Agency

May 2007 finally saw the passage of a long-sought law to establish an independent anti-corruption agency. The National Integrity Agency (ANI) is designed to remedy shortcomings in the monitoring of conflicts of interest and public officials' assets. The law establishing the agency[11] followed a series of drafts: one was written by TI Romania in 2004; a second by the minister of justice in June 2006; and a third was a heavily amended version of the second. The fourth and final version was adopted by the Senate in May 2007.

All four envisioned an institution that would verify asset declarations, and monitor unexplained wealth and possible conflicts of interest.[12] All four provided for a three-tiered structure with a representative council, a management body and a body of inspectors to perform controls. All concurred that the submission of a false declaration of wealth or making false statements would be considered an act of forgery. Another point of convergence was that penalties for illicit enrichment, conflict of interest and incompatibilities were beyond the new agency's competence, so files would be forwarded to the Prosecutor's Office, disciplinary commissions or fiscal authorities. The ANI can impose fines only for failure to submit documents or for overstepping deadlines for submitting declarations.

The system previously in place was seriously fragmented, assigning wealth and conflict of interest control to separate institutions with little capacity for collaboration. This fragmentation prevented any unitary legal approach to corruption prevention. Further inefficiencies derived from the wealth control commission's lack of diligence and the absence of mechanisms to certify that declarations had been submitted. In addition, because conflict of interest complaints were assigned to authorities within the public institutions, there were no guarantees of impartiality or insulation from undue influence.

The law establishing ANI was adopted in a context of mounting pressure both at home and internationally. In 2004 a draft law by TI Romania was sent to parliament and passed the lower house, although the Senate delayed discussion for over a year. With the Second National Anti-Corruption Strategy (2005–7) the deficiencies in corruption prevention were clearly visible, and a proposal was put forth for the creation of 'a single independent body tasked with verifying asset and interest declarations, as well as incompatibility situations'.[13] These domestic efforts were mirrored in pressure from the European Commission.

The adoption of the law establishing the ANI was no easy task. What particularly inflamed public debate were the radical modifications brought to the Ministry of Justice's draft by the Chamber of Deputies. Between 14 August and 11 October 2006 the chamber's legal commission returned with more than ninety-two separate modifications, which effectively left the ANI a highly dependent body with fewer powers. These modifications outraged the ministry and domestic NGOs, and increased the vigour of the debate. In response, TI Romania

10 'TI Romania National Corruption Report 2007, Romania' (Bucharest: TI Romania, 2007).

11 Law no. 144/2007 on the establishment, organisation and function of the National Integrity Agency, or Agenţia Naţională de Integritate in Romanian.

12 Except for the version that resulted from debates within the Chamber of Deputies, which eliminated most of the agency's powers of investigation.

13 Annex 1 to the government decision 231/2005 on the approval of the National Anti-Corruption Strategy for 2005–7.

submitted a second document, 'Basic Principles for an Anti-corruption Public Policy Dedicated to the National Integrity Agency', which won support from civil society organisations.

The principles became the object of intense advocacy. TI Romania had proposed enlarging and improving the legal definition of conflict of interests, achieving a unitary regulatory framework for incompatibilities, and focusing wealth control on assets obtained during the occupation of public office only. It recommended that the ANI have operational independence, access to all public databases, a mandatory character for its decisions (which can, however, be appealed) and the power of dismissal of those in conflict of interest or incompatibility situations.

Applying the current legislation may be problematic. Having administrative jurisdiction, the institution may consider only conflicts of interests as defined by administrative law, which refers to benefits for oneself and immediate relatives solely of a material nature. This ignores non-material benefits and intermediaries. Criminal law contains a much wider definition, meaning that the ANI can effectively do little to combat conflicts of interest despite its mission. Rather, it will be forced to forward findings to the Prosecutor's Office.

The risk of insufficient human or financial resources may also be a problem. The law provides for a maximum of 200 employees and a central office in Bucharest. These employees face the enormous task of checking the wealth and interest declarations of virtually all persons occupying positions in the public sector. Procedures for overcoming capacity constraints are lengthy and beyond the control of the ANI's management.

Anti-corruption agencies can easily become political weapons in the hands of those in power if not sufficiently insulated from pressure. Senate oversight may still allow influence over appointments and dismissals of agency management, which is unsettling because of the political class's inconsistent attitude towards the ANI. It is important to remember that the agency's belated creation was intimately connected to EU pressure, so the degree of genuine political support is difficult to ascertain. The instability of Romania's anti-corruption legislation and inconsistencies in its legal texts will negatively impact the ANI's performance.

Parliamentary disregard for standards of legislative technique make anti-corruption measures vulnerable to abusive interpretation. The law establishing the ANI seems no exception to this: on 30 May 2007, less than one month after its adoption, the government passed an emergency ordinance lowering the financial threshold for wealth control procedures.[14] Although positive in itself, it would have been preferable to have included it in the original defining text for legal clarity.

The law establishing the ANI is one of the most important pieces of anti-corruption policy in Romania – and one of the most thoroughly debated. In the one to two years after the adoption of the law the ANI must demonstrate important successes if it is to make an impact. The chances of such success should be increased by connecting the institution to other preventive instruments, such as public awareness campaigns, anti-corruption education and whistleblower protection, eventually leading to more coherent corruption prevention.

14 The ordinance came after TI Romania expressed criticism regarding the excessively high 'obvious difference' between actual and declared wealth, which can justify the commencement of control procedures.

The Superior Council of the Magistracy's enduring deficiencies

Reform of the judiciary has been a priority since 1990 (see *Global Corruption Report 2005* and *Global Corruption Report 2007*). The prolonged negotiations for accession to the European Union were a powerful impetus for reform and stressed the independence of the judiciary as a central theme. In 2004 an overhaul of the judiciary was initiated through a package of three laws[15] that empowered the Superior Council of the Magistracy as the official representative of the judiciary in its relations with other state authorities and the guarantor of its independence. The CSM consists of nine judges and five prosecutors, elected by their peers, and by law includes the minister of justice, the Supreme Court president, the general prosecutor and two civil society representatives. A number of sensitive issues, such as the appointment of magistrates, career development and disciplinary action, are placed exclusively in the CSM's competence. Three years after passing the three-package law, the CSM continues to be the target of criticism over its efficiency, credibility and integrity. It is illustrative that, of the four benchmarks instituted by the European Commission in September 2006, one explicitly targets the CSM: 'Ensure a more transparent and efficient judicial process notably by enhancing the capacity and accountability of the Superior Council of the Magistracy.'[16]

The CSM made some progress towards implementing key measures within the official reform strategy for the judiciary during the period under review.[17] It increased its administrative capacity, completed and ran new procedures for the promotion, relocation and transfer of magistrates and set up mechanisms to ensure uniform jurisprudence throughout the court system (i.e. a mechanism of periodic consultation among judges and the so-called 'appeal in the interest of law').[18]

Outstanding problems persist regarding CSM's performance as a disciplinary body, however. This is particularly problematic as the judiciary continues to be perceived as one of Romania's most corrupt institutions.[19] In the course of 2006 the Disciplinary Commission received 231 complaints, mostly from litigants, of which 193 were dismissed.[20] In the absence of decisive action by the CSM, the press and civil society have assumed a key role in monitoring the state of the justice system and the performance of magistrates. In response, judges and prosecutors perceive the press as the major factor of pressure on the judiciary.[21]

The CSM also has serious flaws in its integrity standards. The legal framework requires CSM

15 Law no. 303/2004 on the status of judges and prosecutors; Law no. 304/2004 on the organisation of the judiciary; and Law no. 317/2004 on the organisation and functioning of the Superior Council of the Magistracy.

16 European Commission, 'Monitoring Report on the State of Preparedness for EU membership of Bulgaria and Romania', Communication from the Commission (Brussels: European Commission, 2006).

17 Government decision no. 232/2005 for the approval of the reform strategy for the judiciary for 2005–7 and the action plan for the implementation of the reform strategy for the judiciary for 2005–7.

18 A mechanism by which courts of appeal or the general prosecutor can introduce certain cases to the Supreme Court, whose decisions then become obligatory for all courts and can be modified only by law.

19 Data from the Global Corruption Barometer show that the justice system has been ranked as the second most corrupt institution in Romania since 2003, surpassed only by political parties, customs or, alternatively, parliament.

20 According to a report on the activity of the CSM in 2006, published 19 March 2007; for further details, see www.csm1909.ro/csm/linkuri/19_03_2007__9024_ro.doc.

21 According to the 2006 'Study Regarding the Perception of Magistrates on the Independence of the Judiciary', produced by TI Romania at the CSM's request, 50.6 per cent of magistrates consider the press the most important factor for change, compared to 7.6 per cent for the executive and 7.0 per cent for the legislature.

members to be suspended from positions in courts or prosecutors' offices. At the end of 2006 five of fourteen elected members faced potential conflicts of interest as inspectors, since they also held leading positions (albeit suspended) in the judicial system. This not only raised serious ethical issues, it created a capacity deficit.

These conspicuous flaws, coupled with the limited impact of reforms on the judiciary, have further weakened the credibility of the magistracy. According to a TI Romania report, in 2006 only 43 per cent of magistrates thought that the CSM had the ability to guarantee their independence, compared to the 60 per cent who responded the same in 2005. The satisfaction of magistrates with the CSM has also decreased, with only 51 per cent saying that they were satisfied with the institution, compared to 61 per cent a year earlier.[22]

Iulia Cospanaru, Matthew Loftis and Andreea Nastase (TI Romania)

Further reading

V. Alistar (coordinator), I. Coşpănaru *et al.* (authors), 'National Corruption Report 2007' (Bucharest: TI Romania, 2007).

Government of Romania, 'The National Anti-Corruption Strategy 2005–07' (Bucharest: Government of Romania, 2005).

C. Walker (ed.), 'The Anti-corruption Policy of the Romanian Government: Assessment Report' (Washington, DC: Freedom House, 2005).

World Bank, 'Diagnostic Surveys of Corruption in Romania' (Washington, DC: World Bank, 2001).

O. Zabava and C. Vrabie (coordinators), A. Savin and R. Malureanu (authors), 'National Integrity Systems TI Country Study Report of Romania 2005' (Bucharest: TI Romania, 2005).

TI Romania: www.transparency.ro.

22 Ibid.

Slovakia

Corruption Perceptions Index 2007: 4.9 (49th out of 180 countries)

Conventions

Council of Europe Civil Law Convention on Corruption (signed June 2000; ratified May 2003)
Council of Europe Criminal Law Convention on Corruption (signed January 1999; ratified June 2000)
OECD Convention on Combating Bribery of Foreign Public Officials (signed December 1997; ratified September 1999)
UN Convention against Corruption (signed December 2003; ratified June 2006)
UN Convention against Transnational Organized Crime (signed December 2000; ratified December 2003)

Legal and institutional changes

- In November 2006 *Daily SME* published a table showing the political division of public sector jobs in Banská Bystrica, where the **ruling coalition allocated almost 300 posts in state administration to politically affiliated people.**[1] There is a high probability that similar arrangements exist in other regions. The selection of staff is conducted with minimum criteria, jeopardising accountability. One example was the appointment of the daughter of deputy Jozef Ďuračka, deputy chairman of the Committee on Finance, Budget and Currency, to the supervisory board of the Trnava District Heating Plant, even though she had barely completed her university studies.[2]
- An **amendment to the Act on State Service**, adopted in December 2006, increased the potential for political nominations in public agencies, including the Public Procurement Office, the Anti-Monopoly Office and the Statistics Office. The amendment strengthened the government's ability to dismiss the directors of those agencies without showing just cause, and to bypass the existing weak accountability mechanisms.[3]
- Three years after the law on the protection of public interest came into force in May 2004 (see *Global Corruption Report 2005*), **the government has still not introduced more transparent accountability measures.** When in opposition SMER-SD, the strongest party in the coalition, sharply criticised the poor application of the law. Property declarations are carefully buried on the parliamentary website, and require time and expertise to unearth. The Committee on Incompatibility of Functions, a political body

1 *Daily SME* (Slovakia), 20 November 2006.
2 *Pravda* (Slovakia), 6 November 2006.
3 *Daily SME* (Slovakia), 25 November 2006.

made up of coalition and opposition deputies, is supposed by law to shed light on conflicts of interest and asset declarations. Nevertheless, it approved the behaviour of the government's hardman, the justice minister, Štefan Harabin, whose wife used a ministerial car for personal needs. The parliamentary committee imposed no sanction on the minister.[4]

- In an interview with the weekly *Trend* magazine in March 2007, **the minister of economy, L'ubomír Jahnátek, advocated the use of bribery to win arms contracts** if the state hoped to compete with private weapons dealers. 'These are non-traditional forms of business,' he said, 'and we can't afford to close our eyes to them.' The former prime minister, Mikuláš Dzurinda, said the comments 'confirmed that Jahnátek abetted, instigated and endorsed the use of black money and corruption in the arms trade.'[5]
- A year after winning power in mid-2006 **the government has still not clarified its anti-corruption strategy**. In April 2007 the prime minister, Róbert Fico, charged the Interior Ministry with responsibility for submitting legislative proposals concerning the fight against corruption. The Anti-Corruption Department, which worked under the Office of the Government during the last two electoral terms, was incorporated in the Section of Control and Fight against Corruption in May 2007. The current strategy is based on two pillars: not to privatise and to allow the law enforcement agencies to fulfil their duties. Both are crumbling, however. First, without reform the public sector will become ever more vulnerable to corruption, while the Justice Ministry appears bent on destroying the Special Court (see below), an important source of impartial justice in Slovakia.
- In August 2006 **the deputy prime minister and minister of justice, Štefan Harabin, took steps to dissolve the Special Court and Special Prosecutor's Office** established during Dzurinda's second term (see *Global Corruption Report 2005*). With jurisdiction across the entire country, the two agencies were designed to short-circuit the powerful family connections that link citizens, entrepreneurs and administrators at a local level, and prevent corrupt activities from coming to light in local courts (see below).
- In May **the Justice Ministry prepared a small amendment to the Civil Judicious Order to speed up court proceedings**. Henceforth, the appellate court will consider only the appeal itself: all administrative work will be done in the first-degree court. Previously, the appellate court had dealt with time-consuming administrative matters, which prolonged court processes and made it the target of attempted corruption and obstruction.

Turf war weakens Special Court's credibility

Corruption has long been widespread in the public sector and judiciary. The previous government, which made several reforms in this area, managed at least to decrease the level of perceived corruption. In the area of law enforcement, for example, perceived corruption fell from 59 to 47 per cent from 1999 to 2006, according to TI Slovakia surveys.[6]

What did Slovakia do to achieve this result? One important step was to establish a Special Court and Special Prosecutor's Office (see *Global Corruption Report 2006*) insulated from the conventional – and compromised – judicial and investigative authorities. More effective prosecution of corruption requires specialisation and

4 *Pravda* (Slovakia), 19 March 2007.
5 *Daily SME* (Slovakia), 14 April 2007.
6 FOCUS research agency, for TI Slovakia, March 2006, May 2004, March 2002 and October 1999.

better coordination. For this reason, a Special Court was established in Pezinok with national jurisdiction, and powers to hear and decide cases specified in the Criminal Procedure Code. The Special Prosecutor's Office has similar jurisdiction and powers.

The establishment of specialised institutions was driven by pragmatic concerns. First, corruption is a continuing phenomenon and is particularly deep-rooted in the judiciary. Second, in small, provincial communities, relations between officials, citizens and entrepreneurs are tight-knit, placing family loyalty above alien notions of integrity. The 'bumper principle', whereby local power-holders give preference to family and friends, creates a solid foundation for petty corruption. Local courts are simply an extension of this network. With national jurisdiction, the Special Court and Special Prosecutor's Office benefited from disconnectedness from parochial affairs.

The two institutions have been functional for only two years. Nevertheless, the Special Court handed down some significant sentences in highly publicised cases under Chief Justice Igor Králik, head of political and organised crime inquiries. In May 2006 his court sentenced two gangsters to twenty-five years to life for some particularly grisly acid murders; on 10 January 2007 it sentenced two pyramid scheme bosses to eleven and a half years; and on 30 January the former mayor of Rača, Pavol Bielik, was sentenced to five years for corruption.[7]

The court's high point was its investigation of Horizon Slovakia and BMG Invest, two investment companies that went bankrupt in 2002 owing clients more than US$554 million.[8] On 10 January 2007 the court sentenced three of their former managers to seven to eleven years in prison for embezzlement.[9]

These cases were highly publicised and the quality of sentencing reinforced public confidence in the integrity of the Special Court and Special Prosecutor's Office. Since coming to office in July 2006, however, Harabin has persistently attacked the reputation of the court, describing it as expensive, inefficient and unconstitutional.[10] In August 2006 he threatened an audit and later drafted a law that would have dissolved both institutions. After a campaign by civil society, including a nationwide petition, SMER announced in December that the court would be preserved. 'The Court,' conceded Fico with suspiciously faint praise, 'has more advantages than disadvantages.'[11]

But the justice minister had not finished, and his mood became increasingly vindictive. In December the post of head of the Special Court – Králik's job – was put out to tender, with three candidates, Králik included, applying. On 8 February 2007 Harabin selected Justice Michal Truban, then serving at the Special Court's Banská Bystrica branch, to replace Králik, though the latter had beaten him on points in two previous tender rounds. Králik left his post on 1 April 2007.[12]

It is too early to predict the impact of this on the Special Court and prosecutor system. While there appeared to be clear personal animus against Králik, it is more likely that the court's autonomy rankled with the minister.

Harabin's scornful attitude was already having consequences in December 2006. Several of the

7 *Slovak Spectator*, 12 February 2007.
8 *Agence France-Presse* (France), 10 January 2007.
9 *hr online* (Slovakia), 30 May 2007.
10 *Slovak Spectator*, 12 February 2007.
11 *Agence France-Presse* (France), 10 January 2007.
12 *Slovak Spectator*, 12 February 2007.

accused refused to attend hearings, in the expectation that the Special Court would soon be dissolved,[13] and doubt about the future of the institution resulted in the number of citizens' complaints dropping.[14]

Legal double agent or whistleblower?

The legal profession in Slovakia is regulated by the 2003 Advocacy Act, which states that an advocate should be independent in the performance of his or her profession.[15] The degree of this independence has never been established by binding regulation, but the Slovak Bar Association (SBA) is responsible for interpreting the law. The SBA was established by statute and all practising advocates must belong to it.

Slovak law also tightly regulates the activities of undercover agents, whether they are members of the police or other law enforcement agencies. Based on orders from a prosecutor or court to disclose, identify and convict perpetrators of felonies, agents are answerable to clauses within the Criminal Code as to how they pursue their operations. Nonetheless, the Ministry of the Interior is also authorised to recruit individuals who do not belong to the law enforcement agencies to help with investigations. The position of these agents is more anomalous.

In July 2004 Pavol P., an advocate, offered his colleague, Mária Mešencová, a bribe of K100,000 (US$4,035) to influence the testimony of her witness in a case. Had the testimony been altered, the witness stood to gain by the same amount. Mešencová notified the police about the case and agreed to cooperate with them as an undercover agent.[16] In September that year Mešencová met Pavol P. in the Hotel Eden in Piešt'any, where he handed over the bribe. He was arrested by police and charged. On 30 November 2006 Pavol P. was placed on probation for two years. He appealed and is now awaiting a verdict.[17]

On 28 March 2006 the SBA expelled Mešencová, arguing that a lawyer could not collaborate with the police without hampering his or her professional performance. According to the association, such cooperation spelt a loss of independence that could jeopardise the interests of any present or future client.[18]

A number of media and civil society organisations came to Mária Mešencová's support for knowingly risking her career in order to expose corruption in the legal profession. Legal experts questioned the SBA's decision to disbar her on the grounds that it was not right to destroy her career for helping the police. Many felt that the SBA should have focused instead on the corruption among its membership.

Mešencová appealed to the Supreme Court against the decision, arguing that no legal regulation prohibited her from cooperating with the police as an agent in cases other than ones in which she was engaged as a participating party. On 15 February 2007 the court struck down Mešencová's expulsion, though it upheld the 'higher principle' that lawyers cannot work on behalf of the police.[19] Though the outcome is a victory for Mešencová, the court's ruling will undoubtedly make Slovakian lawyers hesitant to report corruption in the future.

Emilia Sičáková-Beblava (TI Slovakia)

13 *Daily SME* (Slovakia), 16 December 2006.
14 *Daily SME* (Slovakia), 26 February 2007.
15 Act no. 586/2003, sect. 2, paragraph 1.
16 *Slovak Spectator*, 29 May 2006.
17 *Slovak Spectator*, 26 February 2007.
18 Ibid.
19 Ibid.

Further reading

E. Sičáková-Beblavá, 'Transparentost' a Korupcia', in M. Kollár *et al.* (eds.), Súhrnná správa o stave spolocnosti (Bratislava: IVO, 2006).

E. Sičáková-Beblavá and P. Nechala, 'Protikorupčné minimum 2006' (Bratislava: TI Slovakia, 2006).

TI Slovakia: www.transparency.sk.

Spain

Corruption Perceptions Index 2007: 6.7 (25th out of 180 countries)

Conventions

Council of Europe Civil Law Convention on Corruption (signed May 2005; not yet ratified)
Council of Europe Criminal Law Convention on Corruption (signed May 2005; not yet ratified)
OECD Convention on Combating Bribery of Foreign Public Officials (signed December 1997; ratified January 2000)
UN Convention against Corruption (signed September 2005; ratified June 2006)
UN Convention against Transnational Organized Crime (signed December 2000; ratified March 2002)

Legal and institutional changes

- **Law 28/2006 for the improvement of public services**, which incorporates a new organisational formula into the general state administration, was passed on 18 July 2006. The law gives government agencies greater autonomy and flexibility in management, but at the same time requires reinforcement of the mechanisms for monitoring effectiveness and promoting accountability. The law creates an agency for monitoring public policies and service quality that specifically includes participation among its behavioural principles, understanding this as a commitment to consulting with stakeholders in carrying out their tasks.[1]
- The post of **special prosecutor for ecology and town-planning offences**, with a brief to act 'directly, strictly and forcefully' against any infraction of this kind, was created in April 2006, though it only began to function

1 Law 28/2006, no. 171.

from summer 2006.[2] At least one prosecutor with specialised knowledge about environmental matters will be posted in each province and another in each autonomous community. They are responsible for conducting investigations, participating in trials, acting in the public interest and writing reports. The Special Prosecutor's Office has become very important in the struggle against corruption in town-planning (see below).[3]

- **Law 5/2006 on the Regulation of Conflicts of Interest of government members and senior administration officials** is also noteworthy, as it contains a range of rules covering, *inter alia*, incompatibility with other activities, the limits of patrimony and after-work activities. The law is complemented by provisions that establish the publication of a record of activities for senior officials, and the publication in the official state bulletin of the goods and property of government members and secretaries of state.[4]
- The **Ethical Code for Public Employees**, incorporated in the Basic Statute for Public Employees, passed on 12 April 2007 and came into force in May. The code starts with a general obligation to perform assigned tasks diligently, to bear in mind the public interest, and to observe the constitution and other laws. It then sets out a series of principles of behaviour, including: conduct shall be based on respect for people's fundamental rights and freedoms; public employees should exclude themselves from matters in which they have a personal interest, as well as any private activity that involves the risk of conflicts of interest with their public service; they should not accept any favour from persons or private bodies that implies privilege or unjustified advantage; and they shall act in accordance with the principles of effectiveness, economy and efficiency.[5]
- From May 2007 the **new Land Law, 8/2007, introduces new incompatibilities for elected members and senior executives, as well as for local comptrollers and city managers**. According to the law, which came into effect on 1 July 2007, two registers, one for officials' goods and the other for their activities, will both be open to public scrutiny. In contrast to the previous system, which affected only locally elected members, the obligation to declare wealth and provide supporting documentation has been extended to unelected members of local government councils and senior executives. For two years after the end of their mandate, representatives who have held executive positions in local government are not allowed to work for companies directly related to their previous activities, in accordance with article 8 of Law 5/2006 on the regulation of conflicts of interest of government members and senior administration officials. The new law also requires councils to set up websites to provide information on town-planning agreements with private developers or landowners, and any reclassifications or increases in building permits. Such decisions were previously taken by the mayor and were subject to little transparency (see below).

Corrupt web of construction spreads

Most of this new legislation reflects government alarm at the scale of corruption in Spain and how best to fight it. The area of greatest concern is local government and, within it, land and town-planning policy. The number of corruption cases in town-planning and land regulations reported or under investigation has shot up from 2,016 in 2004 to 3,279 in 2005 and to 3,846 in 2006.[6] A

2 Law 10/2006.

3 See www.unep.org/labour_environment/PDFs/speech_narbona.pdf.

4 Law 5/2006, no. 86.

5 Law 7/2007, sect. 6.

6 Memoria de la Fiscalía General del Estado, 2004, 2005 and 2006; available at www.fiscal.es.

recent report by Greenpeace estimates that there are 41,000 illegal buildings along Andalusia's 817 kilometres of coastline, and 700,000 new homes are already planned for the autonomous state.[7]

The Andalusian resort of Marbella was the first place where links between town-planning, money-laundering and the country's booming construction sector came to light in 2001, and then again following Operation White Whale in 2005 (see *Global Corruption Report 2006*). Despite all the police and media attention, corruption has proven singularly intractable in Marbella's town hall.

In March 2006 the Marbella courts and prosecutor's office launched Operation Malaya,[8] arresting the mayor, Marisol Yagüe; her deputy, Isabel García Marcos; and José Antonio Roca, the town-planning adviser involved in the first property-related scandal to break under the former mayor, the late Jesús Gil. Judge Miguel Ángel Torres said that Roca was 'the driving force in Marbella City Hall and that the mayor performed a mere symbolic role'.[9] In a second phase of the operation, the police arrested a former chief of police, a deputy mayor, a number of councillors and two construction entrepreneurs.[10] The combined operations yielded €2.4 billion (US$3.2 billion) in confiscated property and led to 1,000 bank accounts being frozen.

A sense of outrage is now beginning to coalesce around corruption in less flashy locations. In November 2006 two former mayors of the small town of Ciempozuelos, south of Madrid, were charged with bribery and money-laundering.[11] Again, the motive was building permits and the enormous profits to be made in the property market. Pedro Antonio Torrejón, who stepped down as mayor when the allegations were made public, was accused of approving the building of 5,600 new homes by the construction company Esprode in exchange for a commission of €40 million (US$54 million), shared with his predecessor as mayor, Joaquín Tejeiro.[12]

In Telde, Grand Canary Island, the mayor and five city councillors resigned on 10 November 2006 for their alleged role in the 'Faycán' case, in which illegal commissions were paid during the tendering of public works projects.[13] An adviser of the public works councillor declared that part of the funds, reportedly as high as 20 per cent of the contract value, went to finance the local People's Party (PP) branch. The party announced that it would sue.

Another reported scam in the same autonomous community was based on the complaint of a businessman, Alberto Santana, about corruption in the granting of concessions to build wind farms in the Canary Islands.[14] According to Santana, José Manuel Soria, president of both the Canary Islands PP and the Grand Canary Council, is alleged to have disallowed a bid from Megaturbinas and passed privileged information to a competing firm, Promotora de Recursos Eólicos, in exchange for financial incentives. The deal was handled by his brother, Luis Soria, industry adviser to the Canary Islands government from 2003 to 2005.[15]

---

7 Greenpeace, *Destrucción a toda costa: Informe sobre la Situación del Litoral Español* (Madrid: Greenpeace, 2007).
8 See www.citymayors.com/politics/marbella_corruption.html.
9 Ibid.
10 See www.spanish-review.com/article654.html.
11 See www.citymayors.com/politics/spain_corruption.html.
12 Ibid.
13 Ibid.
14 *El País* (Spain), 16 June 2006.
15 *El País* (Spain), 10 February 2006.

A similar corruption trend has emerged in the Balearic Islands. In November 2006 police arrested Eugenio Hidalgo, mayor of Andratx, western Mallorca, for cultivating links with officials in the building supervision office in order to obtain illegal construction contracts and change zoning restrictions.[16] Allegedly, he had acquired property on a protected stretch of coast and instructed the town council to approve an application to build 150 units in twenty-six condominiums for an estimated profit of €10 million (US$13.5 million). He was also accused of having accepted bribes. The regional director general of land, Jaume Massot, was also detained for alleged money-laundering.[17]

Fuelling town hall corruption is a spiralling property market and Spain's popularity as a retirement destination, but the origins of the problem go back to the nineteenth century and Spain's traditional land regulations.[18] The latter allowed landowners to retain the benefits of urban expansion, capturing a public policy for private profit. This is because the law allows the virtual land value (for example, 100 new apartments) to be incorporated into the real land value, building high levels of inflation into the process before the first brick is even laid. Add to this the proliferation of inscrutable rules (more than 5,000 pages of norms), which few citizens outside the planning department can understand.

Conflicts of interest office loses first round

Law 5/2006, regulating conflicts of interests of members of the government and high officials in the general administration, has been a positive step forward in preventing corruption in central government. Looked at in closer detail, however, some aspects leave much to be desired.

For example, there is inconsistency in the system of sanctions. Current provisions penalise making a false statement to the Conflict of Interest Office, with a potential sanction of dismissal and disqualification from high office for five to ten years. Failure to present a statement at all is only considered an infraction, however, whose sole outcome is publication in the official gazette. This opens the gates to abuse of position.

The most serious failure, however, is in the regulation of the Conflict of Interest Office. On the one hand, its functional autonomy is recognised, but, on the other, it is organisationally dependent on the Ministry of Public Administration. This dependence will tend to neutralise its autonomy. Spain's administrative model is based on the constitutionally established principle of hierarchy, which is widely incorporated in formal and informal practices, and daily routines. The Conflict of Interest Office was initially situated under the general secretary of public administration as an office that required management by a senior official with the rank of subdirector.

In January 2007 its position within the ministry was changed, concomitant with an extension of its powers to include the incompatibilities of public employees in the general administration.[19] It is now located within the General Direction of Organisation and Inspection of Services. This increase in power comes with serious limitations to effective investigation, however:

- it does not have its own budget or autonomous staff;
- it has no right to access employees' tax and contributions information, which is controlled by the tax agency, so detecting data falsification is very difficult;

16 *Spiegel International* (Germany), 1 March 2007.
17 *La Vanguardia* (Spain), 26 April 2007.
18 See www.elmundo.es/especiales/2006/11/espana/corrupcion_urbanistica/problema.html.
19 Real decreto 9/2007, 12 January 2007.

- it cannot investigate without formal authorisation from the minister; and
- it has no power of sanction.

The baptism by fire for its independence occurred in June 2006, when it was discovered that the ex-president, José Maria Aznar, had been receiving €10,000 (US$13,500) per month through his company, Famaztella S.L., from News Corp, the group owned by media magnate Rupert Murdoch, since September 2004 for 'strategic consultancy' activities. The discovery was made when the Murdoch group reported the payments in a filing to the US Securities and Exchange Commission.[20]

Aznar was not legally entitled to receive this money, since he was a councillor of state at the time, a post that disqualifies holders from accepting any direct or indirect payment for private sector activities. Aznar had neither declared this income nor his consultancy for the Murdoch group, and would have been disallowed from employment with News Corp for two years because he had had direct relationships with it during his presidency. The media reported the matter[21] and the minister of public administration announced an investigation, but the Conflict of Interest Office did not press ahead with it nor propose any further sanction. The matter was filed away and, in so doing, vast amounts of legitimacy drained away from the new office.

Manuel Villoria (TI Spain)

Further reading

J. L. Díaz Ripollés *et al.*, *Prácticas Illícitas en la Actividad Urbanística: Un Estudio en la Costa del Sol* (Valencia: Tirant lo Blanc and Instituto Andaluz de Criminología, 2004).

P. Eigen, *Las Redes de la Corrupción* (Barcelona: Planeta, 2004).

F. Jiménez, 'La Incidencia de la Corrupción en la Democracia', presentation at the Informe sobre la Democracia en España: la Estrategia de la Crispación, Fundación Alternativas, Madrid, 2007.

F. Jiménez and M. Caínzos, 'Political Corruption in Spain', in M. Bull and J. Newell (eds.), *Corruption in Contemporary Politics* (London: Palgrave, 2003).

J. Malem, *La Corrupción: Aspectos Eticos, Económicos, Políticos y Jurídicos* (Barcelona: Gedisa, 2002).

N. Sartorius, 'Informe sobre la Corrupción Urbanística en España' presentation at the Informe sobre la Democracia en España: la Estrategia de la Crispación, Fundación Alternativas, Madrid, 2007.

M. Villoria, *La Corrupción Política* (Madrid: SíntesisSistema, 2006).

'Informe Global sobre la Corrupción en España: La Corrupción Judicial' (Madrid: TI Spain, 2007).

TI Spain: www.transparencia.org.es.

20 *El País* (Spain), 28 June 2006.
21 *El País* (Spain), 29 June 2006.

Switzerland

Corruption Perceptions Index 2007: 9.0 (7th out of 180 countries)

Conventions

Council of Europe Criminal Law Convention on Corruption (signed February 2001; ratified March 2006)

OECD Convention on Combating Bribery of Foreign Public Officials (signed December 1997; ratified May 2000)

UN Convention against Corruption (signed December 2003; not yet ratified)

UN Convention against Transnational Organized Crime (signed December 2000; ratified October 2006)

Legal and institutional changes

- On 1 July 2006, the day that Switzerland ratified the Council of Europe Criminal Law Convention on Corruption and its Additional Protocol, the Federal Council accordingly revised the Swiss Criminal Code and the Federal Law on Unfair Competition.[1] There were **three important changes in anti-corruption law**. First, active bribery of a foreign official was extended to encompass passive bribery, including members of foreign parliaments, international organisations and international courts. Second, under article 4 of the federal law, passive bribery in the private sector became an offence. Third, the criminal responsibility of legal entities was extended to active bribery in the private sector. That effectively means that a private company shall be punished independently of any employee if it is deemed not to have taken all necessary measures to prevent such an offence.
- In a partial break with its secretive past, **the federal government introduced a freedom of information law**[2] on 1 July 2006, thereby catching up with the rest of Europe and some of its own cantons.[3] The Law on Transparency guarantees legal access to official documents of the federal administration, and assures the right to information of all organisations and persons, public or private, with the authority to pass decisions of first instance.[4] Important exemptions from the law, however, include the Swiss National Bank, the Federal Banking Commission, the Federal Assembly, parliamentary commissions and the Federal

1 See www.ejpd.admin.ch/ejpd/de/home/themen/kriminalitaet/korruption/ref_korruption__europarat.html.

2 For further information, see the website of the Federal Data Protection and Information Commissioner (FDPIC), at www.edoeb.admin.ch/index.html?lang=en.

3 M. Pasquier and J.P. Villeneuve, 'Access to Information in Switzerland: From Secrecy to Transparency', *Open Government*, vol. 2, no. 2 (2006).

4 That is to say, decisions based on public law which, *inter alia*, create or modify rights or obligations, or rule on the existence and extent of right or obligations.

Council. Parliament can also effectively withdraw particular administrative units or organisations from the obligations of the law if their mandates require it. Access to documents can also be denied when the requested document is not complete – or if it was completed before the passage of the new law. Nonetheless, there is strong evidence that the public is using its new right. Between 1 July and 31 December, ninety-five requests for information were lodged, of which forty-one were rejected.[5]

- **Whistleblower protection** does not have a proud history in Switzerland, where those who denounce corruption have routinely been dismissed as troublemakers. In 2003 two MPs initiated legislation to improve the legal status of employees who report wrongdoing, but key political and economic stakeholders were hostile to the proposal. Since then, however, the level of awareness has grown, perhaps influenced by international developments. In June 2007 parliament adopted an amended version of the 2003 motion, and mandated the Federal Council to elaborate a draft whistleblowers' act. The council was asked, in particular, to consider: the protection from unfair dismissal or other discrimination of private sector employees who report illegal acts; existing civil legal sanctions on unfair dismissal and modifications to improve their effectiveness; equalising levels of protection for public and private sector employees alike; and the introduction of a legal obligation for public officials to report misconduct.[6] The first draft of the bill will be presented to parliament in 2008.
- Company law is to be modernised to bring it into line with the needs of the economy. A **draft of the Company and Accounting Law Reform**, prepared by the Federal Department of Justice and Police (FDJP), **was favourably received during the statutory consultation process.**[7] The reform affects corporate governance, capital structure, modernisation of the general assembly and accounting and is intended to bring about more effective anti-corruption measures. In the area of corporate governance, the draft recommends an extension of information and voting rights for shareholders, including the right to know the remuneration of senior management in private companies.

The Duvalier case

During his fifteen years in power the former Haitian dictator, Jean-Claude 'Baby Doc' Duvalier, along with his family and entourage, embezzled over US$515 million of public funds, according to evidence collected by the Haitian authorities in 1987.[8] Duvalier transferred revenues from tobacco taxes, the mining industry, vehicle insurance and the national lottery to an extra-budgetary 'social fund' that he alone controlled. He justified the practice as benevolent paternalism, but actually used the fund as a private account. He and his cronies also contracted loans from the national bank without repaying them.

After Duvalier's escape to France in 1986 the new government sent the Swiss Justice Department documentation of all known illegal bank transactions that had occurred during his rule.[9] The Swiss authorities discovered and froze Fr7.6 million (US$6.4 million) of the estimated total in Swiss accounts.[10] Haiti failed to provide the necessary documents in due time, however. In

5 FDPIC, '14th Annual Report 2006/2007' (Bern: FDPIC, 2007).

6 Report of the Commission for Legal Affairs (2006), available at www.parlament.ch/afs/data/d/bericht/2003/d_bericht_s_k25_0_20033212_0_20060221.htm#1.

7 FDJP, 'Reform of Swiss Company Law Approved in Principle', press release, 14 February 2007.

8 Letter from Onill Millet, governor of Banque de la République d'Haiti, 15 January 1987.

9 R. Vogler, *Swiss Banking Secrecy: Origins, Significance, Myth* (Zurich: Association for Financial History, 2006).

10 *SwissInfo*, 22 August 2007.

2002 the Directorate of International Public Law in the Federal Department of Foreign Affairs, in acknowledging the persistent weakness of Haiti's judicial system, extended the freeze of Duvalier's accounts until 31 May 2007.[11] By law, the funds can only be frozen for a limited duration if the government requesting restitution does not provide the relevant evidence.

An initiative to reach an extra-judicial agreement between Haiti and the Duvaliers to share the total amount failed in early 2007. In mid-May the directorate stated its concern that the freeze could not be extended without limit and, under the rule of law, it was bound to return the funds to the Duvaliers as their formal proprietors. It subsequently installed a special commission for the revision of the International Mutual Legal Assistance Act.

International, Haitian and Swiss NGOs were outraged at the plan to return the funds to the Duvaliers and requested a further extension so that Haiti could take steps to accuse the former dictator of human rights violations, as well as seek technical legal assistance. This was in line with Switzerland's own intentions, as the time limit had expired in 2002, and only crimes against *ius cogens* could lead to a new procedure of mutual legal assistance.

Facing public anger, the Federal Council postponed the funds' release for a further three months. The Swiss NGO Aktion Finanzplatz Schweiz (AFS) noted[12] that this was merely a temporary remission and petitioned the government to take steps to freeze such funds until the country of origin was capable of launching a proper case of international mutual legal assistance.

There were parallel initiatives by international agencies. The UN's special rapporteur on impunity, Louis Joinet, a founding member of the movement against impunity, negotiated with the Haitian president, René Préval; the Swiss Directorate of International Public Law sent two delegates to Port-au-Prince; and the World Bank president, Robert Zoellick, offered technical legal assistance, should Haiti request it.

On 13 August 2007 President Préval wrote to Federal President Micheline Calmy-Rey of Switzerland stating that Haiti intended to set up a procedure of mutual legal assistance, but that this would entail at least twelve months' further delay. The Swiss government judged this sufficient to prolong the freezing of the Duvalier accounts for a further year.[13]

The Swissair case

Before 2000, when the government ratified the OECD Convention on Combating Bribery of Foreign Public Officials, it was legal to bribe in the interests of doing business abroad and to declare the amount as an extraordinary tax expense. Since the revision of the Criminal Code only one conviction has been handed down, however, and that was for a trivial matter involving a mere €500 (US$675).[14]

The collapse in October 2001 of the national airline, Swissair, is worthy of analysis in this respect. Thousands of employees lost jobs and pensions, and anger at the demise of one of the

11 Ibid.

12 AFs, press release, 13 June 2007.

13 *Tax-news.com* (British Virgin Islands), 27 August 2007.

14 In 2001 a court in Ticino convicted a foreigner who had offered an Italian customs official Fr800 (US$675) in exchange for a false stamp on his passport. The court handed down a suspended prison term of thirty days and expelled the defendant from Switzerland for three years.

nation's most prestigious companies resulted in a cry for the punishment of those responsible. The trial, which started in January 2007, sought to discover to what extent the former board and management of Swissair had been responsible for the airline's bankruptcy. In addition to the main charge of mismanagement, a further accusation was made alleging unjustified payments.

The latter referred to unjustified payments to the management of the Polish airline, LOT, by Swissair's former CEO, Philippe Bruggisser. In 1999 the Polish government had announced it would privatise 37 per cent of the state-run airline. Three airlines responded to the offer: British Airways (BA), Lufthansa and Swissair. A Swissair stake was strategically important, because Austrian Airlines had recently cancelled its partnership arrangement, limiting Swissair's access to the East European market. Bruggisser needed the LOT deal to stuff the huge hole in the east.[15]

In the summer of 1999 a Finnish employee of Swissair made two interesting statements. First, that Swissair was slightly ahead of Lufthansa in the competition for LOT, and streets ahead of BA. But, second, that BA had offered LOT's CEO, Jan Litwinski, a salary of €150,000 (US$202,500) on top of the share price. 'In such a competition,' Bruggisser is reported to have said, 'all weapons will be used.'[16]

At the end of an intense bidding competition, Swissair emerged as the victor in November 1999. Bruggisser offered Litwinski a consulting contract to develop Swissair's strategy for Eastern Europe at a monthly fee of €9,000 (US$12,150) for three years. In total, Swissair paid Litwinski €170,000 (US$230,000).[17]

In March 2000 the Polish parliament passed the Kapp Law, which placed an upper limit of €41,000 on all CEO salaries. The SAir Group, Swissair's holding company, agreed to top up his salary through a consulting arrangement.[18] As a result, the Swiss attorney general accused Litwinski, Bruggisser and another Swissair employee of making false certification claims.[19]

In the eyes of the public, Swissair had simply bribed the LOT manager in exchange for the shares, a theory supported by the absence of written board approval for his consultancy services.[20] But it was not an open-and-shut case. In 1999, when the payments were made, the revised Criminal Code had not become binding. In Switzerland, an act of making public officials compliant is considered to be 'giving an advantage',[21] and is liable. The payments to Litwinski satisfied the requirement of giving undue advantage, but the Criminal Code refers only to Swiss, and not foreign, public officials.

All the accused were acquitted on 7 June 2007. The court found that the district attorney had not proved that the defendants had deliberately caused damage to the SAir Group, and that intention was therefore missing. The charges against

15 *Tagesanzeiger* (Switzerland), 30 January 2007.
16 Ibid.
17 District Court Bülach, Overview of the Charges, available at www.bezirksgericht-buelach.ch/zrp/buelach.nsf/wViewContent/A5BC0C0446C2701DC1257259002BB307/$file/SAirgroup%20Anklage.pdf.
18 *Tagesanzeiger* (Switzerland), 29 January 2007.
19 *Tagesanzeiger* (Switzerland), 30 January 2007.
20 Ibid.
21 Article 322 *quinquies* of the Criminal Code states: 'Any person who offers, promises or gives any undue advantage to a member of a judicial or other authority, a state employee, an expert, translator or interpreter employed by any authority, an arbitrator or a member of the armed forces so that he accomplishes the duties of his position shall be liable to imprisonment or a fine.'

Bruggisser and Litwinski of disloyal management were also dropped because the alleged activity occurred not in Switzerland, but Poland.[22]

The Swissair case illustrates the shortcomings of existing law in the domains of bribery, corruption and disloyal management. TI Switzerland strongly favours modifying the law on 'giving an advantage' so that it is also made applicable to foreign public officials.

TI Switzerland

Further reading

O. Hafner, 'Korruption und Korruptionsbekämpfung in der Schweiz' (Berne: TI Switzerland, 2003).

D. Jositsch, 'Das Schweizerische Korruptionsstrafrecht' (Zurich: Schulthess Juristische Medien, 2006).

Z. Ledergerber, *'Whistleblowing' unter dem Aspekt der Korruptionsbekämpfung* (Berne: Stämpfli Verlag, 2005).

M. Pieth, 'Korruptionsgeldwäsche', in J.-B. Ackermann *et al.* (eds.), *Wirtschaft und Strafrecht: Festschrift für Niklaus Schmid zum 65 Geburtstag* (Zurich: Schulthess, 2001).

PricewaterhouseCoopers AG/SA, 'Economic Crime Survey, Switzerland' (Pricewaterhouse-Cooper AG/SA, 2005).

N. Queloz, 'Prävention und Sanktion der Korruption als Beitrag zum Schutz der Menschenrechte', in A. Donatsch *et al.* (eds.), *Strafrecht, Strafprozessrecht und Menschenrechte: Festschrift für Stefan Trechsel zum 65 Geburtstag* (Zurich: Schulthess Juristische Medien, 2002).

Swiss State Secretariat for Economic Affairs (SECO), 'Preventing Corruption: Information for Swiss Businesses Operating Abroad' (Berne: SECO, 2003).

TI Switzerland: www.transparency.ch.

22 *International Herald Tribune* (US), 7 June 2007.

Ukraine

Corruption Perceptions Index 2007: 2.7 (118th out of 180 countries)

Conventions

Council of Europe Civil Law Convention (signed November 1999; ratified September 2005)
Council of Europe Criminal Law Convention (signed January 1999; not yet ratified)
UN Convention against Corruption (signed December 2003; not yet ratified)
UN Convention against Transnational Organized Crime (signed December 2000; ratified May 2004)

Legal and institutional changes

- **'On the Way to Integrity', a presidential decree on anti-corruption policy issued on 11 September 2006**, was intended as a follow-up to the 'Concept of the Fight against Corruption 1998–2005', introduced by the former president, Leonid Kuchma. Kuchma's programme was only partially implemented and had few appreciable results. The latest 'concept' identifies reform of the civil service structure, civil service salaries, the rotation of public officials, administrative procedure, public procurement, the judiciary and the activities of elected bodies. Strengthening the media through guarantees on freedom of information and increased NGO involvement in formulating anti-corruption policy are also urgently required. The concept paper was criticised as too vague, with few deadlines and institutional responsibilities for implementation or monitoring.[1]
- The Justice Ministry and the parliamentary committee on the fight against corruption and organised crime are preparing the **second reading of a package of anti-corruption laws**, including drafts on changes concerning corruption offences, the prevention and countering of corruption and the responsibilities of legal entities for the commission of corrupt acts.[2] Introduced by President Viktor Yushchenko in October 2006, the package is aimed at harmonising Ukraine's anti-corruption legislation with international standards and implementing the norms of the UN Convention against Corruption, the EU's Criminal Convention on the Fight against Corruption and Additional Protocol ratified by the Verkhovna Rada (parliament) on 18 October 2006. The package was approved on its first reading on 12 December 2006.
- Work on amending the procurement law was approved on 15 August 2007 – the eleventh occasion of its amending since 2000. The current system has serious transparency defects, according to international analysts.[3] The new amendment **identifies an anti-monopoly committee as the most appropriate body to supervise public procurement and creates an inter-agency commission** to regulate procurement, but it still allows for the presence of non-audited mediators in the process. These companies secure high profits from monopolies on legal consultation, information about tenders and the provision of documentation. As a result, tender participants incur additional costs and compensate by boosting their offer prices. President Yushchenko created a working group in January 2007 to draft a new law on public procurement by August 2007. Public control of procurement is vested in the Tender Chamber of Ukraine (TCU), which has been criticised for non-transparency and insider trading with mediators. In response to repeated challenges by journalists,[4] the TCU submitted more than fifty court appeals in 2006–7 intended to protect the 'honour, dignity and business reputation' of its members. Tenders are published online, but with only limited, prepaid access that provides partial information, and without proper

1 OECD ACN, 'Istanbul Anti-Corruption Action Plan for Armenia, Azerbaijan, Georgia, the Kyrgyz Republic, the Russian Federation, Tajikistan and Ukraine: Ukraine Monitoring Report' (Paris: OECD ACN, 2006).

2 See the draft Law on Introducing Changes to Legislative Acts Concerning Responsibility for Corruption at gska2.rada.gov.ua/pls/zweb_n/webproc4_1?id=&pf3511=28736; the draft Law on the Basics of Preventing and Counteracting Corruption at gska2.rada.gov.ua/pls/zweb_n/webproc4_1?id=&pf3511=28741; and the draft Law on the Responsibility of Legal Entities for Committing Corrupt Offences at gska2.rada.gov.ua/pls/zweb_n/webproc4_1?id=&pf3511=28745.

3 OECD Support for Improvement in Governance and Management (SIGMA), 'Ukraine Governance Assessment' (Paris: OECD SIGMA, 2006).

4 See www.cripo.com.ua.

search systems for tracing separate tender histories or monitoring prices.[5]

- In October–December 2006 the **OECD's Anti-Corruption Network for Transitional Economies conducted an assessment of Ukraine's progress in complying with the 2004 Istanbul recommendations.**[6] The report notes that current coordination mechanisms for anti-corruption policy do not appear strong enough, and that there is a need to identify and support with the necessary mandate and resources a lead institution to take charge of implementing the anti-corruption agenda, and ensure effective coordination of the various institutional components. Of the twenty-four Istanbul recommendations, Ukraine was recognised as non-compliant with twelve and only partly compliant with nine. As for compliance with GRECO's guiding principles, the first and second evaluation rounds were completed in March 2007, but Ukraine chose not to publicise the results.
- **Two notable long-term anti-corruption projects started in 2006 and 2007**: the US$45 million Millennium Challenge Corporation programme and the Council of Europe's Ukrainian Project Against Corruption (UPAC), valued at €1.75 million (US$2.4 million). The MCC programme has ambitious goals concerning improved measurable indicators of corruption in Ukraine. Around 95 per cent of its funding is to be allocated to reforms in priority spheres, including the judiciary; increased monitoring and enforcement of ethical and administrative standards; streamlining and enforcing regulations; and combating corruption in higher education. About 5 per cent of the funding will be distributed to NGOs to monitor the activities of short-term government projects. The UPAC includes helping the authorities increase their ability to prevent corruption and contributing to the formulation of laws that bring legislation into conformity with the Council of Europe's Criminal and Civil Law Conventions against Corruption, and the UNCAC.

Revolution in turmoil as elections approach

The Orange Revolution of November–December 2004 was a landmark in Ukrainian history. Massive public protests led to the cancellation of the presidential election results, and a new election brought the current president, Viktor Yushchenko, to power. The decision to repeat the second round of the election was accompanied by an amendment to the constitution that gives the parliamentary majority the right to appoint the prime minister, whom the president cannot remove. Since implementation in January 2006, this law has strengthened a political crisis, which has dogged the government through much of 2007.

After the revolution the new government declared the fight against corruption a top priority, and sought to harmonise Ukraine's anti-corruption legislation with international conventions. This broad political consensus regarding corruption was undermined, however, by tensions between President Yushchenko and the prime minister, Viktor Yanukovych, and their respective parties, Our Ukraine, the Party of Regions, the Socialist Party of Ukraine and the Communist Party of Ukraine, and oppositional bloc of Yulia Tymoshenko. The situation is complicated further by tensions between the political and economic elites at national and regional levels, and the absence of traditions of good governance.

In spring 2007 a new threat to the government emerged. A number of deputies from the bloc of

5 See www.zakupivli.com.
6 OECD ACN, 2006.

Yulia Tymoshenko and Our Ukraine party changed their affiliations and crossed the floor to join the Party of Regions, the largest in parliament. One of their aims was to increase the Party of Regions' majority to 300, thereby creating a sufficient majority to change the constitution should the occasion arise. The president responded on 2 April 2007 by dissolving parliament. According to a second presidential decree, a fresh election should have been held on 24 June, but by common agreement between the president, prime minister and speaker of parliament the date was pushed back further to 30 September 2007.

With no sitting government the anti-corruption agenda was bound to suffer, but, even before these events, there was little stomach to tackle corruption. One recent scandal concerned the tender for NGOs interested in becoming implementing partners under the MCC. On 22 May the Ministry of Finance announced that the winner of the H3.5 million (US$717,000) contract was Anticorruption Forum, which, it turned out, had been founded by the minister of finance, Mykola Azarov, when he was head of the state tax administration.[7] President Yushchenko wrote to Yanukovych demanding an official investigation of allegations that the Ministry of Finance and Control and Revision Department had spent H1.8 billion (US$368.7 million) for auditing without adequate transparency.[8]

Lack of clarity as to how future anti-corruption policy might be consolidated is aggravated by selectivity in the fight against corruption. Most legal proceedings are instituted against junior public officials or political opponents of the governing majority. In November 2006, the defence minister, Anatoliy Hrytsenko, the foreign minister, Boris Tarasyuk, and the interior minister, Yuriy Lutsenko – all Cabinet appointees as part of the presidential quota – were ordered to 'report on their work to parliament', frequently a prelude to allegations of corruption.[9] Anatoliy Hrytsenko, Ukraine's first civilian defence minister, has fought effectively against corruption in the security sector and claimed to have sent numerous cases of defence corruption to the prosecutor general, without achieving any result.[10] Earlier, the speaker of parliament, Oleksandr Moroz, was accused of taking a US$300 million bribe for promoting the Coalition of National Unity.[11] The prosecutor general found no evidence to confirm the allegation.

In this context, attention should be paid to the uproar over Suzanna Stanik, a Constitutional Court judge, who was appointed rapporteur on a case arising out of the presidential decree to dissolve parliament on 2 April 2007. Two weeks later she was accused of corruption;[12] that same day, the prosecutor general's office said it had investigated the allegations and found them groundless. This did not prevent President Yushchenko from asking the court to reconsider the accusations and to relieve Stanik of her duties for 'breaching the oath'.[13] As it turned out, she had been the focus of a corruption scandal in 2001, but the case was dropped because of her

7 *Ukrayinska Pravda* (Ukraine), 1 June 2006; *Zerkalo Nedeli* (Ukraine), 31 August 2007.
8 *Unian* (Ukraine), 3 August 2007.
9 *Voice of America* (US), 15 November 2006.
10 *Moldova.com* (Moldova), 9 November 2006.
11 *Ukraina Moloda* (Ukraine), 26 July 2007. Transcripts of tape recordings that allegedly confirm the bribery were submitted to the general prosecutors' office by Oleh Lyashko, a member of the bloc of Yulia Tymoshenko parliamentary faction.
12 Valentyn Nalyvaychenko, head of Ukraine's Security Service, or SBU, announced on April 16 that a relative of Stanik had illegally received property worth US$12 million, apparently from the government. See www.whatsonkiev.com/index.php?go=News&in=view&id=2240.
13 *Ukrainian Echo* (Ukraine), 12 June 2007.

close relations with the former president, Leonid Kuchma.[14]

The issues of more transparent electoral regulations became relevant as the newest voting round approached. 'On the Way to Integrity', the presidential decree on fighting corruption issued in September 2006, pointed to the need for improved election-financing mechanisms, the gaps in regulation of political party financing, the limited nature of legal sources of party financing and the absence of proper state supervision of political party activities.

The main tasks for legislators are to improve the procedures of pre-term elections and implement the law on the state registration of voters, which was envisaged under the common agreement between the president, prime minister and speaker in May 2007. A new electoral law regulating the pre-term elections,[15] however, contains a number of tactics that many experts consider deleterious to Ukraine's fledgling democracy, including the cancellation of absentee voting cards or permission to vote at home.[16] Meanwhile, there are serious problems with setting up the electronic database required under the voter registration law,[17] given the short lead-in time.

Instilling integrity in the education sector

Having signed up to the Bologna Declaration in May 2005, Ukraine is obliged to reform its higher education system to bring it into line with European standards. Integrating it is complicated not only by the scale of required reforms, however, but by the fact that Ukraine's educational system is riddled with corruption.

Bribery is widespread in higher education facilities, from college entry and exam results to marking doctoral or master's theses. According to the Management Systems International (MSI) sociological survey, corruption in higher education takes fifth position in the list of the most corrupt spheres (43.6 per cent), after vehicle inspection (57.5 per cent), the police (54.2 per cent), health care (54 per cent) and the courts (49 per cent).[18] Between a third and a half of all Ukrainians who dealt with the government in the previous twelve months experienced extortion, but the worst offenders were universities, where 47.3 per cent of respondents said a bribe had been demanded and 29 per cent said they had given it freely.[19]

On the one hand, teachers and professors may be tempted by their low salaries[20] to seek additional payments while, on the other, students may see corruption as a fast and convenient way of gaining diplomas. The situation is aggravated by public awareness of impunity for bribers, a tolerant social attitude to corruption in general and the absence of initiatives by the authorities to uproot graft in education. Financial bribes range from US$10–50 for an exam pass to several thousand for entry to a prestigious university.

14 *Aratta Ukraine* (Ukraine), 19 April 2007.

15 The Law on Introducing Changes to the Law on Elections of People's Deputies and some other Legislative Acts of Ukraine (concerning the order of conducting pre-term elections to the Verkhovna Rada, and the replacement of people's deputies whose authority was cancelled before the appointed time) was adopted on 1 June 2007.

16 See, for instance, *Unian* (Ukraine), 5 June 2007.

17 OSCE/ODIRM, International Election Observation Mission, 'Ukraine Pre-Term Parliamentary Elections 30 September 2007' (Vienna: OSCE/ODIHR, 2007).

18 MSI and Kyiv International Institute of Sociology (KIIS), 'Corruption in Ukraine: 2007 Baseline National Survey for the MCC Threshold Country Program' (Kiev: MSI and KIIS, 2007).

19 Ibid.

20 Monthly salaries in education vary from H800 to H1500 (US$165 to US$310), according to the Ministry of Education. On 1 July 2007 an average teacher's monthly salary was H1,245 (US$255). See www.mon.gov.ua/newstmp/2007/20_08.

According to the Ministry of Interior, bribes vary from H400 to H100,000 (US$80 to US$21,500).[21]

'On the Way to Integrity' does not mention the word 'education' once, nor 'health care' either, for that matter. In June 2007 President Yushchenko wrote to the Ministry of Interior and prosecutor general urging them to investigate, expose and prevent corruption in institutes of higher education.[22] But the most noticeable activities are being provided within the framework of the MCC Threshold Country Plan,[23] one of whose components specifically targets corruption in education entrances.

The goal is to establish a legal framework requiring a minimum test score for admission to universities, develop a functioning security system for exam results and ensure that 100 percent of students are tested and that test centres are fully operational. A system of independent testing is being implemented by the Ukrainian Centre of Evaluation of Education[24] and the Ministry of Education. In 2006 more than 40,000 pupils participated in external tests, and in 2007 116,000 (26 per cent) of the country's 448,000 graduating pupils took part.[25]

Most were tested on Ukrainian, mathematics and history, and were selected for entry to university on the basis of their results. Participants were generally positive, though many still believed they would not be able to gain a college place without leveraging their connections.[26] The Ministry of Education is more optimistic and has adopted a decree that all universities will accept only the independent test certificates as of next year.

Anna Yarovaya and Olga Mashtaler
(NGO 'Anticorruption Committee', TI national contact in Ukraine)

Further reading

J. Anderson and C. Gray, *Anti-corruption in Transition 3: Who Is Succeeding . . . and Why?* (Washington, DC: World Bank, 2006).

OECD SIGMA, 'Ukraine Governance Assessment' (Paris: OECD SIGMA, 2006).

USAID, 'Corruption Assessment: Ukraine' (Washington, DC: MSI, 2006).

USAID, MCC and MSI, 'Promoting Active Citizen Engagement (PACE) in Combating Corruption in the Ukraine' (Washington, DC: USAID, MCC and MSI, 2007); available at www.pace.org.ua/content/view/24/1/lang,en.

TI Ukraine: www.transparency.org.ua.

21 *Khreshchatyk* (Ukraine), 7 July 2007.

22 'Pres. sends letters to Medvedko, Korniyenko', press statement, press office of President Viktor Yushchenko, 27 June 2007.

23 Agreed and signed in early December 2006.

24 Supported by Soros Foundation Network's International Renaissance Foundation at www.irf.kiev.ua; see also www.mon.gov.ua (Ministry of Education of Ukraine).

25 See *Ukraina i Svit* (Ukraine), 25 June 2007; *Gazeta po Ukrainski* (Ukraine), 14 June 2007; *Dzerkalo Tyzhnya* (Ukraine), 18–24 August 2007.

26 See www.osvita.org.ua/ukrtest/news/2007-05-04.

United Kingdom

Corruption Perceptions Index 2007: 8.4 (12th out of 180 countries)

Conventions

Council of Europe Civil Law Convention on Corruption (signed June 2000; not yet ratified)
Council of Europe Criminal Law Convention on Corruption (signed January 1999; ratified December 2003)
OECD Convention on Combating Bribery of Foreign Public Officials (signed December 1997; ratified December 1998)
UN Convention against Corruption (signed December 2003; ratified February 2006)
UN Convention against Transnational Organized Crime (signed December 2000; ratified February 2006)

Legal and institutional changes

- In July 2006 the prime minister, Tony Blair, asked Hilary Benn, the secretary of state for international development, to lead the government's work on combating overseas corruption. In the same month the government published its **first annual action plan for combating international corruption**. It focused on investigating and prosecuting bribery overseas; eliminating money-laundering and recovering stolen assets; promoting responsible business conduct in developing countries; and supporting international efforts to fight corruption. The actions set out in the plan include the establishment by November 2006 of a dedicated 'overseas anti-corruption unit' in the United Kingdom, staffed by City of London and Metropolitan Police staff, to investigate allegations of bribery and money-laundering.
- In March 2007 the Home Office published a response to the consultation it had launched in December 2005 on **reform of the UK corruption/bribery law**.[1] The government acknowledged that there was broad support for reform of the Prevention of Corruption Acts of 1906 and 1916, and that there had been influential opposition to its 2003 Corruption Bill.[2] The government said, however, that, because there was fundamental disagreement on which of a number of approaches should be adopted, it had asked the Law Commission to undertake a review of

1 UK Home Office/National Offenders Management Service, 'Consultation on Bribery: Reform of the Prevention of Corruption Acts and Serious Fraud Office Powers in Cases of Bribery of Foreign Officials – Summary of Responses and Next Steps' (London: Home Office, 2007); House of Commons Hansard, 5 March 2007.

2 The all-party Joint Parliamentary Committee (JPC), which scrutinised the government's draft Corruption Bill, concluded, *inter alia*, that the bill did not state clearly in language that could be readily understood by the police and prosecutors what types of conduct were punishable as corrupt. The JPC therefore proposed what it believed would be a simpler approach to the definition of corrupt behaviour.

corruption law and produce a new draft bill for consideration. The Law Commission is expected to make its recommendations only in 2008.

- **The Fraud Act 2006**, which came into force early in 2007, established a new general offence of fraud, which can be committed in three ways: by false representation; failing to disclose information; and abuse of position. It established a number of specific offences to assist in the fight against fraud. According to the government, the new offences were expected to simplify the law and better equip police and prosecutors to face the challenge of combating complex fraud in the twenty-first century.
- In March 2007 Sir Hayden Phillips published his **review of the funding of political parties**, which the prime minister had asked him to carry out in the wake of the 'loans for peerages' affair in March 2006. He recommended, *inter alia*, limits on donations and expenditure, but pointed out that the two principal obstacles to reaching a consensus on reforms were the specific design of these limits.[3] He noted that the immediate imposition of a cap on donations (say £50,000 from any one source) would place the Labour Party at a peculiar disadvantage, because it continues to depend heavily on funding from a relatively small number of large organisations. Also, it was difficult to attain consensus on whether existing caps on expenditure on general election campaigns (introduced in 2000) should be lowered.

Anti-corruption strategy falters

The government's Action Plan for Combating International Corruption, published in July 2006, called for the strengthening of anti-corruption efforts across several areas, ranging from the investigation and prosecution of foreign bribery to encouraging resource-rich countries to implement the Extractive Industries Transparency Initiative. With three-year funding of £6 million (US$11.95 million), a dedicated Overseas Anti-Corruption Unit, staffed by City of London and Metropolitan Police, was established in November 2006 to investigate allegations of bribery and money-laundering. This was a response to criticism that, in the nine years since the United Kingdom had ratified the OECD Anti-Bribery Convention, not a single company or individual has been indicted for the offence of bribing a foreign public official, though several UK companies have been implicated.

This positive move was undermined in December 2006 by the decision of the Serious Fraud Office (SFO) to terminate its investigation into the activities of British Aerospace Systems Plc (BAeS) in relation to the Al Yamamah defence contract with Saudi Arabia.[4] The SFO cited 'representations' made to the attorney general and the director of the SFO by ministers concerning the need to safeguard 'national and international security', which made it 'necessary to balance the need to maintain the rule of law against the wider public interest'.[5]

3 H. Phillips, 'Strengthening Democracy: Fair and Sustainable Funding of Political Parties', Review of the Funding of Political Parties (London: HMSO, 2007).

4 The SFO, investigation, which began in November 2004, focused on suspected false accounting in the Al Yamamah defence contract, which provided for the sale of combat aircraft and related equipment and services worth some £40 billion. The media reported the existence of a secret fund established by BAeS to channel benefits to Saudi agents in the contract. In 2006 Saudi-owned Swiss bank accounts were said to be under investigation. A year later, the UK media alleged that payments exceeding £1 billion had been paid to a senior Saudi official in relation to the contract, and that the UK Defence Ministry was involved in covering up such payments. BAeS has repeatedly denied any wrongdoing. See *BBC News* (UK), 7 June 2007 and 11 June 2007.

5 Serious Fraud Office, press release, 14 December 2006.

The SFO said that it had given no weight to 'commercial interests or to the national economic interest' in reaching its decision. Article 5 of the OECD convention requires that, when considering prosecutions in respect of offences under the convention, contracting states should give no weight to three factors: national economic interest, relations with other states and the identity of those involved. The SFO statement addressed the first of these, but not the other two. In his statement to the House of Lords, however, the attorney general claimed that he and the SFO were precluded 'from taking into account considerations of the national economic interest or the potential effect upon relations with another state, and we have not done so'.[6]

The termination of the Al Yamamah investigation was severely criticised in the United Kingdom and abroad.[7] Concerns were expressed about the United Kingdom's commitment to enforcement of the OECD convention. It was also pointed out that the United Kingdom was guilty of adopting a double standard: while it expected other countries to uphold the rule of law and observe their obligations under international anti-corruption conventions, it reserved the right to ignore its obligations when this was politically or commercially expedient. A new UK–Saudi defence deal, reportedly worth £20 billion (involving the sale by BAeS of new combat aircraft to the Saudi air force), had been agreed in principle in 2006.[8]

Two NGOs, the Corner House and Campaign against the Arms Trade, instituted legal proceedings against the government, arguing that:

- the decision to terminate the Al Yamamah investigation was based on considerations of potential damage to relations with Saudi Arabia, although this is expressly forbidden under article 5 of the OECD anti-bribery convention;
- the prime minister had improperly taken into account considerations of damage to diplomatic relations; and
- the advice the prime minister gave to the attorney general amounted to a direction to discontinue the investigation, which constituted an unlawful interference into the independence of prosecutors.[9]

In March 2007 the OECD working group on bribery repeated its serious concerns about the suspension of the Al Yamamah investigation and outlined further shortcomings in the United Kingdom's anti-bribery legislation. It decided to conduct a supplementary review of UK compliance with the OECD convention, focusing on progress in enacting a new foreign bribery law and in broadening liability of legal persons for foreign bribery.

The United Kingdom's failure to enact new corruption legislation continued to be of particular concern. The current law of corruption rests on a mix of common law and statutes, principally those enacted in 1889, 1906 and 1916. In March 2007 the Home Office announced that the definition of bribery would be referred back to the Law Commission, an exercise expected to be completed only in 2008, with no assurance that parliamentary time would be found to enact legislation.

The Al Yamamah affair also raised questions about the role of the attorney general, whose consent continues to be required for all foreign bribery prosecutions. The attorney general is the senior law officer (with a supervisory role in respect of all criminal prosecutions in England and Wales), as well as a member of the government and the

6 House of Lords Hansard, 14 December 2006.
7 OECD, press release, 14 March 2007.
8 Timesonline, 19 August 2006.
9 The Corner House and Campaign against the Arms Trade, press release, 16 December 2006.

Cabinet. The government has not implemented an assurance it gave the OECD working group on bribery in 2005 that it would replace the statutory requirement for the attorney general's consent with a requirement for the consent of the director of public prosecutions, or a nominated deputy.[10]

Anti-money-laundering measures strengthened

The Serious Organised Crime Agency established new structures to ensure that intelligence was prioritised and linked to targeted activity. It also coordinated regular meetings of the enforcement agencies, bringing together the Metropolitan Police, the City of London Police, the Financial Services Authority, the Asset Recovery Agency and the SFO to facilitate the exchange of financial intelligence. These measures were intended to improve the effectiveness of the United Kingdom's efforts to address the money-laundering threat posed by so-called 'politically exposed persons'. The Metropolitan Police responded to requests from the Nigerian government relating to two former state governors, Joshua Dariye and Diepreye Alamieyeseigha. In one case, £1 million was returned. In total, about £35 million (US $70.5 million) of Nigerian assets were reportedly under restraint.[11]

In January 2007 the Treasury initiated consultations on draft regulations to implement the Third EU Money Laundering Directive with a view to presenting final regulations in mid-2007, which would come into effect by the end of the year.[12] The draft regulations provide for the introduction of a supervisory, registration and enforcement regime for trust and company service providers (TCSPs), holding out the possibility over time of some effective measures being put in place to stem the abuse of anonymous corporate entities and trusts for money-laundering and terrorist financial activity. The initial regime foreseen by the draft regulations was one of 'light touch', however, and did not seem capable of ensuring that the persons who effectively direct the business of TCSPs and the beneficial owners of those entities were fit and proper persons. Furthermore, the agency assigned to supervise TCSPs, Her Majesty's Revenue and Customs, does not have a proven capacity in this area and may be inappropriate for the role of regulator.

Weaknesses in political party funding are exposed

The 'loans for peerages' affair further reinforced public cynicism about the integrity of political parties. In March 2006 the House of Lords Appointments Commission rejected the Labour Party's nomination for peerages of four individuals who had provided it with funds before the 2005 general election.[13] Such funds had allegedly not been disclosed under the Political Parties Elections and Referendum Act 2000 on the grounds that they were 'commercial' loans.[14] The Labour Party treasurer claimed he had not been informed of these loans.[15]

10 OECD, 'UK: Phase 2 Follow-up Report on the Implementation of Phase 2 Recommendations' (Paris: OECD, 2007).
11 DfID, 'Combating International Corruption: UK Action Plan for 2006–07, Interim Progress' (London: DfID, 2007).
12 HM Treasury, 'Implementing the Third Money Laundering Directive: Draft Money-Laundering Regulations' (London: HMSO, 2007).
13 Six hundred of the 732 members of the House of Lords are life peers. Some are appointed as non-party political peers by the House of Lords Appointments Commission. Others, who are working or party political peers, are nominated by party leaders. The latter are vetted for propriety by the House of Lords Appointments Commission.
14 *Telegraph* (UK), 14 March 2006.
15 *Sunday Telegraph* (UK), 15 October 2007; *BBC News* (UK), 20 July 2007.

Questions were raised about the loans, as well as others made to the opposition Conservative Party. These questions included whether the funds were intended as donations or soft loans, but were listed as commercial loans so as to circumvent the disclosure requirements of the Electoral Commission.

The Metropolitan Police launched an investigation on 21 March 2006 into whether crimes had been committed by one or both of the two main political parties. Crimes would have been committed if there had been deception in concealing the true nature of donations or loans, and/or if the loans or donations were offered in exchange for peerage nominations, which would breach the Honours (Prevention of Abuses) Act 1925. This law was introduced following the sale of honours when David Lloyd George was in office as prime minister.

The 'loans for peerages' affair highlighted the need for rules governing the funding of political parties and elections to be strengthened and procedures for the appointment of peers in the House of Lords made less susceptible to possible abuse. It remains to be seen whether Sir Hayden Phillip's recommendations for improving the system for political party funding will be acted on.

TI UK

Further reading

Chartered Institute of Building, 'Corruption in the UK Construction Industry Survey' (London: Chartered Institute of Building, 2006).

TI UK, 'Corruption in the Official Arms Trade' (London: TI UK, 2002).

'Corruption and the Funding of UK Political Parties' (London: TI UK, 2006).

'Corruption Bill' (London: TI UK, 2006).

'"Loans for Peerages" and other Connected Issues: Analysis of Possible Crimes' (London: TI UK, 2006).

'Project Anti-corruption Systems: Consultative Edition' (London: TI UK, 2007).

TI UK: www.transparency.org.uk.

Part three
Research

Introduction

Dieter Zinnbauer[1]

Each year the *Global Corruption Report* presents selected highlights of recent research on corruption that can help to design and target anti-corruption interventions more effectively. The breadth of initiatives and insights shows that our understanding of both the scope and dynamics of corruption is advancing steadily.

The big picture: measuring corruption and benchmarking progress in the fight against corruption

Several large-scale research initiatives measure and compare corruption within and across countries, providing the empirical basis to target and benchmark anti-corruption efforts more effectively. These studies paint a rather bleak picture, with some facets of hope. Johann Graf Lambsdorff summarises the main results from the TI Corruption Perceptions Index 2007, which is based on fourteen different surveys to assess corruption in 180 countries. He finds that some rather middle-performing countries have made progress and are catching up with the group of best performers, while at the bottom some of the countries most affected by corruption have seen a deterioration of their situation.

Juanita Riaño reports on the TI Global Corruption Barometer 2007, TI's annual survey of public opinion on corruption. The Barometer 2007 covered more than 60,000 households in sixty countries. It finds that households are most often confronted with bribery when dealing with the police and judiciary – and that the poorest households have to pay bribes most often. Parliaments and political parties continue to be widely viewed as the most corrupt public institutions, according to the Barometer 2007. A majority of respondents, most pronounced in Asia, expect corruption to get worse in their countries. Jonathan Werve and Nathaniel Heller corroborate some of these findings in their summary of the Global Integrity Report, an assessment of more than 290 governance indicators in forty-three countries. Their research identifies political finance and legislatures as two of the weakest links in governance systems, alongside insufficient freedom of information.

Mitchell Seligson and Dominique Zéphyr add a regional perspective with their large-scale household survey in the Americas. They construct a victimisation index, which clearly documents the everyday burden of bribery: across twenty countries in the region, more than one in five respondents had been asked for a bribe over a twelve-month period when dealing with public institutions. Verena Fritz and colleagues take a somewhat different approach, and focus

1 Dieter Zinnbauer is the editor-in-chief of the *Global Corruption Report*.

in their assessment of national governance systems on the views of ten key local stakeholder groups. Corruption emerges as by far the most pressing issue across the ten countries that they have studied. Finally, Sarah Repucci presents a new measurement tool that will bridge the gap between qualitative study and comparative assessment of integrity systems. By scoring the National Integrity System country studies, TI will build on the in-depth assessment of national laws, institutions and practices that underpin governmental accountability and use a locally adaptable methodology to document progress over time and enable cross-country comparisons.

Sectoral insights: capturing corruption risks and performance in key sectors

A second group of research initiatives included in the *Global Corruption Report 2008* focuses on measuring corruption – and reporting on transparency – in high-risk sectors. Juanita Olaya presents the Revenue Transparency Project, which explores the strength of integrity mechanisms in extractive industries. The most recent phase of the project assesses revenue transparency in more than forty oil and gas companies, and reveals that disclosure on a country-by-country basis is being practised by a few leaders, but is still too limited to enable the benefits of transparency to be felt in resource-rich countries, many of which remain poor despite their mineral wealth. Bruno Speck and Silke Pfeiffer focus on another pivotal sector: political finance. They employ a range of methods, including innovative field tests, to assess the transparency of political party financing in eight countries in Latin America. Their results show that party financing lacks meaningful public oversight and transparency, a finding that dovetails with the perception of parliaments and parties as the most corrupt institutions in the Barometer 2007 survey.

Understanding the details: investigating the dynamics of corruption

A third strand of corruption research included in this report focuses on understanding individual motivations, drivers and attitudes towards corruption. Richard Rose, using data from a nationwide household survey in Russia, explains the gap between the reported experience and perception of corruption that many surveys detect. Johann Graf Lambsdorff sets up an experimental game to examine how individuals engage in collusive partnerships, and draws important conclusions for anti-corruption policy design. Benjamin Olken and Patrick Barron document in detail the patterns of corruption along trucking routes and show how insights from industrial organisation theory can help understand and tackle this type of corruption. Ray Fisman and Edward Miguel study the parking behaviour of diplomats in New York, shedding light on the relative importance of norms versus sanctions in deterring corrupt behaviour.

The final two contributions deliver a robust scientific blow to lingering prejudices that corruption may somehow grease the wheels of local administrations. Rema Hanna and colleagues conduct a field study and experiment with regard to obtaining driving licences in Delhi, uncovering convincing evidence that corruption produces social inefficiencies and distortions in this area. Finally, Emmanuelle Lavallée explores a large survey dataset from Africa and powerfully rebuts the assumption that petty corruption to 'get things done' can lead to more confidence in administrative service delivery – showing that the opposite is the case.

8 The big picture: measuring corruption and benchmarking progress in the fight against corruption

Corruption Perceptions Index 2007

Johann Graf Lambsdorff[1]

The Corruption Perceptions Index (CPI) ranks countries in terms of the degree to which corruption is perceived to exist among public officials and politicians. Now in its thirteenth year, it is a composite index, making use of surveys of business people and assessments by country analysts. The statistical work is carried out at the University of Passau, Germany, and the CPI 2007 was published by Transparency International in September 2007.

It ranks 180 countries (an increase from 163 countries last year), and draws on fourteen different polls and surveys from twelve independent institutions, using data published or compiled between 2006 and 2007. Data from the following sources were included.

- Country Performance Assessment Ratings by the Asian Development Bank
- Country Policy and Institutional Assessment by the African Development Bank
- Bertelsmann Transformation Index by the Bertelsmann Foundation
- Country Policy and Institutional Assessment by the IDA and IBRD (World Bank)
- Economist Intelligence Unit
- Nations in Transit by Freedom House
- Global Insight (formerly World Markets Research Centre)
- International Institute for Management Development (in Lausanne)
- Merchant International Group Limited
- Political and Economic Risk Consultancy (in Hong Kong)
- African Governance Report by the United Nations Economic Commission for Africa
- World Economic Forum

Since the first publication of the CPI in 1995 a growing body of scientific literature has made use of the results, contributing to our evolving understanding of corruption around the world.[2]

1 Johann Graf Lambsdorff holds the chair in economic theory at the University of Passau, Germany, and is a research adviser for Transparency International, for which he has coordinated the CPI since 1995.

2 For a summary of related contributions, see J. Lambsdorff, *The Institutional Economics of Corruption and Reform: Theory, Policy and Evidence* (Cambridge: Cambridge University Press 2007).

The strength of the CPI lies in the combination of multiple data sources in a single index, lowering the probability of misrepresenting a country's level of corruption. For 2007 the CPI includes data from three new sources: the Asian Development Bank, the African Development Bank and the Bertelsmann Foundation.

The CPI primarily provides a snapshot of the views of business representatives and country analysts, with less of a focus on year-to-year trends. To the extent that changes can be traced to sources that enter on a consistent basis, however, trends can be identified. Countries with a CPI 2007 score that decreased significantly relative to the CPI 2006, and where this deterioration is not the result of technical factors, are Austria, Bahrain, Belize, Bhutan, Jordan, Laos, Macao, Malta, Mauritius, Oman, Papua New Guinea and Thailand. Significant improvements can be observed for Costa Rica, Croatia, Cuba, the Czech Republic, Dominica, Italy, Macedonia, Namibia, Romania, Seychelles, South Africa, Suriname and Swaziland.

The CPI is not capable of answering whether the world as a whole is improving or not in terms of perceived corruption. But it can reveal whether regions or clusters of countries are improving relative to each other. The analysis shows that poorly scoring countries tend to have difficulties escaping a downward trend. Likewise, the best performers increasingly face competition from others that are catching up. In order to better capture this trend, it was decided to change the methodology slightly in 2007. The modification had virtually no effect on the ranking of countries, but a slight impact on the way the scores are displayed. For example, countries scoring between 4 and 6 improved relative to the best-scoring countries. The best- and the worst-scoring countries, to the contrary, deteriorated slightly. The modification ensures that scores are consistent across time and better reveal whether countries have improved or deteriorated.

A more detailed description of the methodology is available at www.transparency.org and at www.icgg.org.

Table 5 Corruption Perceptions Index 2007

Country rank	Country/territory	2007 CPI score[a]	Surveys used[b]	Confidence range[c]
1	Denmark	9.4	6	9.2–9.6
	Finland	9.4	6	9.2–9.6
	New Zealand	9.4	6	9.2–9.6
4	Singapore	9.3	9	9.0–9.5
	Sweden	9.3	6	9.1–9.4
6	Iceland	9.2	6	8.3–9.6
7	Netherlands	9.0	6	8.8–9.2
	Switzerland	9.0	6	8.8–9.2
9	Canada	8.7	6	8.3–9.1
	Norway	8.7	6	8.0–9.2

(*Continued*)

Table 5 (continued)

Country rank	Country/territory	2007 CPI score[a]	Surveys used[b]	Confidence range[c]
11	Australia	8.6	8	8.1–9.0
12	Luxembourg	8.4	5	7.7–8.7
	United Kingdom	8.4	6	7.9–8.9
14	Hong Kong	8.3	8	7.6–8.8
15	Austria	8.1	6	7.5–8.7
16	Germany	7.8	6	7.3–8.4
17	Ireland	7.5	6	7.3–7.7
	Japan	7.5	8	7.1–8.0
19	France	7.3	6	6.9–7.8
20	United States	7.2	8	6.5–7.6
21	Belgium	7.1	6	7.1–7.1
22	Chile	7.0	7	6.5–7.4
23	Barbados	6.9	4	6.6–7.1
24	Saint Lucia	6.8	3	6.1–7.1
25	Spain	6.7	6	6.2–7.0
	Uruguay	6.7	5	6.4–7.0
27	Slovenia	6.6	8	6.1–6.9
28	Estonia	6.5	8	6.0–7.0
	Portugal	6.5	6	5.8–7.2
30	Israel	6.1	6	5.6–6.7
	Saint Vincent and the Grenadines	6.1	3	4.0–7.1
32	Qatar	6.0	4	5.4–6.4
33	Malta	5.8	4	5.3–6.2
34	Macao	5.7	4	4.7–6.4
	Taiwan	5.7	9	5.4–6.1
	United Arab Emirates	5.7	5	4.8–6.5
37	Dominica	5.6	3	4.0–6.1
38	Botswana	5.4	7	4.8–6.1
39	Cyprus	5.3	3	5.1–5.5
	Hungary	5.3	8	4.9–5.5
41	Czech Republic	5.2	8	4.9–5.8
	Italy	5.2	6	4.7–5.7
43	Malaysia	5.1	9	4.5–5.7
	South Africa	5.1	9	4.9–5.5
	South Korea	5.1	9	4.7–5.5

Table 5 (continued)

Country rank	Country/territory	2007 CPI score[a]	Surveys used[b]	Confidence range[c]
46	Bahrain	5.0	5	4.2–5.7
	Bhutan	5.0	5	4.1–5.7
	Costa Rica	5.0	5	4.7–5.3
49	Cape Verde	4.9	3	3.4–5.5
	Slovakia	4.9	8	4.5–5.2
51	Latvia	4.8	6	4.4–5.1
	Lithuania	4.8	7	4.4–5.3
53	Jordan	4.7	7	3.8–5.6
	Mauritius	4.7	6	4.1–5.7
	Oman	4.7	4	3.9–5.3
56	Greece	4.6	6	4.3–5.0
57	Namibia	4.5	7	3.9–5.2
	Samoa	4.5	3	3.4–5.5
	Seychelles	4.5	4	2.9–5.7
60	Kuwait	4.3	5	3.3–5.1
61	Cuba	4.2	4	3.5–4.7
	Poland	4.2	8	3.6–4.9
	Tunisia	4.2	6	3.4–4.8
64	Bulgaria	4.1	8	3.6–4.8
	Croatia	4.1	8	3.6–4.5
	Turkey	4.1	7	3.8–4.5
67	El Salvador	4.0	5	3.2–4.6
68	Colombia	3.8	7	3.4–4.3
69	Ghana	3.7	7	3.5–3.9
	Romania	3.7	8	3.4–4.1
71	Senegal	3.6	7	3.2–4.2
72	Brazil	3.5	7	3.2–4.0
	China	3.5	9	3.0–4.2
	India	3.5	10	3.3–3.7
	Mexico	3.5	7	3.3–3.8
	Morocco	3.5	7	3.0–4.2
	Peru	3.5	5	3.4–3.7
	Suriname	3.5	4	3.0–3.9
79	Georgia	3.4	6	2.9–4.3
	Grenada	3.4	3	2.0–4.1

(Continued)

Table 5 (continued)

Country rank	Country/territory	2007 CPI score[a]	Surveys used[b]	Confidence range[c]
	Saudi Arabia	3.4	4	2.7–3.9
	Serbia	3.4	6	3.0–4.0
	Trinidad and Tobago	3.4	4	2.7–3.9
84	Bosnia and Herzegovina	3.3	7	2.9–3.7
	Gabon	3.3	5	3.0–3.5
	Jamaica	3.3	5	3.1–3.4
	Kiribati	3.3	3	2.4–3.9
	Lesotho	3.3	6	3.1–3.5
	Macedonia	3.3	6	2.9–3.8
	Maldives	3.3	4	2.3–4.3
	Montenegro	3.3	4	2.4–4.0
	Swaziland	3.3	5	2.6–4.2
	Thailand	3.3	9	2.9–3.7
94	Madagascar	3.2	7	2.5–3.9
	Panama	3.2	5	2.8–3.4
	Sri Lanka	3.2	7	2.9–3.5
	Tanzania	3.2	8	2.9–3.4
98	Vanuatu	3.1	3	2.4–3.7
99	Algeria	3.0	6	2.7–3.2
	Armenia	3.0	7	2.8–3.2
	Belize	3.0	3	2.0–3.7
	Dominican Republic	3.0	5	2.8–3.3
	Lebanon	3.0	4	2.2–3.6
	Mongolia	3.0	6	2.6–3.3
105	Albania	2.9	6	2.6–3.1
	Argentina	2.9	7	2.6–3.2
	Bolivia	2.9	6	2.7–3.2
	Burkina Faso	2.9	7	2.6–3.4
	Djibouti	2.9	3	2.2–3.4
	Egypt	2.9	7	2.6–3.3
111	Eritrea	2.8	5	2.1–3.5
	Guatemala	2.8	5	2.4–3.2
	Moldova	2.8	7	2.5–3.3
	Mozambique	2.8	8	2.5–3.1
	Rwanda	2.8	5	2.3–3.3
	Solomon Islands	2.8	3	2.4–3.1

Table 5 (continued)

Country rank	Country/territory	2007 CPI score[a]	Surveys used[b]	Confidence range[c]
	Uganda	2.8	8	2.5–3.0
118	Benin	2.7	7	2.3–3.2
	Malawi	2.7	8	2.4–3.0
	Mali	2.7	8	2.4–3.0
	São Tomé and Príncipe	2.7	3	2.4–3.0
	Ukraine	2.7	7	2.4–3.0
123	Comoros	2.6	3	2.2–3.0
	Guyana	2.6	4	2.3–2.7
	Mauritania	2.6	6	2.0–3.3
	Nicaragua	2.6	6	2.3–2.7
	Niger	2.6	7	2.3–2.9
	Timor-Leste	2.6	3	2.5–2.6
	Vietnam	2.6	9	2.4–2.9
	Zambia	2.6	8	2.3–2.9
131	Burundi	2.5	7	2.0–3.0
	Honduras	2.5	6	2.3–2.6
	Iran	2.5	4	2.0–3.0
	Libya	2.5	4	2.1–2.6
	Nepal	2.5	7	2.3–2.7
	Philippines	2.5	9	2.3–2.7
	Yemen	2.5	5	2.1–3.0
138	Cameroon	2.4	8	2.1–2.7
	Ethiopia	2.4	8	2.1–2.7
	Pakistan	2.4	7	2.0–2.8
	Paraguay	2.4	5	2.1–2.6
	Syria	2.4	4	1.7–2.9
143	Gambia	2.3	6	2.0–2.6
	Indonesia	2.3	11	2.1–2.4
	Russia	2.3	8	2.1–2.6
	Togo	2.3	5	1.9–2.8
147	Angola	2.2	7	1.8–2.4
	Guinea-Bissau	2.2	3	2.0–2.3
	Nigeria	2.2	8	2.0–2.4
150	Azerbaijan	2.1	8	1.9–2.3
	Belarus	2.1	5	1.7–2.6

(Continued)

Table 5 (continued)

Country rank	Country/territory	2007 CPI score[a]	Surveys used[b]	Confidence range[c]
	Congo, Republic	2.1	6	2.0–2.2
	Côte d´Ivoire	2.1	6	1.7–2.6
	Ecuador	2.1	5	2.0–2.3
	Kazakhstan	2.1	6	1.7–2.5
	Kenya	2.1	8	1.9–2.3
	Kyrgyzstan	2.1	7	2.0–2.2
	Liberia	2.1	4	1.8–2.4
	Sierra Leone	2.1	5	2.0–2.2
	Tajikistan	2.1	8	1.9–2.3
	Zimbabwe	2.1	8	1.8–2.4
162	Bangladesh	2.0	7	1.8–2.3
	Cambodia	2.0	7	1.8–2.1
	Central African Republic	2.0	5	1.8–2.3
	Papua New Guinea	2.0	6	1.7–2.3
	Turkmenistan	2.0	5	1.8–2.3
	Venezuela	2.0	7	1.9–2.1
168	Congo, Democratic Republic	1.9	6	1.8–2.1
	Equatorial Guinea	1.9	4	1.7–2.0
	Guinea	1.9	6	1.4–2.6
	Laos	1.9	6	1.7–2.2
172	Afghanistan	1.8	4	1.4–2.0
	Chad	1.8	7	1.7–1.9
	Sudan	1.8	6	1.6–1.9
175	Tonga	1.7	3	1.5–1.8
	Uzbekistan	1.7	7	1.6–1.9
177	Haiti	1.6	4	1.3–1.8
178	Iraq	1.5	4	1.3–1.7
179	Myanmar	1.4	4	1.1–1.7
	Somalia	1.4	4	1.1–1.7

[a] ‘2007 CPI score’ relates to perceptions of the degree of corruption as seen by business people and country analysts, and ranges between 10 (highly clean) and 0 (highly corrupt).
[b] ‘Surveys used’ refers to the number of surveys that assessed a country’s performance. Overall, fourteen surveys and expert assessments were used, and at least three were required for a country to be included in the CPI.
[c] ‘Confidence range’ provides a range of possible values of the CPI score. This reflects how a country’s score may vary, depending on measurement precision. Nominally, with 5 per cent probability the score is above this range and with another 5 per cent it is below. Particularly when only a few sources are available, however, an unbiased estimate of the mean coverage probability is lower than the nominal value of 90 per cent.

Global Corruption Barometer 2007

Juanita Riaño[1]

Transparency International's Global Corruption Barometer 2007 ('the Barometer') seeks to understand how and in what ways corruption affects ordinary people's lives, providing a snapshot of the scope and scale of corruption from the view of citizens around the world.

The Barometer is a public opinion survey carried out for TI by Gallup International as part of its Voice of the People Survey. It has been published annually since 2003. In 2007 the Barometer polled 63,199 men and women aged fifteen and older in a sample weighted according to overall population structure in sixty low-, middle- and high-income countries and territories.[2] The fieldwork was carried out between July and September 2007.

The Barometer explores the experience of citizens with petty bribery when interacting with different institutions and public services. It also examines how members of the public expect the corruption problem to evolve in their country and how they rate their government's performance in fighting it.

Experience of bribery

The Barometer asked respondents about their contact with different service organisations and whether they had to pay bribes in their dealings with them.

According to the 2007 results, low- and middle-income citizens are the most affected by bribery when dealing with the eleven core institutions (presented in figure 3): 10 per cent of the high-income citizens who had contact with any of the included institutions reported paying a bribe, while 14 per cent of the low-income citizens reported the same.

Figure 3 shows that the experience of paying bribes differs greatly between the different organisations covered in the Barometer. Across the entire sample the police are by far the institution to which bribes were most commonly paid, followed by the judiciary. These findings raise significant concerns about corruption in the overall system of law enforcement. In contrast, utilities such as telephone services and gas providers are reported as the least affected by bribery.

Figure 4 shows that the extent of corruption in law enforcement varies enormously across regions when the data is disaggregated. Only a very small proportion of respondents from North America and the EU+ regional groupings have paid a bribe to the police and judiciary. In contrast, about a half of the respondents in Africa who had contact with the police in the past twelve months had paid a bribe. In between these two extremes, the other regional groupings all

1 Juanita Riaño is a research coordinator at Transparency International.

2 Countries included in the Global Corruption Barometer, as well as their regional classification, can be seen in table 6.

Table 6 Regional classification

Region	Countries
Africa	Cameroon, Ghana, Nigeria, Senegal, South Africa
Asia-Pacific	Hong Kong, India, Indonesia, Japan, Malaysia, Pakistan, Philippines, Singapore, South Korea, Thailand, Vietnam
EU+	Austria, Bulgaria, Czech Republic, Denmark, Finland, France, Germany, Greece, Iceland, Ireland, Italy, Lithuania, Luxembourg, Netherlands, Norway, Poland, Portugal, Romania, Spain, Sweden, Switzerland, United Kingdom
Latin America	Argentina, Bolivia, Colombia, Dominican Republic, Guatemala, Panama, Peru, Venezuela
South-east Europe	Albania, Bosnia and Herzegovina, Croatia, Kosovo, Macedonia, Serbia and Montenegro, Turkey
Newly independent states	Moldova, Russia, Ukraine
North America	Canada, United States

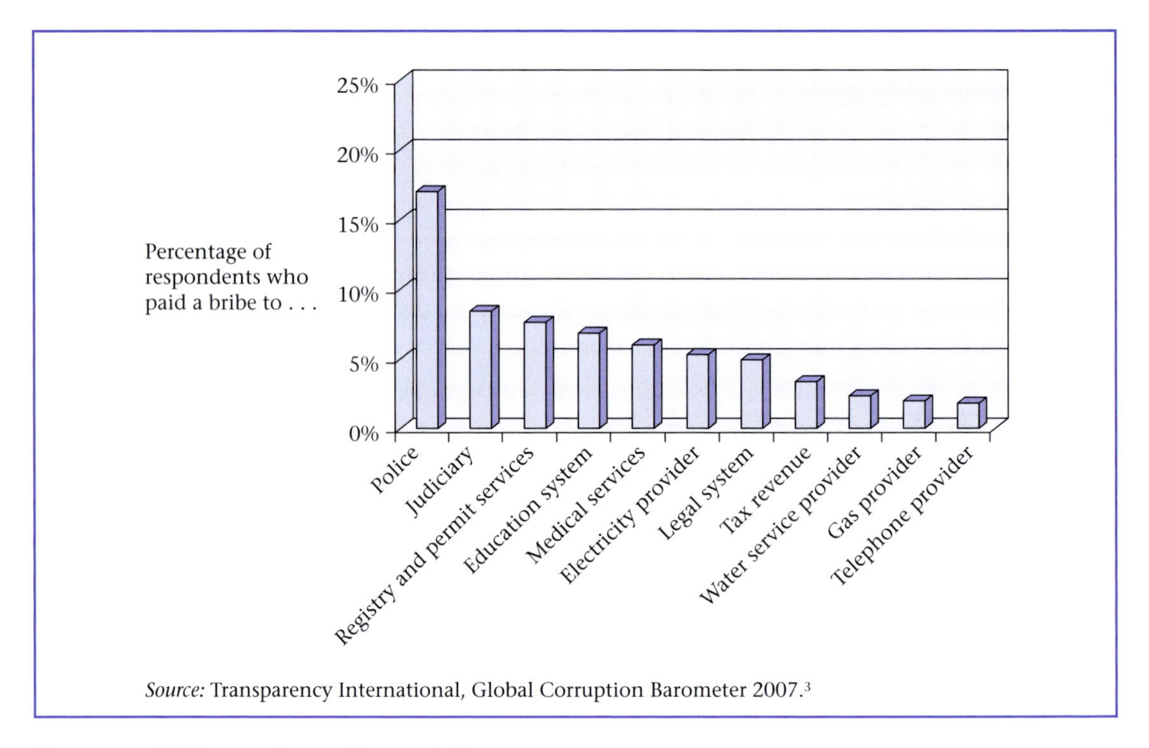

Source: Transparency International, Global Corruption Barometer 2007.[3]

Figure 3 Worldwide experience with petty bribery

3 Percentages are calculated for citizens who contacted the agencies seeking attention and weighted.

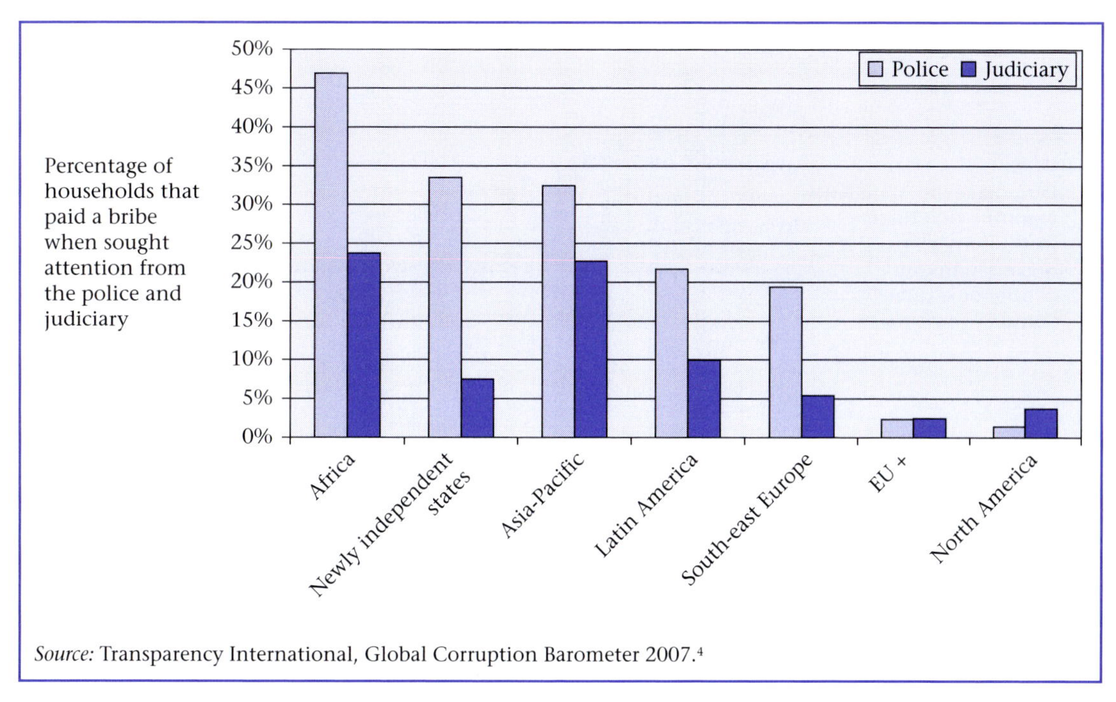

Source: Transparency International, Global Corruption Barometer 2007.[4]

Figure 4 Bribery to the police and judiciary by region

exhibit significant levels of corruption in the police and the judiciary. In Asia-Pacific and in the newly independent states one in three respondents who had contact with the police paid a bribe. In Latin America and South-east Europe the number varies between 15 and 20 per cent. Regional differences are equally prominent when analysing bribery in the judiciary: while 20 per cent of the citizens in the Asia-Pacific and Africa regions paid a bribe when in contact with the judiciary, only 2 per cent of respondents from the European Union and other Western countries in contact with the judiciary did the same.

Which sectors are most affected by bribery?

The Barometer asked households the extent to which they considered corruption to affect fourteen key institutions and services in their countries.

According to citizens' assessments, these services and institutions can be grouped into three categories:

- *very corrupt (considered as corrupt by more than a half of respondents)*, including political parties, parliaments/legislatures and the police;

4 Percentages are weighted and calculated for citizens who had contact with the agencies.

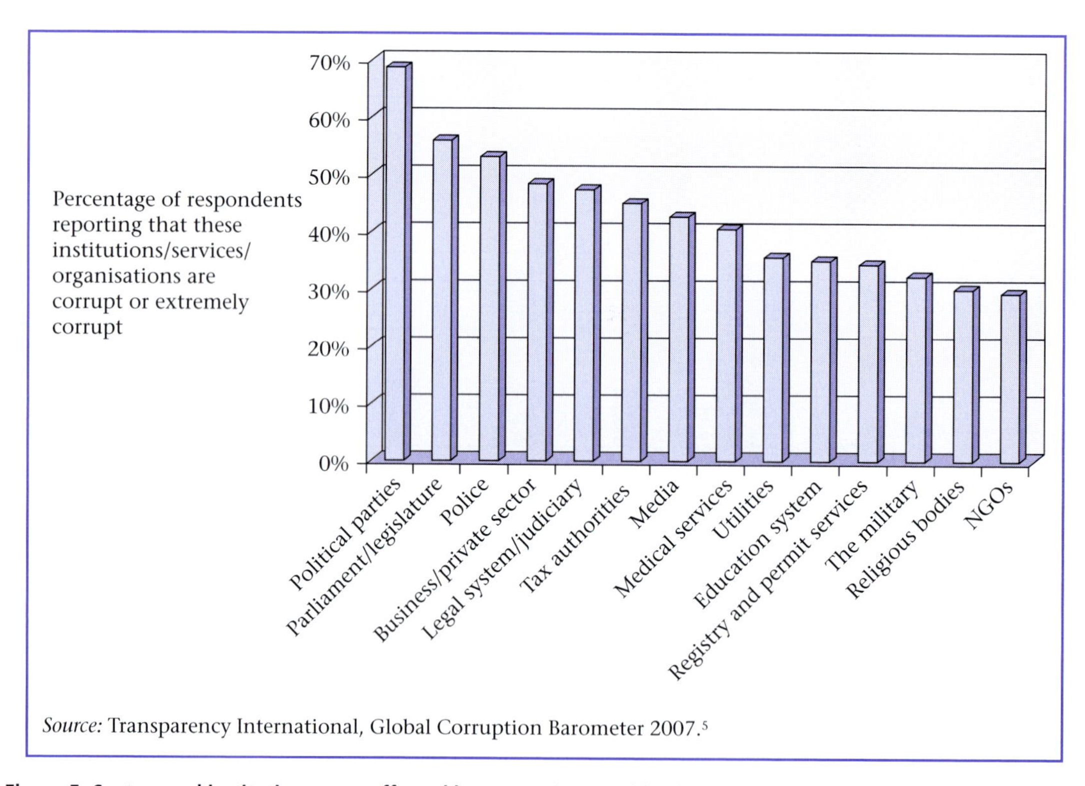

Source: Transparency International, Global Corruption Barometer 2007.[5]

Figure 5 Sectors and institutions most affected by corruption, worldwide perceptions

- *corrupt (considered as corrupt by more than a third of respondents)*, including the private sector, the media, the judiciary and legislative system, the medical services and the tax authorities; and
- *less affected by corruption (considered as corrupt by fewer than a third of respondents)*, including non-governmental organisations and religious bodies.

Do households around the world expect the level of corruption to change in the near future?

According to the 2007 results, 54 per cent of all interviewed households expect the level of corruption in their country to increase in the next three years. A closer look at these results reveals important differences across regions, however: 64 per cent of households in the Asia-Pacific region expect corruption to increase, whereas roughly a half of the interviewees in the European Union, Western Europe and the Americas foresee an increase in the level of corruption in their countries.

5 Percentages are weighted.

Generally speaking, citizens do not give their governments good marks in the fight against corruption. Only one-third of the interviewed households reported that their governments were being effective in this regard.

Global Integrity Report

Jonathan Werve and Nathaniel Heller[1]

Global Integrity is an independent non-profit organisation that tracks governance and corruption trends around the world. Using on-the-ground local research teams and blind peer review panels, the Global Integrity Report provides an expert assessment of the strengths and weaknesses of national governance and anti-corruption efforts by combining qualitative journalism with almost 300 data points for each country.

Approach

In 2006 Global Integrity conducted fieldwork in forty-three countries/territories including Argentina, Armenia, Azerbaijan, Benin, Brazil, Bulgaria, Cambodia (journalistic reporting only), the Democratic Republic of Congo, Egypt, Ethiopia, Georgia, Ghana, Guatemala, India, Indonesia, Israel, Kenya, Kyrgyzstan, Lebanon, Liberia, Mexico, Montenegro, Mozambique, Nepal, Nicaragua, Nigeria, Pakistan, the Philippines, Romania, Russia, Senegal, Serbia, Sierra Leone, South Africa, Sudan, Tajikistan, Tanzania, Uganda, the United States, Vietnam, the West Bank (journalistic reporting only), Yemen and Zimbabwe.

Our team of more than 200 journalists and researchers around the world systematically examined the laws, institutions and practices that prevent abuses of power and ensure that governments are responsive, and responsible, to their citizens.[2] Our methodology identifies the 'implementation gap' that exists when a country has anti-corruption laws or institutions on the books, but in practice suffers from poor implementation and ineffective enforcement of anti-corruption mechanisms. It allows researchers to unpack the index into distinct indicators, addressing both law and legal provisions and the practical application of the law.

Each country report contains the following:

- an integrity scorecard comprising 290 indicators that analyse the strengths and weaknesses of a country's integrity framework across six broad dimensions of governance;
- a journalistic narrative explaining how corruption manifests itself in everyday life for the average citizen;

1 Jonathan Werve is director of operations for Global Integrity, an NGO monitoring governance and corruption worldwide. Nathaniel Heller is the managing director of Global Integrity (www.globalintegrity.org).

2 For a list of researchers and journalists involved in the 2006 report, see www.globalintegrity.org/whoweare/team.cfm.

- a timeline of major corruption scandals and political milestones; and
- critical commentary and outside perspectives on both the qualitative reporting and quantitative data from a peer review panel of three to five local and international governance experts.

For a scorecard example, see annex 1.

Additionally, the Global Integrity Index aggregates the integrity indicators into an accessible cross-country comparative table, promoting active public discussion on governance and anti-corruption progress. The Global Integrity Index publishes each of its 12,000 data points – including scorer comments, references, and critical peer review feedback for every indicator – on the organisation's website.[3]

Score data is controlled for cross-country variation by an initial international review, followed by a peer review panel of three to five local and international experts, and finally an international scoring committee that reviews indicators across countries, ensuring that the narrative descriptions and peer review commentary correspond correctly with the numerical scores. Additionally, scorers and reviewers are given detailed scoring criteria that establish what specific situations trigger each score on any given indicator.

Highlights

Key findings from the 2006 Global Integrity Report (published in January 2007) include the following.

(1) Political financing is the biggest anti-corruption challenge facing many countries. Recent scandals in both poor and wealthy nations confirm that weak political finance regulation is a central driver of corruption. Global Integrity found that many developing countries seem destined to repeat the mistakes of more developed nations when it comes to regulating the flow of money into the political process. Thirty-five of the forty-one countries assessed earned the lowest possible rating in this area, with a median score of 38 out of 100.

(2) Weak legislatures threaten to undermine other crucially needed long-term anti-corruption reforms. Only lawmakers can pass the necessary anti-corruption legislation. Global Integrity found that legislative accountability at the national level is uniformly weak around the world compared to other branches of government.

(3) Vietnam, one of the world's fastest-growing economies, has the second weakest overall anti-corruption framework of the 2006 group of countries. This should be a cause for concern to prospective investors, particularly since the findings suggest that governance and corruption challenges in Vietnam are deeply rooted and systemic.

(4) Russia has made minimal progress in establishing and enforcing effective anti-corruption mechanisms compared to other Soviet Union successor states. The data confirm recent

3 See www.globalintegrity.org.

anecdotal evidence that the consolidation of power and the crackdown on the media in Russia have negatively affected overall governance.

(5) New EU countries Romania and Bulgaria display a gap in overall anti-corruption performance, with Romania exceeding the performance of Bulgaria. Both countries' relatively strong assessments compared to other 2006 countries, however, suggest that the 'carrots and sticks' EU accession process has been effective in promoting institutional reform in both countries.

(6) Weak access to information mechanisms and whistleblower protection threaten government accountability in almost every country. Laws ensuring citizens' right to access government information and protecting citizens who speak out against corruption are non-existent in some countries and poorly implemented or simply ignored in many others. A majority of the countries assessed earned the lowest possible rating in the 'Whistleblowing measures' subcategory; the same was true in the 'public access to information' subcategory. Armenia, Bulgaria, Lebanon and Serbia each received the worst possible assessment in all eight indicators addressing 'Whistleblowing measures'. A summary of the findings is presented graphically in annex 2.

Global Integrity was expected to publish the next Global Integrity Report in late 2007, covering fifty-five countries.

Annex 1

Example

Global Integrity Report 2006: Kenya scorecard

The Kenya scorecard highlights a lack of official accountability mechanisms in Kenya. Six subcategories earn the lowest rating, all of which relate to government accountability.

Category I Civil society, public information and media: 68 / weak
I-1 Civil society organisations: 81 strong
I-2 Media: 68 weak
I-3 Public access to information: 56 very weak

Category II Elections: 64 / weak
II-1 Voting and citizen participation: 89 strong
II-2 Election integrity: 86 strong
II-3 Political financing: 15 very weak

Category III Government accountability: 56 / very weak
III-1 Executive accountability: 49 very weak
III-2 Legislative accountability: 57 very weak
III-3 Judicial accountability: 42 very weak
III-4 Budget processes: 75 moderate

Category IV Administration and civil service: 70 / weak
IV-1 Civil service regulations: 49 very weak
IV-2 Whistleblowing measures: 66 weak
IV-3 Procurement: 89 strong
IV-4 Privatisation: 78 moderate

Category V Oversight and regulation: 89 / strong
V-1 National ombudsman: 93 very strong
V-2 Supreme audit institution: 87 strong
V-3 Taxes and customs: 88 strong
V-4 Financial sector regulation: 93 very strong
V-5 Business licensing and regulation: 84 strong

Category VI Anti-corruption and rule of law: 79 / moderate
VI-1 Anti-corruption law: 100 very strong
VI-2 Anti-corruption agency: 83 strong
VI-3 Rule of law: 67 weak
VI-4 Law enforcement: 65 weak

All scores range from 0 to 100. Data reflect conditions from July 2005 to June 2006.

Annex 2

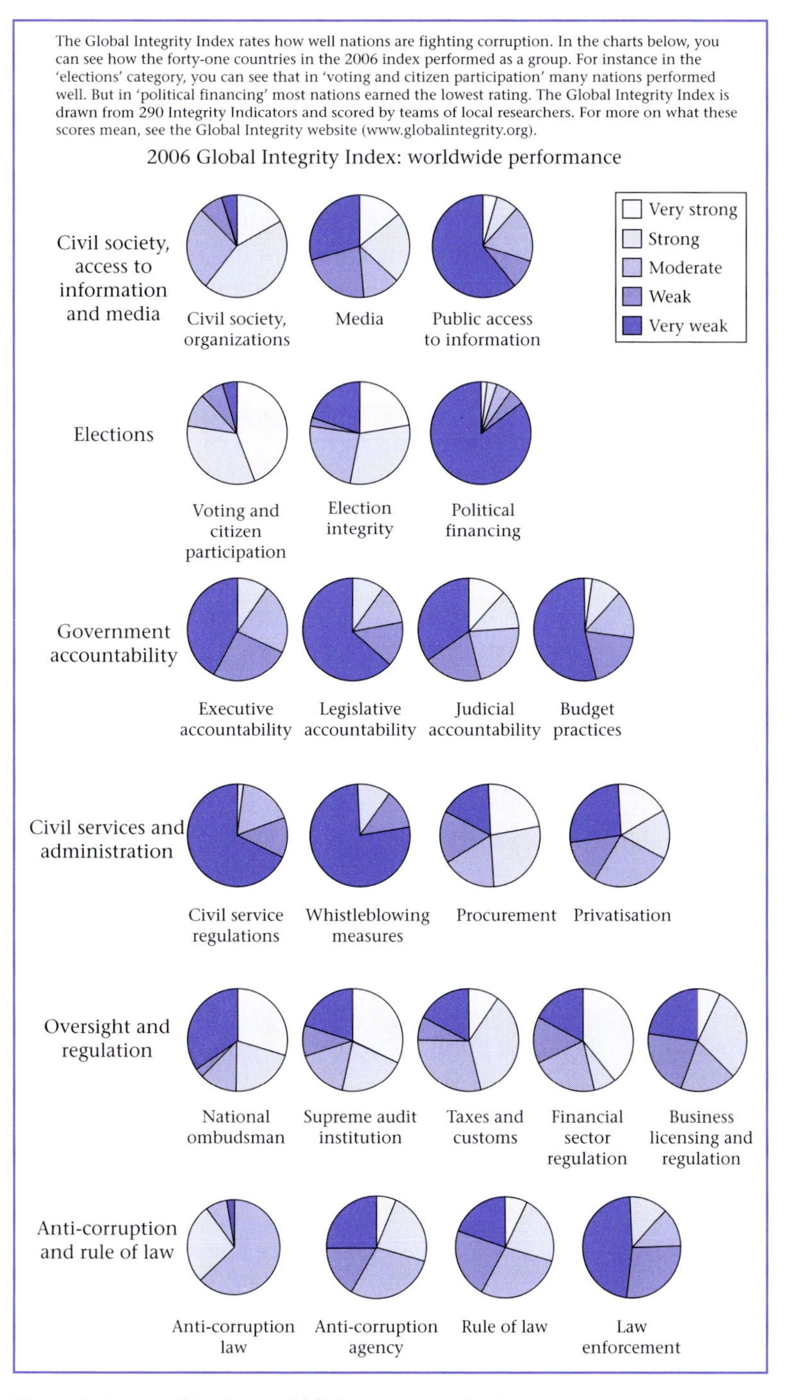

Figure 6 How well is the world fighting corruption?

The Americas Barometer 2006: report on corruption

Mitchell A. Seligson and Dominique Zéphyr[1]

The Americas Barometer (*El Barómetro de las Américas*), the survey effort of the Latin America Public Opinion Project, focuses on street-level corruption as experienced in the daily lives of citizens. In 2006 a total of 31,477 people were included in the sample, generally around 1,500 per country, in face-to-face interviews lasting an average of forty-five minutes. Interviews were conducted in the local languages; numerous indigenous languages were included in the Andes and Guatemala, while Creole was used in Haiti and French among the French-speakers of Canada. The 2006 study was expanded to include twenty countries, with representation of North America and the Caribbean now included.[2]

The focus on direct experience with corruption rather than on the perception of corruption has been the hallmark of the LAPOP studies published in past *Global Corruption Reports*. The wisdom of that decision is reinforced by the latest findings. For the twenty countries as a whole, the correlation between individual perception of corruption and individual reports of having been solicited for a bribe are extremely low, not rising above an r of 0.06. Indeed, Bolivia and Haiti, both in the group of countries with the highest recorded levels of actual corruption, have perceptions of corruption that are lower than any other country except Canada, a nation that scores at or near the very low end of corruption experience in the Americas.

While the study asks an entire battery of questions on corruption experience, direct comparisons of the signature item in the series are revealing. We asked: 'Have you been asked to pay a bribe by a public official in the last year?' The results are shown in figure 7. The chart includes an 'I' at the end of each bar to show the range of the confidence intervals of the samples. The yawning gap between the United States and Canada, on the one hand, and the high-level corruption countries on the other is striking; a person from Bolivia is fifty times more likely to be asked for a bribe by a public official than a person from the United States. Even in countries that are moderate in their levels of corruption, such as Costa Rica, where only 6.1 per cent of the sample reported being asked by a public official to pay a bribe in the last year, the rate is twenty times higher than in the United States.

1 Mitchell A. Seligson is Centennial Professor of Political Science at Vanderbilt University and director of the Latin American Public Opinion Project (LAPOP). Dominique Zéphyr is research coordinator and data analyst at LAPOP.

2 Because of the high costs of face-to-face interviews, in the United States and Canada alone surveys were conducted via random-digit-dialling phone calls, and samples there were around 600. All other samples were based on national sample frames, stratified by region and sub-stratified by urban/rural residence. Full details can be found at www.AmericasBarometer.org.

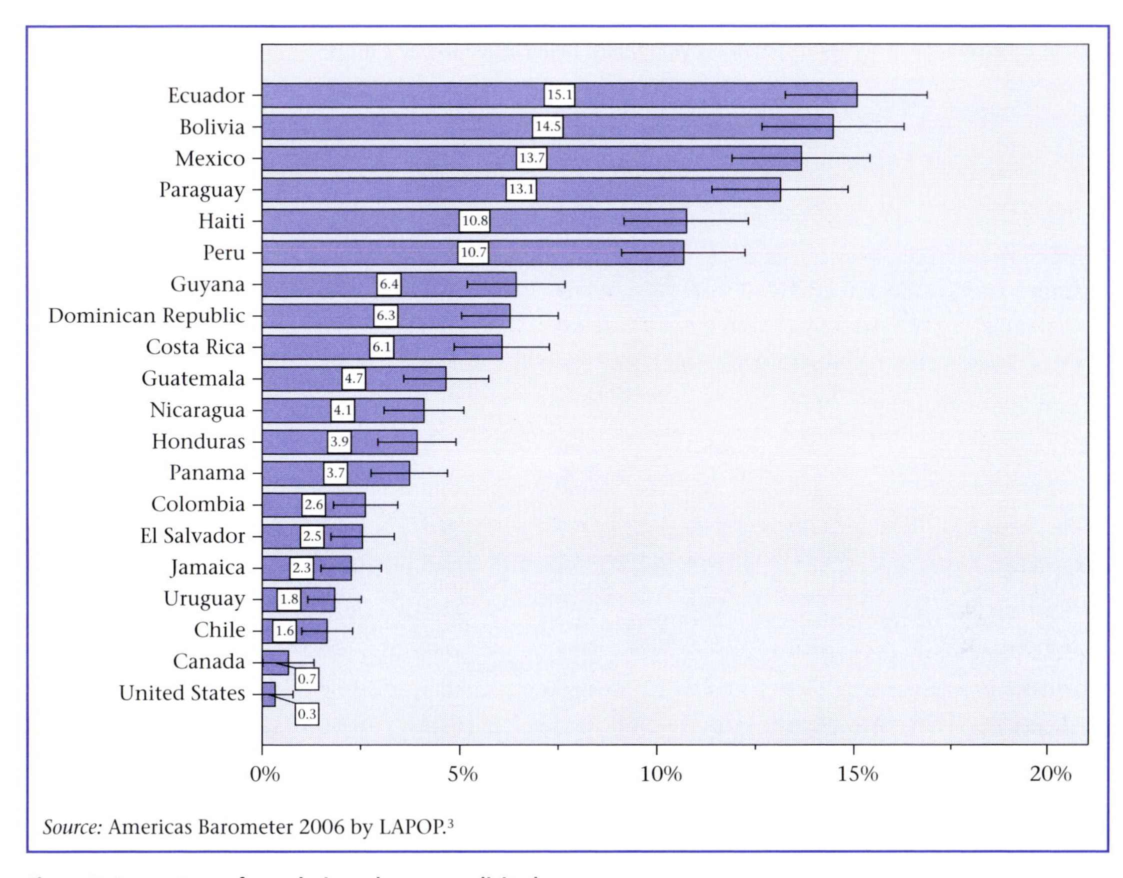

Source: Americas Barometer 2006 by LAPOP.[3]

Figure 7 Percentage of population who were solicited

For all countries in the sample apart from the United States and Canada, where the full series of questions was not asked, an overall index of corruption victimisation was constructed, based on the number of different ways in which a person was requested to pay a bribe in the twelve months preceding the survey. In total, seven different possible venues of corruption were measured, including bribe requests by police, in the courts, in local government, in the public health service, in the public school system, at work and by public officials. In Uruguay, which represents the best case, fewer than 1 per cent of the population were asked to pay a bribe in the twelve months preceding the interview. Haiti emerges as an extreme case, with one out of every two adults reporting being victimised. The average for the region was 22.5 per cent of a country's population being asked to pay a bribe. The results for the

3 The error bars represent 95 per cent confidence intervals.

Table 7 Percentage of survey respondents who report being asked to pay a bribe[a]

	Police bribery	Public employee bribery	Municipal bribery	Bribery work	Court bribery	Health service bribery	School bribery
Bolivia	20.5	14.5	24.1	12.5	19.0	10.2	10.2
Canada		0.7					
Chile	2.3	1.7	5.6	6.5	5.3	3.0	3.5
Colombia	4.5	2.6	4.4	3.6	3.3	3.7	1.8
Costa Rica	8.7	6.1	5.9	4.9	3.0	4.5	4.4
Dominican Republic	10.7	6.3	19.5	3.2	12.5	5.1	3.6
Ecuador	11.6	15.1	14.8	7.4	22.9	8.7	13.2
El Salvador	6.6	2.5	6.0	3.3	2.8	6.7	3.5
Guatemala	11.0	4.6	6.4	9.0	6.3	7.6	7.4
Guyana	11.8	6.4	13.4	16.7	10.1	13.6	
Haiti	10.2	10.8	61.9	51.1	50.2	57.7	59.6
Honduras	11.0	3.9	10.4	2.7	7.8	3.7	3.9
Jamaica	7.0	2.3	16.0	35.4	16.8	35.7	30.1
Mexico	22.8	13.7	24.0	13.4	25.0	13.7	12.7
Nicaragua	7.3	4.1	12.5	9.9	22.7	10.2	9.3
Panama	6.6	3.7	16.2	2.8	14.3	3.9	4.1
Paraguay	11.6	13.1	13.0	10.0	17.0	3.9	3.1
Peru	18.8	10.7	14.9	9.2	11.6	3.9	8.2
United States		0.3					
Uruguay	2.3	1.9	1.8	4.0	0.0	1.4	1.6

[a] Among those who used the public service described, except for 'public employees', which was a generic category without the 'filter' for users. Pre-tests revealed extremely low levels of corruption in the United States and Canada, and thus, to economise on precious interview time, for these countries the rest of the series was eliminated.

individual items in the series are shown in table 7, above, and the summary graph is shown in figure 8.

As in prior studies, the 'hot spots' of corruption are the cities, where more public officials are present to extract bribes from their victims. The data also reveal that males are far more likely to be victims than females, no doubt because of their greater dealings with public life than females in the Latin American and Caribbean environment. Finally, even though the poor may pay a higher percentage of their incomes in bribes, it is the wealthier who have the 'deep pockets' and are more likely to be seen as good targets for those who have bribery in mind.

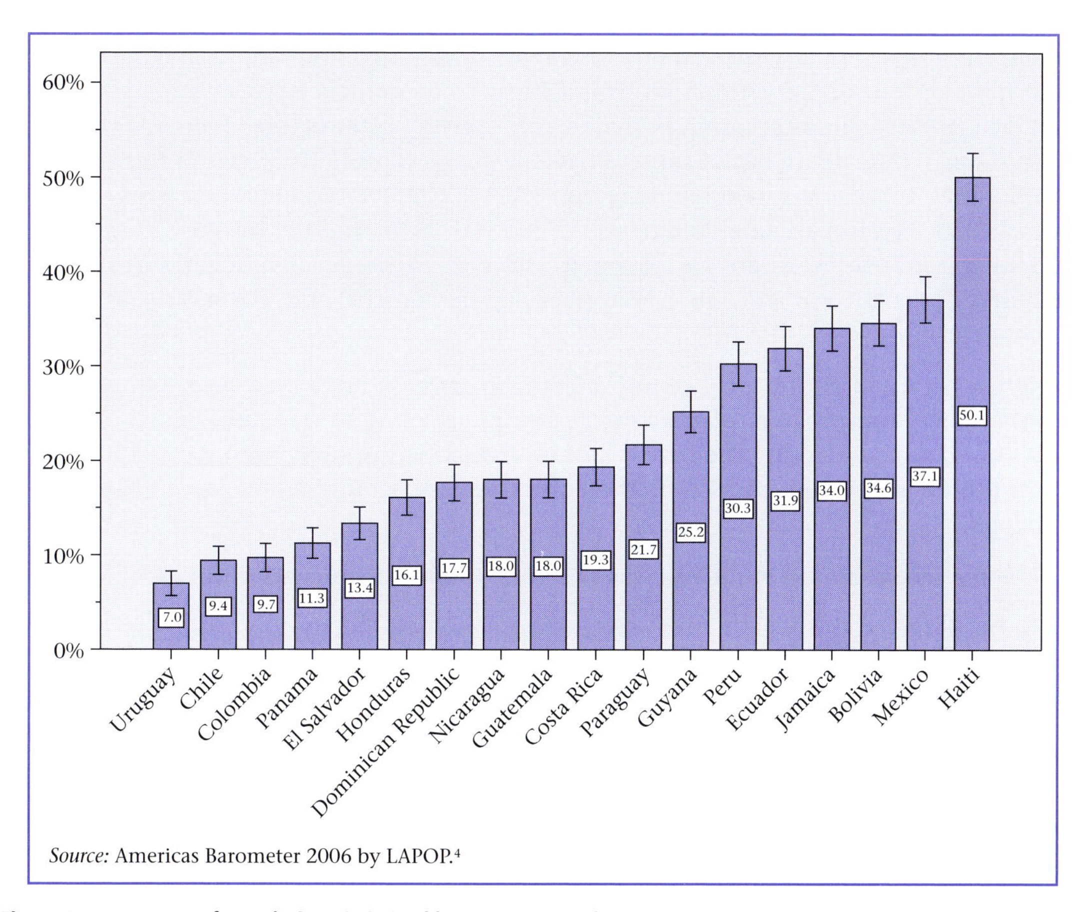

Source: Americas Barometer 2006 by LAPOP.[4]

Figure 8 Percentage of population victimised by corruption at least once in past year

4 The error bars represent 95 per cent confidence intervals.

The World Governance Assessment: corruption and other dimensions of governance

Verena Fritz, Ken Mease, Marta Foresti and Goran Hyden[1]

The fundamental idea of the World Governance Assessment (WGA), which was first developed in 2000, is to assess governance by capturing the views of local stakeholders. Local

1 Verena Fritz and Marta Foresti are research fellows at the Overseas Development Institute (ODI), London. Goran Hyden is a Professor of Political Science at the University of Florida and Kenneth Mease is a member of the Department of Economics at the University of Florida.

stakeholders may perceive governance problems differently from those making a judgement from the outside. Their opinion is less likely to be influenced by existing surveys – a problem that governance indices based on expert assessments increasingly face.[2] Furthermore, while a number of surveys have begun to capture the views of local business communities, the WGA is special in reflecting the views of ten stakeholder groups – ranging from civil society, to parliamentarians to the judicial sector and the business community.[3] Respondents to the WGA are also divided into two broad groups: 'regime incumbents' (members of the government, the civil service and parliament) and 'governance guardians' (all other groups).

The WGA is carried out by local coordinators who are based in a local research institution or NGO. Local coordinators play a crucial role in selecting respondents based on predefined criteria, in conducting the survey and in disseminating survey results locally. Table 8 shows the questions/indicators that the survey covers, arranged into six arenas and six principles.

In 2006 the second round of surveys was carried out in ten countries/territories.[4]

While only one of the WGA's thirty-six indicators refers directly to corruption, there are several others that are pertinent. This is particularly true with regard to 'grand' forms of corruption, such as the transparency of political parties and the degree to which policy reflects public preferences and legislators are accountable to the public.

Two important observations emerge from the WGA. First, among the ten countries/territories covered by the survey, some 'traditional' governance concerns are, in fact, relatively positively rated. Freedom of association and freedom of expression as well as freedom of the media are the three most highly rated indicators on average (and generally for all ten countries/territories individually as well). Peaceful competition for power is also highly rated, except in Kyrgyzstan and Uganda. Even government protection of private property rights is positively rated in nine out of the ten countries/territories, the exception being Kyrgyzstan.

Second, it is striking that the most negatively rated indicators all relate to corruption and associated concerns: the question with the lowest average score is the one directly addressing corruption. The next lowest average ratings are for a merit-based system for recruitment, the transparency of political parties and the efficiency of the judicial system. With some variation, these are seen as the main problem areas across the countries/territories surveyed. Figure 9 shows the single most highly rated indicator (freedom of expression – question 3) and the

---

2 Author conversation with Bertelsmann Transformation Index team.

3 For further information on the WGA, see www.odi.org.uk/wga_governance/. The ten stakeholder groups are: representatives of the academic community, business, international organisations, the judiciary, the media, non-governmental organisations, religious groups, civil servants, government/executive branch and parliament.

4 The countries/territories are Argentina, Bulgaria, Indonesia, Kyrgyzstan, Mongolia, Namibia, Palestine, Peru, Trinidad and Tobago, and Uganda. The survey was conducted with a minimum of seventy respondents per country.

Table 8 Governance arenas and principles covered by the WGA

Principle/ arena	*Participation*	*Fairness*	*Decency*	*Accountability*	*Transparency*	*Efficiency*
Civil society	1. Freedom of association	2. Society free from discrimination	3. Freedom of expression	4. Respect for governing rules	5. Freedom of the media	6. Input in policy-making
Political society	7. Legislature representative of society	8. Policy reflects public preferences	9. Peaceful competition for political power	10. Legislators accountable to public	11. Transparency of political parties	12. Efficiency of legislative function
Government	13. Intra-governmental consultation	14. Adequate standard of living	15. Personal security of citizens	16. Security forces subordinated to civilian government	17. Government provides accurate information	18. Efficiency of executive branch
Bureaucracy	19. Civil servants shape policy	20. Equal opportunities to public services	21. Civil servants respectful towards citizens	22. Civil servants accountable	23. Civil service decision-making transparent	24. Merit-based system for recruitment
Economic society	25. Private sector consulted on policy	26. Regulations applied equally	27. Governments respect private property rights	28. Regulating private sector to protect workers	29. Transparency in international trade policy	30. Interventions free from corruption
Judiciary	31. Non-formal processes of conflict resolution	32. Equal access to justice for all citizens	33. Human rights incorporated in national practice	34. Judicial officers held accountable	35. Clarity in administering justice	36. Efficiency of the judicial system

lowest-rated indicator (control of corruption – question 30) on average. (The top bar is the average for all countries/territories; the bars below show the averages for all seventy or so respondents per country/territory.)

A further interesting dimension of the WGA is the comparison between views of 'regime incumbents' and 'governance guardians'. For example, in Mongolia, 'governance guardians' have a far less favourable view than 'regime incumbents' about the accountability of legislators to the public.

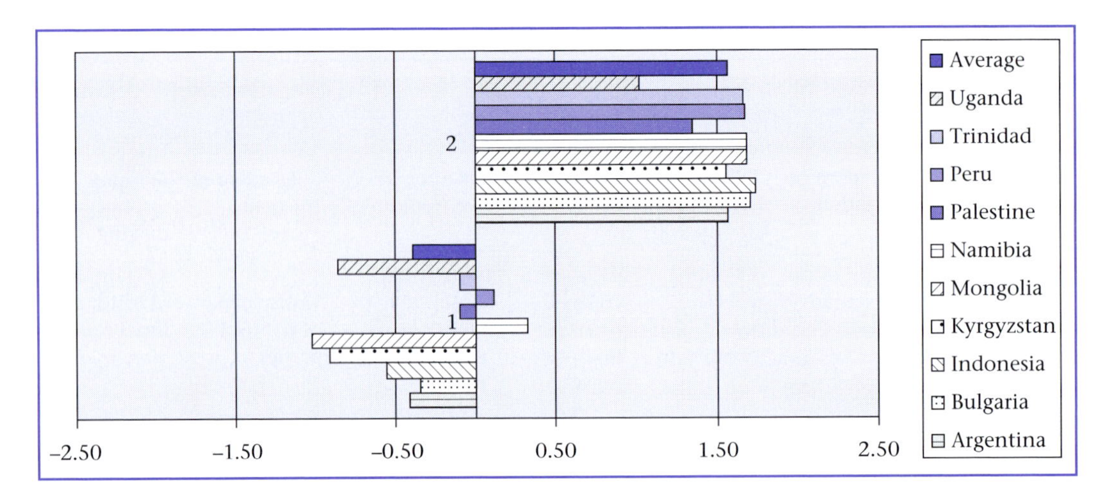

Figure 9 Aggregate ratings for freedom of expression (top) and for control of corruption (bottom)[a]

When interpreting these results, we need to keep in mind the fact that the countries/territories included in the survey are non-representative of the 'universe' of low- and middle-income countries. In a number of countries that are not covered, freedom of expression and freedom of association continue to be real concerns (e.g. Uzbekistan, Ethiopia, Vietnam). What is important is that, in those countries/territories where freedoms to raise issues ('voice') are rated rather highly by local stakeholders, the same stakeholders nonetheless express a rather negative view of public sector efficiency and accountability.

Donors and 'democracy promoters' have invested considerable efforts in establishing and improving democratic governance in low- and middle-income countries. Since the 1990s in particular they have sought to strengthen civil society and, more recently, to improve governmental transparency (through freedom of information acts and so on). These efforts are certainly worthwhile and appear to have achieved some effect in the countries/territories surveyed, but they do not seem to generate the expected results in terms of effective accountability and associated improvements in governance. This may be a matter of time. It is also possible, however, that there are other important factors (such as the prevalence of patronage, or high levels of inequality) that make citizens' voices less effective.

Given the limited and non-representative country/territory sample, we have to be careful in drawing generalised conclusions from the 2006 WGA. We hope that its findings on corruption and wider governance issues can provide interesting evidence, however, as well as some leads on issues worthy of further research and policy thinking.

a Average ratings for all respondents per country/territory on a five-point scale. Negative ratings indicate an average rating below the midpoint, positive ratings above the midpoint.

National Integrity System scoring system

Sarah Repucci[1]

The National Integrity System (NIS) is a concept developed by Transparency International to describe the sum total of laws, institutions and practices in a country that maintains governmental and private sector accountability and integrity. Since 2001 TI has used the NIS concept to produce more than seventy NIS Country Studies, which compare the on-the-ground situation in a country to standards for integrity and accountability. NIS Country Studies are a powerful tool for analysing the effectiveness of a country's integrity system and to facilitate the formulation of targeted and effective national anti-corruption strategies.[2]

NIS Country Studies currently offer only qualitative analysis. In order to evaluate strengths and weaknesses more effectively, as well as facilitate comparisons within a country over time, TI is developing an NIS scoring system to complement the current NIS qualitative methodology. NIS Country Studies and scores will be produced concurrently in order to be mutually reinforcing.

Methodology

Over the last two years TI has been working with its national chapters, external experts and consultants from Pact – an NGO committed to capacity-building with regard to local leaders and organisations – to design a scoring system that can be implemented by TI partners and other stakeholders in nearly any country in the world. Actors from different sectors working on promoting transparency and fighting corruption will be able to use the newly developed indicators to help develop strategies for advocacy and improve the ability to monitor and evaluate results over time, while government reformers can use the results for targeting resources and decision-making.

The scoring system will build on the sixteen areas or 'pillars' of the NIS.[3] These will be grouped into seven main elements: executive functions, the legal framework, the judicial system, independent bodies, civil society, the private sector and media/information. A final score will be compiled for each main element. In addition, sub-scores on accountability, transparency, integrity and complaints mechanisms, resources, and the role of each pillar in the system will be calculated.

Local ownership and adaptation

The NIS scoring system is based on in-country development and production. The starting point is the global scoring model, which specifies the fundamental qualities of integrity, accountability and transparency that are important for any governance system under consideration. To

1 Sarah Repucci is a programme officer at Transparency International.

2 More information on the NIS and completed Country Studies are available at www.transparency.org/policy_research/nis.

3 Executive, legislature, political parties, electoral commission, supreme audit institution, judiciary, civil service/public sector agencies, law enforcement agencies, public contracting system, ombudsman, government anti-corruption agencies, media, civil society, business sector, regional and local government, international institutions.

account for local differences, NIS scoring implementers in each country determine how the global criteria will be measured within the national context, creating a country scoring model. Through a participatory process, led by the NIS implementer, stakeholders will work together to translate global criteria into country-specific indicators and, in the case of adaptations, identify the most appropriate data sources and data collection strategies. Data gathered for the country model will then be aggregated according to the global criteria, in order to generate the final NIS scores.

To enable comparison among different NIS elements, all country-specific indicators are scored using the same scale. Thus, differences in scores among pillars of the system can help identify the weakest elements and target anti-corruption interventions.

Opportunities and challenges

The NIS programme is well established and respected, but the value of adding a scoring component has been emphasised by many stakeholders.

One challenge that TI has faced in the design of the NIS scoring system is how to balance the desire for some standardisation across countries with the wide variation that exists between different countries' information sources, systems and contexts. For example, a country with strong traditional authorities might want to add questions to take this potentially important source of governance into account. Similarly, a country where many integrity-related evaluations have already been carried out might want to incorporate these data rather than generating new information. The solution for respecting this diversity without sacrificing comparability is a framework methodology with room for some country-specific adaptation in data collection and analysis. While this means that the full set of country scores will not be strictly comparable with those of another country, the same core indicators will be published for all countries. Users interested in comparing results across countries can rely on this core dataset, while reform advocates in the country will benefit from the full, country-specific dataset.

A second challenge has been how best to complement existing indicators. Other organisations already produce quantitative indicators on governance at the country level, and TI is not seeking to compete with them. Rather, TI's global network of national chapters and its focus on anti-corruption measures create a unique niche in this field. The NIS scoring system will be generated by and for the countries under consideration, looking at the governance system through an anti-corruption lens. Leaving aside issues such as security or economic governance, it concentrates on political institutions, oversight bodies and other actors in society that have the potential and obligation to resist and fight corruption.

The role of the NIS scoring system in the field of governance measurement is therefore twofold. First, it enables national ownership and tailored application to bridge the divide between standardised global assessment and qualitative in-depth case study. Second, it allows a strong focus on anti-corruption policies, contributing much-needed detail to frameworks and initiatives that aim to measure the broader performance of governance systems.

The draft model will be piloted in two countries in 2008. A review process will follow, and the revised model will be launched later in the year.

9 Sectoral insights: capturing corruption risks and performance in key sectors

Promoting Revenue Transparency Project: from resource curse to resource blessing?

Juanita Olaya[1]

Oil, gas and minerals, or the 'extractive industries', generate great wealth. Oil export revenues alone were estimated to be US$866 billion in 2006.[2] This represents approximately 1.8 per cent of the world's GDP for that year and more than a half of the combined GDP for the world's fifty-three lowest-income nations.[3] While much of this wealth comes from developing countries, it does little to reduce high levels of poverty. In a perverse phenomenon that has been dubbed the 'resource curse', the high revenues generated by extractive industries undermine economic growth and fuel corruption, inequality and conflict. Poverty worsens as social investments are misappropriated or mismanaged. This in turn can weaken political cohesion and the rule of law.

Transparent resource governance is necessary to transform this curse into a blessing. Ensuring access to information about the money that companies pay to governments and that governments receive for oil, gas and mineral resources empowers citizens to hold governments and companies accountable, to monitor how the money is spent and to lobby for responsible public spending. If properly managed, revenues from natural resources provide a basis for poverty reduction, economic growth and sustainable development.

TI, collaborating with the Revenue Watch Institute and other partners, currently manages the Promoting Revenue Transparency (PRT) Project, which is grounded in the belief that transparency can help to reverse the resource curse, ensuring that extractive industry revenues benefit society directly. The project measures and compares the degree of extractive industry revenue transparency in oil and gas companies, host countries where production is taking place and home countries where the companies are based.

1 Juanita Olaya is the Promoting Revenue Transparency programme manager at Transparency International.

2 Nominal billions of dollars. Based on author's calculations from US Energy Information Agency – OPEC Revenues Fact Sheet and Major Non-OPEC Revenues Fact Sheet. For updated figures see www.eia.doe.gov/cabs.

3 Current world GDP in billions of dollars for 2006 is US$48,245 and the GDP for low-income countries is US$1,612. *World Development Indicators 2006* (Washington, DC: World Bank, 2006); calculations are ours.

In the current phase of the project TI has evaluated forty-two oil and gas companies and their upstream operations in twenty-one countries. Of these, twenty-three are state-owned companies (or nationally owned companies). This evaluation, published in a report, uses a questionnaire to assess the revenue transparency policies, practices and management systems of companies on the basis of publicly available information that is company sourced. Information was examined at both company headquarter level and in the countries of operation.

More specifically, the questionnaire covers the following areas.

- *Revenue payments transparency*: questions directly assessing the transparency of revenue payments. These cover the disclosure by companies of payments to the governments of the countries in which they operate through production entitlements, royalty payments, taxes, bonuses and fees, and whether these are disclosed on a country-by-country basis.
- *Transparency of operations*: questions on the disclosure of other details of company operations, regarding subsidiaries, contract details, current and future production volumes and the value of reserves, and financial information regarding revenues, production costs and profits. This enables citizens and investors to put revenue information into context and use it effectively.
- *Anti-corruption procedures*: questions on companies' anti-corruption procedures. Companies need to provide an environment that generally supports transparency and good governance in order for public disclosure and revenue transparency to be sustainable.

Early insights

Preliminary findings show how leaders in the sector are disclosing revenue data on a country-by-country basis, which should serve as a model for other companies that have already started to increase disclosure. Further efforts need to be made, however. The results also indicate that regulations by home governments that mandate revenue transparency are important in securing a level playing field and ensuring that voluntary efforts by leading companies are mainstreamed.

The PRT methodology is cutting-edge in the realm of research performed for advocacy purposes. The participatory engagement of all key stakeholders is one of the most essential principles. Multi-stakeholder engagement and consultation with companies, governments and CSOs is critical in the production of the reports and the advocacy aspects of the project. This approach produced a notable impact long before the *Companies Report* was actually published.

Company engagement has been a main component of the methodology and was sought in every step of the process, starting with the development of the questionnaire and including data verification. As part of the data-checking process, each of the companies covered in the report received its individual results for review. Many companies responded positively and worked constructively with TI to verify results. Establishing and sustaining engagement with stakeholders can be an extra challenge in the research process, but we are convinced that this is the right approach to achieving meaningful change – and to ending the resource curse.

Further information, including the full text of the *Companies Report* in various languages, can be found at www.transparency.org/policy_research/surveys_indices/promoting_revenue_transparency.

Crinis: measuring accountability, disclosure and oversight on who finances whom in politics

Bruno W. Speck and Silke Pfeiffer[1]

Context

Resources to fund elections and political parties, either private or public, are a necessary means to enable political competition. At the same time, political financing represents a risk for distorting the fairness of elections. This can pave the way for corrupt arrangements between elected candidates and their campaign donors that can lead to policy capture and policy distortions.

In addition to regulatory frameworks and state-enforcement, public control by political opponents, journalists, watchdog institutions and individual citizens plays an increasingly important role in overseeing how political finance influences the political process. Public access to comprehensive, reliable information in a timely manner is an important precondition to enable this kind of public control.

Crinis ('ray of light' in Latin) has been developed as a joint project by Transparency International and the Carter Center to assess (i) the accountability of political actors to enforcement agencies and the wider public, (ii) the quality of the information they provide and (iii) the performance of official oversight bodies and public watchdog groups. The assessment is then used as a basis for targeted advocacy for more transparent and accountable political finance systems. In its first round, in 2006, the *Crinis* project has been implemented in the following eight Latin American countries: Argentina, Colombia, Costa Rica, Guatemala, Nicaragua, Panama, Paraguay and Peru.

Concept and methodology

The project covers ten different dimensions of transparency and accountability with regard to political finance. These include the internal accountability of parties (dimension 1), reporting to state oversight institutions (2) and the disclosure of information on party and campaign

1 Bruno Wilhelm Speck is Professor of Political Science at the State University of Campinas, Brazil. Silke Pfeiffer is the director of the Americas Department at Transparency International.

finance to the citizens (3). A second set of dimensions assesses the quality of data in terms of comprehensiveness (4), the level of detail (5) and the reliability of reporting to state oversight institutions (6). A third group of indicators assesses the design and efficiency of preventive mechanisms (7) and sanctions (8), as well as the performance of enforcement agencies (9) and civil society oversight (10).

One of the innovative aspects of the project is that it moves beyond legal obligations for disclosure and assesses the effectiveness and usefulness of disclosure measures from a citizen's viewpoint. Would a citizen be able to find and understand specific information on political finance he or she may be interested in? Would citizens be able to research the amount of private donations and names of major donors that supported the incumbent president in the elections? We designed a field test, in which fifteen citizens, ten students and five journalists were given eight specific questions concerning the sources, recipients and amounts of funding for political parties and the most recent election campaign. The participants were drawn from volunteers. They were invited to use any channel of information at their disposal (internet, library, phone, personal visit) and to report back with their results five days later.

Research findings

Figure 10 shows the result of these field tests for all countries, aggregating three questions on regular party financing and five questions on campaign finance. The score is based on an average of all responses, where 0 means no information at all was obtained and 10 means full transparency. The result revealed significant shortcomings across all countries, with most countries achieving less than a third of available points, signalling unsatisfactory access to information on political financing.

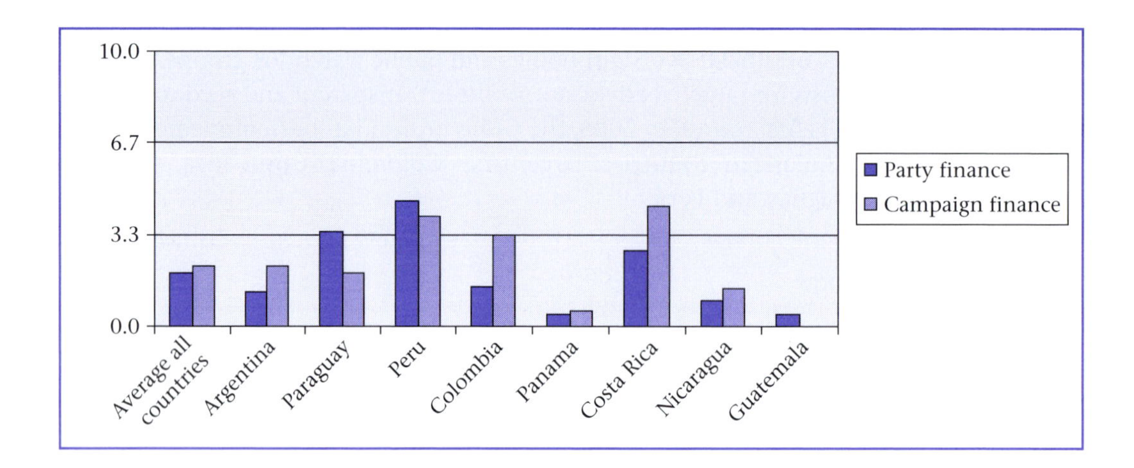

Figure 10 Field tests on access to information: citizens, students and journalists answering eight specific questions, on financing parties and elections in a time frame of five days

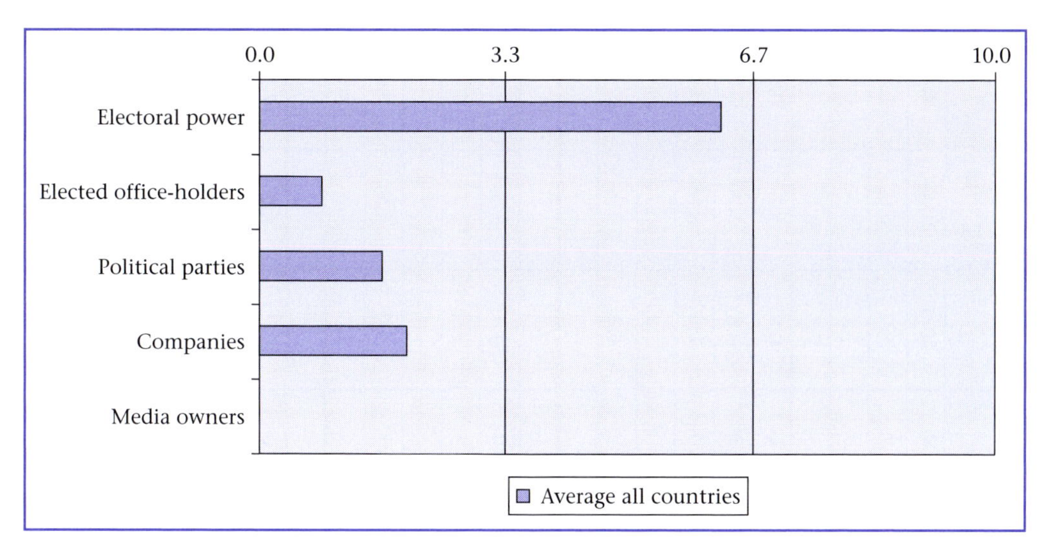

Figure 11 Responsiveness of different stakeholders: letters requesting information on party and campaign finance in a time frame of sixty days

A similar test assessed the responsiveness of different stakeholders to requests for information from the citizenry. The research team in each country sent out letters to parties, elected office-holders, private companies, owners of newspapers, radio and TV stations and the state agencies controlling party and campaign finance requesting detailed information on political finance issues. This field test was run for sixty days, and the research team followed up by phone to make sure the letters had reached their intended recipients. Figure 11 contains the results of this test.

The average results for all eight countries demonstrate the reluctance of all stakeholders except the electoral power (the official agency overseeing the elections) to disclose information on political finance to the citizenry. None of the media companies in any of the eight assessed countries replied to the request.

These tests, together with additional indicators, provide data for scoring a political finance system along the third *Crinis* dimension, its effective disclosure of information on political finance to the citizenry. For the overall *Crini*s assessment, research teams gathered additional information for each country from a broad range of sources, including analyses of pertinent laws and regulations and stakeholder surveys. This yielded a database with 140 indicators for each country, to inform the assessment of political finance and accountability along the ten *Crinis* dimensions mentioned earlier.

Crinis as a diagnostic tool

Crinis as a benchmarking tool makes it possible to identify both best practice and the specific shortcomings of a country's system of political finance (see table 9).

Table 9 Main results of *Crinis*

Dimensions[5]		Regional average	Argentina	Paraguay	Peru	Colombia	Panama	Costa Rica	Nicaragua	Guatemala
(1) Internal bookkeeping	Law	8.6	10.0	7.9	10.0	9.1	8.8	10.0	10.0	3.1
	Practice	5.1	4.5	4.2	4.3	5.3	5.7	5.0	6.0	5.4
(2) Reporting to the electoral management body	Law	4.0	4.7	2.8	5.7	4.6	7.0	3.6	2.7	0.9
	Practice	8.0	0.0	10.0	8.6	8.5	10.0	9.8	6.0	2.8
(3) Disclosure of information to the citizenry	Law	2.0	7.2	0.0	0.0	3.3	3.9	0.0	1.7	0.0
	Practice	2.8	3.8	1.9	5.0	3.8	2.3	3.6	1.3	1.0
(4) Comprehensiveness of reporting	Law	5.4	7.4	8.0	7.2	6.8	2.3	7.0	3.8	0.7
	Practice									
(5) Depth of reporting	Law	5.4	7.8	6.0	6.8	7.6	3.6	6.4	3.9	1.1
	Practice									
(6) Reliability of reporting	Law									
	Practice	4.8	5.1	5.1	4.9	5.0	5.0	4.4	4.3	4.3
(7) Preventive measures	Law	5.1	7.9	2.9	3.7	4.7	7.3	5.0	5.7	4.0
	Practice									
(8) Sanctions	Law	3.4	3.6	1.4	4.2	6.7	0.0	4.7	6.8	0.0
	Practice	4.1	6.9	3.8	3.0	6.8	4.9	1.5	3.3	2.6
(9) State control	Law									
	Practice	6.5	7.4	4.8	6.3	6.7	7.0	6.7	5.7	7.1
(10) Public control	Law									
	Practice	6.1	7.5	4.8	7.7	4.3	4.1	6.6	7.9	5.8

5 0.0–3.3 = insufficient; 3.4–6.6 = regular; 6.7–10.0 = satisfactory.

The assessment shows that a large gap exists between regulatory frameworks and their effective implementation, and that several dimensions of accountable political finance have not yet been sufficiently addressed through legal rules and regulations.

The results from this diagnostic exercise will inform specific advocacy activities that aim to bring on board local stakeholders, to strengthen democratic parties and to empower journalists and civil society activists.

The *Crinis* results underscore the fact that it is imperative to make political financing a priority on the region's agenda for political reform. The governance failures and irregularities caused in this area have a direct and lasting negative impact on democracy, policy-making in the public interest and the quality of life of citizens.

After the first round of application in Latin America *Crinis* went through a process of methodological refinement, and it is anticipated that it will be adapted and replicated in several countries in Africa, Asia and Eastern Europe.

10 Understanding the details: investigating the dynamics of corruption

Bridging the gap between the experience and the perception of corruption

Richard Rose and William Mishler[1]

Corruption has material impacts but there are big differences in how corruption is assessed. The media focus on cases of elite corruption involving contracts worth billions of dollars for defence procurement or exploiting natural resources, but these forms of corruption affect few citizens directly. Transparency International's Corruption Perceptions Index draws principally on the opinions of experts focusing on elite corruption. Perception is not the same as experience, however.

Corruption in the everyday delivery of health, education and social services can affect the mass of the population. A nationwide survey can turn attention from the perception of elite corruption to the first-hand experience that citizens have when contacting public officials in their community. This section draws on evidence from the New Russia Barometer (NRB) survey organised by the Centre for the Study of Public Policy (CSPP), University of Aberdeen, and conducted by the Levada Centre, Russia's oldest not-for-profit survey institution. A nationwide random sample of 1,606 adults was interviewed in their homes between 12 and 23 April 2007.[2]

Public officials are widely perceived as corrupt

When Russians are asked 'How widespread do you think bribe-taking and corruption are?' the results are unambiguous: five-sixths see officials as corrupt. The only difference is between those who see *almost all* public officials as corrupt (35 per cent) and those who see *most* as corrupt (51 per cent). Only 9 per cent see *fewer than a half* as corrupt and just 5 per cent think that *very few* officials are corrupt. The views of NRB respondents are consistent with TI's elite-oriented Corruption Perceptions Index.

1 Richard Rose is a professor at the University of Aberdeen. William Mishler is a professor at the University of Arizona. This section has been prepared with the assistance of a grant from the British Economic and Social Research Council, RES-062-23-0341, and also draws on the experience of working with the Transparency International 2006 Global Corruption Barometer data.

2 See www.abdn.ac.uk/cspp.

When asked about public services with which they are likely to be familiar, two-thirds or more perceive a majority of the police, doctors and hospitals, military service institutions, education, offices issuing permits and tax collectors as corrupt. Even social security, where the qualification for a pension is based on publicly available bureaucratic records, is usually seen as corrupt.

The chief sources of information about corruption are second-hand rather than through direct experience. For 86 per cent of Russians, television and newspapers are the important sources of information about corruption, and talking with friends and neighbours comes second in importance. For the media, instances of gross elite corruption are news, whereas petty corruption is not, and the media can provide the stuff of informal conversations with friends and neighbours.

If people see corruption as widespread this could encourage a 'race to the bottom', in which people accept it as part of their way of life. For example, bribes can be regarded as desirable if they produce what a person wants, such as a free place for a child at a good school or a secure public sector job. All the same, notwithstanding the widespread perception of corruption, 71 per cent of Russians do not think it acceptable to give officials a bribe, even if it is the only way to get something they want.

To determine their experience of corruption, Russians were asked whether anyone in their household had actually paid a bribe to the institutions that are usually perceived as corrupt. The result is strikingly clear-cut. Overwhelming majorities report no experience of bribery for each of seven services. For the median service, getting a permit or registering an activity, only 5 per cent had experienced corruption in the past two years (see figure 12). The gap between experience and perception found in Russia is consistent with findings of the 2006 Global Corruption Barometer published by Transparency International.

The explanation for limited experience is simple: contact with public officials is a necessary condition of paying a bribe, and in a two-year period most households do not use a given public service. For example, education and military service is not continuing but restricted to a particular phase of the life cycle. The one exception is the health service: three-quarters of households have used this service in the past two years.

Calculating the experience of corruption among those dealing with a given public service leads to the same conclusion: big majorities do not have to pay a bribe to use a particular service. More than nine-tenths of encounters with social security offices are honest, and so are three out of four contacts with the health system, education, permit offices and police.

The vulnerability of individuals to corruption is increased, however, by the fact that Russians are entitled to multiple services. For example, an individual may claim a social security pension and health care. In the past two years, 84 per cent had a household member contact at least one public service; the median household had dealt with two services; and one in ten had used five or more services. When all of a household's contacts with public services are taken into account, first-hand experience of corruption rises. A total of 23 per cent have paid a bribe for a public service in the past two years; 61 per cent have contacted public services

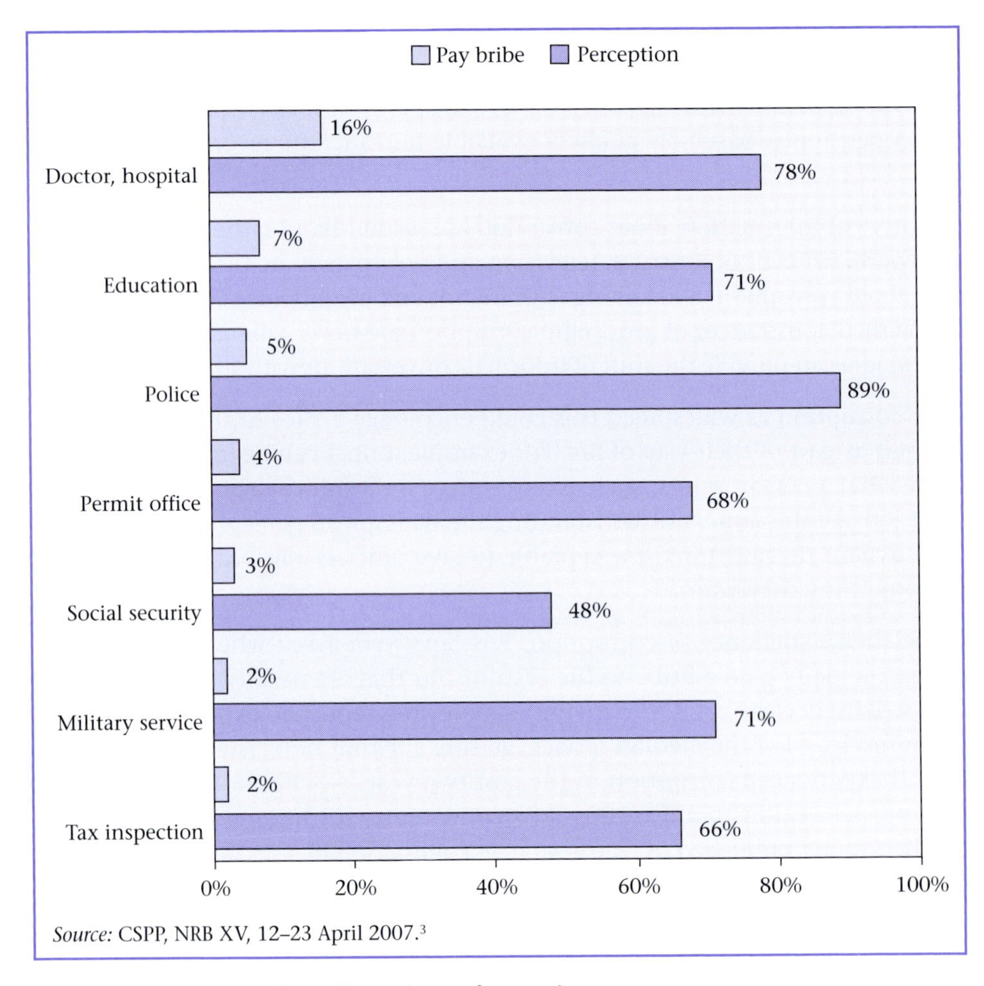

Source: CSPP, NRB XV, 12–23 April 2007.[3]

Figure 12 Gap between perception and experience of corruption

without paying a bribe; and 15 per cent have avoided being asked for a bribe by having no contacts with public officials.

In Russia, the evidence clearly shows that, contrary to popular perceptions and anecdotal examples, 'everybody' is *not* paying bribes to public officials. The services that offer the best opportunities for collecting bribes are a small fraction of all public services – for example, discretion in the issuance of big procurement contracts for military supplies, roads and civil engineering projects or the privatisation of valuable parts of Russia's state-owned economy. Services that are closely supervised or involve computers, such as the payment of social

3 The figure is based on two survey questions: 'To what extent do you think the following institutions are affected by corruption?' and 'In dealing with any of these institutions in the past two years, was it necessary for anyone in your household to give a bribe?'. Number of respondents = 1,606.

Table 10 Experience of contacts and of corruption

	No contact	Contact with no bribe *(percentage of respondents)*	Contact with bribe
Doctor, hospital	24	60	16
Police	75	20	5
Education	74	19	7
Permit office	78	19	4
Social security	67	30	3
Army recruiting	84	14	2
Tax inspections	87	11	2
Total	(23%)	(54%)	(23%)

Source: CSPP, NRB xv, 12–23 April 2007.[4]

security pensions, offer much less scope for corruption than services by street-level bureaucrats such as the police. Moreover, most public services are delivered by professionals. A requirement for delivering health care or education is that an individual must be a trained doctor, nurse, teacher or specialist, and a part of that training emphasises the ethic of helping people. For such professionals, the object of employment is not to maximise income through collecting bribes but to deliver services in accord with their training.[5]

Corrupt contracts can have a durable and pervasive impact. In the course of time most people will have contacts with most public services. Hence, sooner or later a public official will demand a bribe. Survey evidence indicates that, over half a dozen years, most Russian households will have paid a bribe. Since most Russians think it wrong to pay a bribe, having to do so can have a big impact on how people evaluate government, 'locking in' the belief that most officials are corrupt. There is a positive correlation between having paid a bribe and seeing most officials as corrupt.

Experience of corruption is spread throughout all categories of Russian society. Analysis of the New Russia Barometer survey and the Global Corruption Barometer 2006 shows that being prosperous or poor does not protect people from the clutches of extortionate officials. Educated or uneducated, people can be advised to 'get smart' and pay off an official if they want a public service. Young and old and men and women are equally at risk of being asked to pay a bribe.

4 The table is based on two survey questions. 'In the past two years have you or anyone in your household contacted any of the following public institutions?' and, if the answer is 'Yes', 'In dealing with this institution, was it necessary to give a bribe?'.

5 D. Galbreath and R. Rose, 'Fair Treatment in a Divided Society: A Bottom Up Assessment of Bureaucratic Encounters in Latvia', forthcoming in *Governance*, vol. 21 (2008).

Even though the annual incidence of bribery is limited, its pervasiveness means that every Russian household is vulnerable to being asked to pay a bribe. The combination of the public's moral aversion to corruption and vulnerability to being compelled, sooner or later, to pay a bribe creates a popular demand for government to take action to reduce bribery. While corrupt governments tend to be less responsive to citizens than governments high in integrity, they pay a price in diminished political support. The longer the Russian government tolerates corruption among its officials, the more this leads citizens to become dissatisfied with how their country is ruled.[6]

6 R. Rose *et al.*, *Russia Transformed: Developing Popular Support for a New Regime* (New York: Cambridge University Press, 2006).

Corrupt reciprocity

Johann Graf Lambsdorff[1]

In addition to being deterred by penalties, corrupt actors are perhaps even more influenced by such other factors as the expected opportunism of their counterparts. This suggests a novel strategy for fighting corruption – the 'invisible foot' – whereby the unreliability of corrupt counterparts induces honesty and good governance even in the absence of good intentions.[2]

To test this proposition, an experimental corruption game was carried out with first-year economics students at the universities of Clausthal and Passau, Germany.[3] Students from Clausthal were assigned the role of businesspeople who requested being awarded a contract despite offering low-quality work. In the first round, 180 valid questionnaires were collected from participants containing some personal information. In the second round, 176 students at the University of Passau assumed the role of public servants and chose between whistleblowing, opportunism (refusing to provide the favour in spite of acceptance of a payment) and reciprocity (by awarding the selected contract to the anonymous bribe-payer). In the third round, students from Clausthal (businesspeople) could decide on whether or not to blow the whistle on the behaviour of their counterparts. Participants were presented with the following pay-off matrix:

All the participants in Clausthal and in Passau were shown figure 13. Starting from an endowment of €25, the businessperson gives €20 (as a gift or bribe) to the public servant, resulting in an initial endowment of €5. He or she would win a further €35 as a profit from the contract in case of reciprocity and lose €5 if someone blew the whistle. The public servant obtains a

1 Johann Graf Lambsdorff holds the chair in economic theory at the University of Passau, Germany, and is a research adviser with Transparency International.

2 See J. Lambsdorff, *The Institutional Economics of Corruption and Reform* (Cambridge: Cambridge University Press, 2007).

3 See J. Lambsdorff and B. Frank, 'Corrupt Reciprocity', Economics Faculty Discussion Paper no. 51-07 (Passau: University of Passau, 2007).

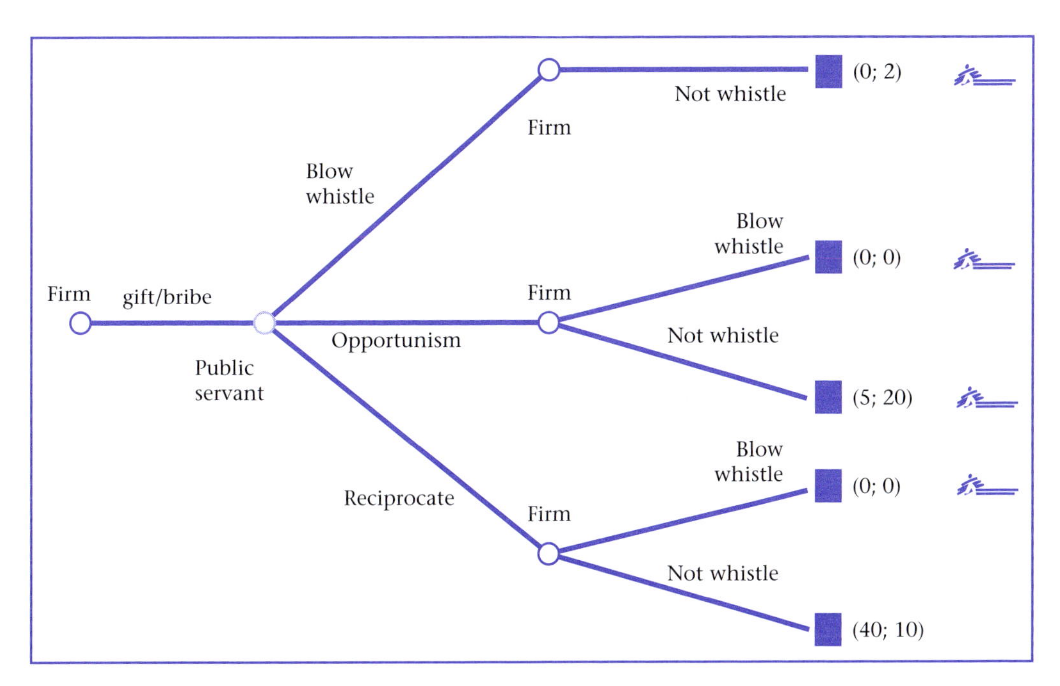

Figure 13 Corrupt reciprocity: the pay-offs to students[4]

pay-off of €20 (gift or bribe) from the businessperson. He or she would have to pass on €10 for arranging the awarding of the contract (reciprocity). Upfront whistleblowing induces confiscation of the gift or bribe but a bonus of €2. If the corrupt transaction were incomplete (either due to opportunism or whistleblowing) no damage would be imposed on society. This was considered in the game by an €8 donation to Médecins sans Frontières.

Forty-nine *public servants* out of 176 in Passau preferred to blow the whistle upfront. As figure 14 shows, a considerable number of public servants reciprocated the bribe, although this goes along with a lower individual pay-off than opportunistic behaviour. The apparent reason is the risk of retaliatory behaviour by *businesspeople* in the final step, who were observed to blow the whistle in twenty-one cases when being confronted with an opportunistic public servant. This behaviour is a departure from income maximisation. It may be motivated by the desire to take revenge for having been cheated by an opportunistic public servant (negative reciprocity). This prospect of retaliation also means that there is a risk for *public servants* to engage in opportunism. Either this risk or positive reciprocity ('Be kind to those who are kind to you') may thus motivate public servants to complete the corrupt transaction, rather than act opportunistically.

The choice of *public servants* was considerably different between male and female participants. Female students were considerably more likely to behave opportunistically and less likely (at a 1 per cent error level) to reciprocate. There is a related finding at the University of

4 The numbers in parenthesis indicate pay-offs in euros (businessperson; public servant). The logo of Médecins sans Frontières indicates an €8 donation.

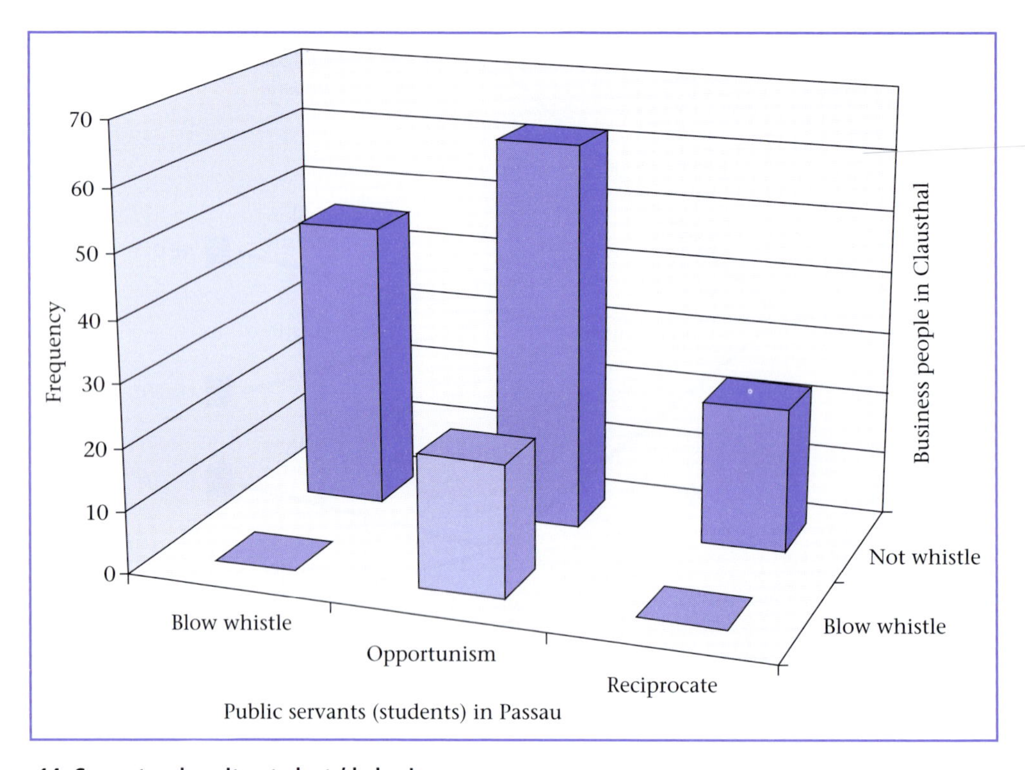

Figure 14 Corrupt reciprocity: students' behaviour

Clausthal, where female students were significantly less likely to blow the whistle – that is, to retaliate after they had been cheated by an opportunistic public servant.

Finally, students in Clausthal were at the outset asked whether they would prefer their payment to be called a 'gift' or a 'bribe': twenty-five questionnaires with a 'gift' wording and twenty-five with a 'bribe' wording were auctioned to students. Those with a lower willingness to pay were assigned a randomly chosen questionnaire. While some preferred the milder 'gift' wording, others expressed (also in written questionnaires) preference for a 'clearer language'. As revealed in figure 16, students who preferred the 'bribe' wording were more willing to retaliate when they were cheated.

We thus observe two different approaches to bribing public servants. While transferring a 'gift' is preferred because it appears less offensive and demanding, a bribe is chosen precisely for the opposite reason: it is more demanding and clearer that reciprocity is expected, including the threat to retaliate in the case of opportunistic behaviour. We found empirical evidence on these differences.

In July 2007 the same game was repeated at a summer school, with forty senior prosecutors and fraud investigators from various continents. The findings were consistent with the ones reported here, making it plausible that they are of general validity beyond the calculus of sophomore students.

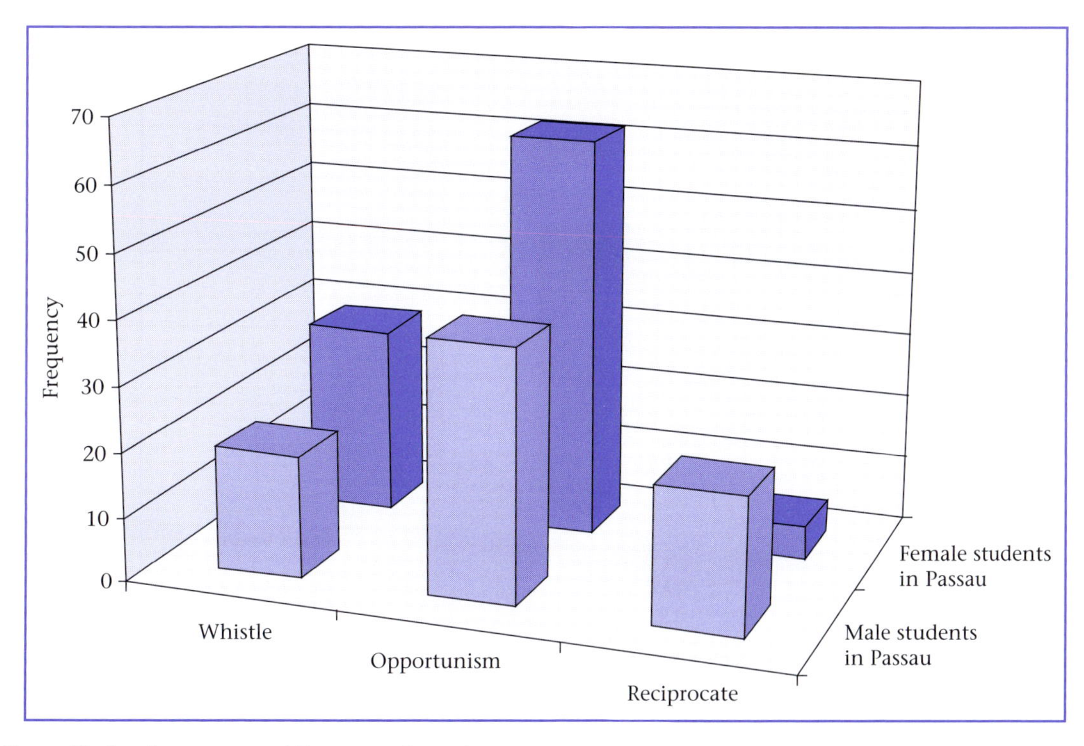

Figure 15 Gender matters: public servants' reaction

Corrupt transactions require trust and cooperation among the criminal partners, because the hidden agreements are not enforceable by courts. Women appear less inclined to engage in this type of trusted corrupt cooperation. This is also corroborated by related empirical evidence, which states that countries with more women in parliament and in the labour force are less affected by corruption.[5]

Corrupt actors must be deterred from their criminal actions. But deterrence involves more than just the threat of suffering from legal sanctions. It also encompasses the risk of being cheated by one's counterpart. There are a number of implications for anti-corruption policies, including:

- strong incentives for the 'good' whistleblowers (those who act upfront or after having completed a corrupt transaction);
- measures to deter 'bad' whistleblowers (those who threaten to retaliate after being cheated, and thus exert pressure on their counterpart to complete the corrupt deal);
- lenience for public servants who report their misconduct (threats of penalties may force them to reciprocate otherwise);

5 For a review of the related evidence, see J. Lambsdorff, 2007, pp. 34–5.

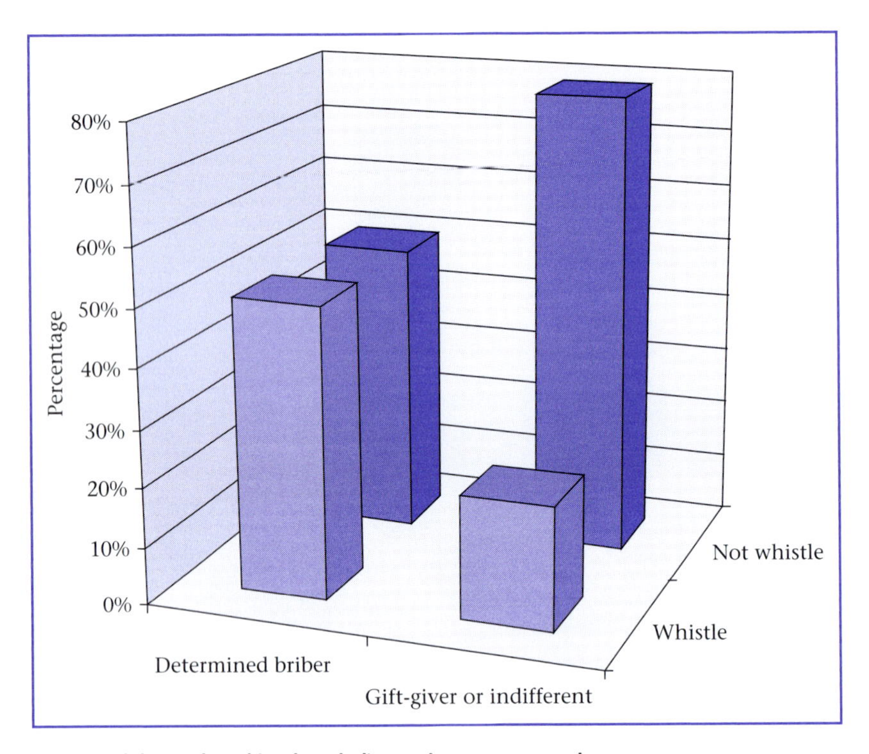

Figure 16 Businesspeople's (students' in Clausthal) reaction to opportunism

- penalties for all businesspeople, and those who were cheated by public servants; and
- the stronger involvement of women in teams, which can be expected to limit the trust in corrupt collusion in male-dominated networks.[6]

The power inherent in economic thinking first became apparent with Adam Smith's notion of the *invisible hand*. Competition substituted for benevolence by guiding self-seeking actors to serve the public. Individual morality lost relevance as a guiding principle for directing behaviour in private markets. May this also be true for politics and administration? Can anti-corruption policies flourish without good intentions? Will anti-corruption policies come to a standstill if they focus on moral sanctions, which may be detrimental to civil liberties? With respect to fighting corruption we may not have a mechanism as powerful as the *invisible hand*.

The experiment reveals a substitute, however. Corrupt actors cannot credibly promise reciprocity. The sphere of illegality may act as a powerful deterrent to engage in corrupt activities in the first place. The risk of betrayal may operate like an *invisible foot*, making life hard for

6 A formal treatment of the game and resulting policy recommendations can be found in J. Lambsdorff and M. Nell, 'Fighting Corruption with Asymmetric Penalties and Leniency', CeGE Discussion Paper no. 59 (Göttingen: University of Göttingen, 2007).

those who fail to adhere to honesty. The intention to cheat the public is kept in check by the fear of being cheated by the partner in the corrupt transaction, presenting honesty as a viable strategy for the self-interested individual.

The simple economics of extortion: evidence from trucking in Aceh

Benjamin A. Olken and Patrick Barron[1]

Can the behaviour of market participants and their pricing strategies under different market conditions also shed light on the strategies of corrupt officials for extorting bribes? A unique empirical study of bribery along trucking routes in Indonesia offers fascinating evidence on how insights from industrial organisation theory can help explain corrupt exchanges, with important implications for the design of anti-corruption policies.[2]

Research design

The study focused on two major long-distance transport routes in Aceh, Indonesia: the Meulaboh road and the Banda Aceh road. From November 2005 to July 2006 surveyors accompanied truck drivers on 304 trips on these two routes to and from Aceh and recorded the frequency, amount and type of corrupt exchanges that took place along the way. As indicators of bargaining power, the surveyors also recorded whether corrupt officials were equipped with guns or could have drawn on support from colleagues in their vicinity in case trouble arose.

During the course of the survey, the Indonesian government withdrew over 30,000 police and military personnel from Aceh. The removal of these officials led to a reduction of over a half in the number of checkpoints along one road, creating an exogenous change in the 'market structure' for illegal payments in the affected area and making it possible to study the effect on bribery prices.

Research findings

The researchers identified three forms of illegal payments along the routes: payments at checkpoints operated by police and military officials; payments to avoid a fine for carrying excess cargo; and protection payments made to criminal organisations to avoid trucks being hijacked

1 Benjamin A. Olken is a fellow at Harvard University. Patrick Barron works with the World Bank in Indonesia.
2 For the full study, see B. Olken and P. Barron, 'The Simple Economics of Extortion: Evidence From Trucking in Aceh', Working Paper no. 13145 (Cambridge, MA: National Bureau of Economic Research, 2007).

or having cargo stolen, or to lower the costs of bribes paid to avoid fines for carrying excess cargo. The surveyors recorded over 6,000 of such illegal payments. On average twenty illegal payments were recorded per trip, amounting to 13 per cent of the costs for the trip and exceeding the combined wages of the truck driver and his assistant. These findings underscore the fact that corruption is rampant in trucking goods throughout this part of Indonesia, adding a significant premium to transportation costs.

The study also reveals that the corrupt behaviour of officials along the way is not only the product of individual incentives and sanctions, but responds to overall changes in market structures (e.g. the number of bribe-takers along the way) in accordance with what industrial organisation theory would predict. From the latter perspective, the sequence of checkpoints along the way can be viewed as a chain of vertical monopolies. All these monopolies are the sole suppliers of different inputs (free passage at all checkpoints) that are indispensable for producing a specific economic value (shipping cargo from A to B).

With over a half of the checkpoints closed on one part of the road, the average price of bribes for the remainder of the road was found to increase. In total it did not reach the previous overall price tag, however. These results are consistent with pricing behaviour in a chain of decentralised monopolies whose total number drops. Distance to destination also mattered, consistent with the theory of hold-up. The closer the checkpoint to the destination and therefore the higher the potential loss from not being able to complete the journey, the higher the price of the bribe was found to climb.

In line with standard market behaviour, corrupt officials were also found to adjust the price of the bribes according to their perceived bargaining power. If the officer had a gun visibly displayed, payments increased by 17 per cent on average, while each additional officer present at the checkpoint drove bribes up by 5 per cent. These factors also increased the likelihood of active price negotiation rather than a hand-off of a bribe without discussion. In addition, the estimated willingness of truck drivers to pay also influenced the bribe discrimination. Drivers of trucks older than twelve years or carrying cargo of lower value paid lower bribes than those of newer trucks or trucks with more precious freight.

Bribe-taking that follows all these principles of market-pricing behaviour has important implications for anti-corruption policies. First, where bribes are decentralised, reducing the number of corrupt officials may be an effective way to reduce the overall price for bribes. This is not obvious, given the possibility that the remaining corrupt agents could simply increase their prices and extract the same overall corruption charge. Second, stamping out a centralised system of corruption without ensuring that corruption does not reappear in a decentralised manner may make matters even worse than before, since agents that cannot coordinate their activities may end up producing a higher bribery burden than if they could coordinate their behaviour. Increased attention to the context or 'market structures' of corruption can therefore offer valuable guidance for those attempting to dismantle such market places of corruption.

Corruption, norms and legal enforcement: evidence from diplomatic parking tickets

Ray Fisman and Edward Miguel[1]

Is corruption mainly a matter of weak legal enforcement or one of social norms? Since many societies that collectively place less importance on rooting out corruption will have both weak anti-corruption social norms and also less effective legal enforcement, it is difficult to disentangle the relative effects of law versus norms of behaviour. But understanding the relative importance of these potential causes of corruption is of central importance in reforming public institutions to improve governance: if corruption is predominantly controlled through social norms, interventions that focus exclusively on boosting legal enforcement may not be sufficient. At the same time, the effectiveness of legal enforcement in rooting out corruption bears on current debates in foreign aid policy for development economics, including the recent World Bank focus on legal measures to improve governance.

Diplomats and parking as a natural experiment[2]

From November 1997 to the end of 2002 diplomats stationed at the United Nations in New York City accumulated over 150,000 unpaid parking tickets – more than US$18 million in outstanding fines. The unruly parking behaviour of diplomats (including the remarkable performance of one Kuwaiti diplomat, who garnered over 1,000 tickets during that period) is a clear example of abuse of public office.

But this natural experiment is also an ideal setting to disentangle the roles of legal enforcement and cultural norms in controlling corruption. Diplomatic immunity, originally intended to protect diplomats and their families from mistreatment abroad, is now more commonly viewed as the 'best free parking pass in town'. Thus one immediate implication of diplomatic immunity – not just in New York, but also in national capitals worldwide – has been that it allows diplomats to park illegally but never suffer any threat of legal punishment, leaving a 'paper trail' of illegal acts with no hard consequences (see table 11).

We use this paper trail to evaluate the importance of norms in a country's government bureaucracy in explaining corruption. In the absence of any legal enforcement, the decision to obey the law and park legally is left to the conscience of each diplomat, which in turn may be directly affected by the norms of the diplomat's home country. But, if cultural norms do not

1 Ray Fisman is Professor of Economics and Finance at Columbia University, New York. Edward Miguel is Professor of Economics at the University of California, Berkeley.

2 For more details, see R. Fisman and E. Miguel, 'Corruption, Norms and Legal Enforcement: Evidence from Diplomatic Parking Tickets', Working Paper no. 12312 (Cambridge, MA: National Bureau of Economic Research, 2006).

Table 11 Average unpaid annual New York City parking violations per diplomat for selected countries, November 1997 – November 2005

Parking violations rank	Country	Violations per diplomat, pre-enforcement (11/1997–11/2002)	Violations per diplomat, post-enforcement (11/2002–11/2005)	TI Corruption Perceptions Index 2006
1	Kuwait	246.2	0.15	4.8
2	Egypt	139.6	0.33	3.3
3	Chad	124.3	0	2.0
4	Sudan	119.1	0.38	2.0
5	Bulgaria	117.5	1.67	4.0
10	Pakistan	69.4	1.23	2.2
24	Indonesia	36.1	0.75	2.4
27	South Africa	34	0.51	4.6
28	Saudi Arabia	33.8	0.53	3.3
30	Brazil	29.9	0.23	3.3
47	Italy	14.6	0.81	4.9
67	China	9.5	0.07	3.3
69	Venezuela	9.1	0.1	2.3
82	India	6.1	0.56	3.3
125	Guatemala	0.1	0.07	2.6
126	Switzerland	0.1	0	9.1
128	United Kingdom	0	0.01	8.6
145	Norway	0	0	8.8
149	Turkey	0	0	3.8

matter, we would expect all diplomats to abuse their parking privileges in the absence of legal punishment.

We find a striking pattern in the parking data: diplomats from high-corruption countries (based on Transparency International surveys) have significantly more parking violations in New York. We establish this relationship in a way that quantitatively accounts for the importance of cultural norms, as our measure of corruption (average unpaid tickets per year per diplomat for each country) is a result of home-country norms, not law enforcement.[3]

3 It is possible that diplomats from high-corruption countries also face stronger internal sanction for minor offences committed under diplomatic immunity. We do not find that diplomats from such countries who accrue unpaid violations early in their careers have shorter tenure, however. While this is not conclusive evidence, it suggests that the behaviour we observe is the result of different norms, not different rules.

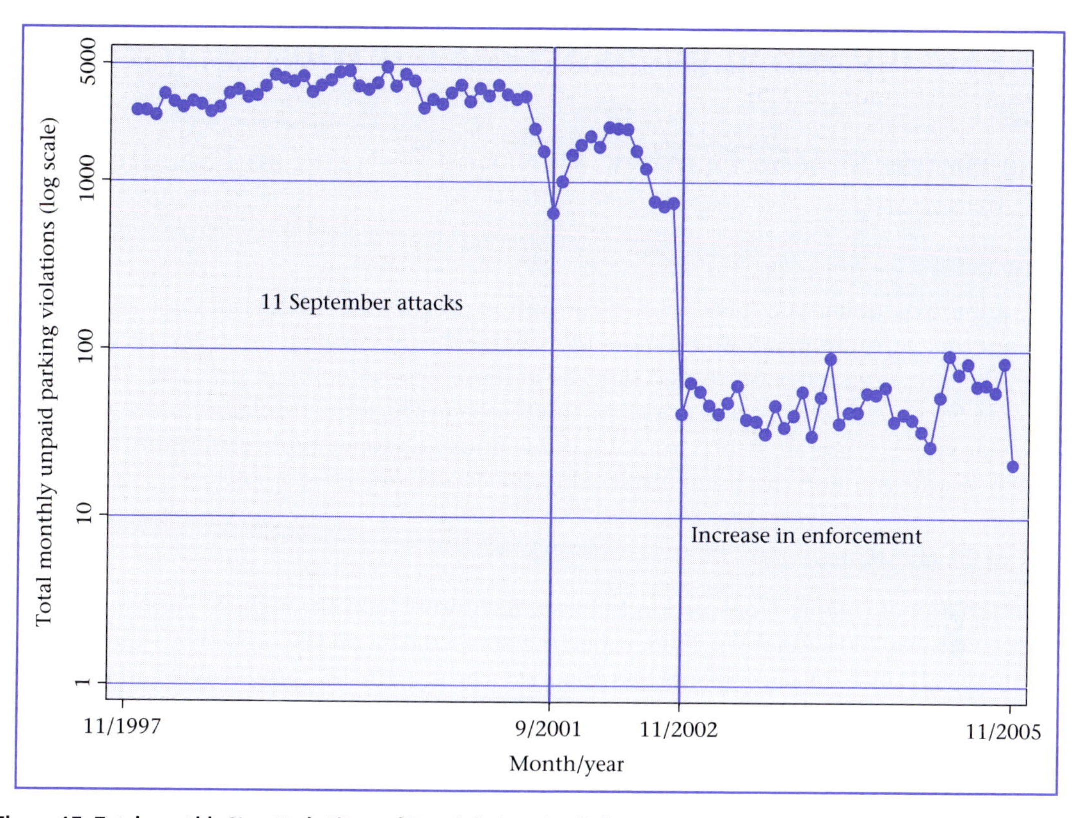

Figure 17 Total monthly New York City parking violations by diplomats, 1997–2005

The New York parking natural experiment also allows us to assess the role of legal enforcement by exploiting a sharp increase in the punishment for parking violations. After October 2002 the New York City government gained permission to seize the diplomatic plates of any vehicle with three or more unpaid violations. This credible increase in enforcement – a number of vehicles were actually made examples of by having their plates stripped in October 2002 – led to immediate and massive declines of approximately 98 per cent in parking violations (see figure 17). Yet our previous conclusion (that high-corruption countries produce more law-breaking diplomats) remains true even in this high-enforcement regime, just at lower average levels of parking tickets.

What are the policy implications? First, enforcement does work: legal sanctions are effective against corrupt government officials. It is worth noting, however, that most countries will not be able to increase legal enforcement instantaneously; the police and other enforcers may themselves have adopted norms of corrupt behaviour. Hence, while we do find that legal enforcement works, reforming attitudes and norms should also be an important element in efforts to reduce corruption and improve the rule of law.

Petty corruption in public services: driving licences in Delhi

Rema Hanna, Simeon Djankov, Marianne Bertrand and Sendhil Mullainathan[1]

While millions of dollars are spent on anti-corruption programmes each year, some analysts still maintain that corruption is nothing more than a tax: the process may be unjust or frustrating but, in the end, it provides goods and services to those who value them the most. Corruption may even 'grease the wheels' and speed up an all too cumbersome regulatory process. A study on how driving licences are issued in Delhi, India, finds this view highly misleading and shows precisely how corruption can dramatically alter the consequences of a policy.[2]

Research design

The International Finance Corporation followed 822 individuals through the licensing process between October 2004 and April 2005, collecting detailed data on the procedures and expenditures involved. Once the participants had obtained a licence the IFC administered an independent – and surprise – driving test to determine how well these individuals could actually drive. An experimental design was included in the study to reveal the efficiency implications of corruption.

Specifically, a randomly selected group of licence candidates were offered a bonus if they obtained a licence within the minimum legal time frame, thirty-two days (the 'bonus group'). A second randomly selected group (the 'lesson group') were given driving lessons. The remaining third served as a comparison group. This design allows for the evaluation of whether individuals with higher willingness to pay, or better qualifications, can obtain licences more easily than the 'average person'.

Research results

Table 12 presents the results, by experimental group. Individuals with a greater need to get a licence (the bonus group) were most likely to do so. They obtained the licence at the highest rate (71 per cent versus 48 per cent for the comparison group), and obtained it quickly (thirty-two days on average, against forty-eight days for the comparison group). While having a higher willingness to pay speeds up the licensing process, it does so at a social cost. Only a small fraction of the licence-getters in the bonus group (38 per cent) took the legally required driving test at the Regional Transport Office (RTO) and nearly 65 per cent of them failed the driving test independently administered by the IFC. This suggests a socially inefficient bureaucracy that allows unqualified drivers with high willingness to pay to obtain licences.

1 Rema Hanna (New York University); Simeon Djankov (World Bank); Marianne Bertrand (University of Chicago); Sendhil Mullainathan (Harvard University).

2 For more details, see M. Bertrand *et al.*, 'Obtaining a Driving License in India: An Experimental Approach to Studying Corruption', *Quarterly Journal of Economics*, vol. 122, no. 4 (2007).

The experience of the lesson group suggests that social considerations play some role in the allocation process. Under an extreme view of a corrupt bureaucracy, the allocation of licences would not depend on applicant qualifications, but only on willingness to pay. We find this is not the case: the lesson group is twelve percentage points more likely than the comparison group to obtain a permanent licence.

All groups spent much more than the official cost to obtain their licences. Individuals in the comparison and bonus groups paid, on average, about twice the official amount to obtain theirs. The lesson group did not pay much less than the other groups, suggesting that even the 'good drivers' had to resort to extra-legal payments to obtain their licences. Corruption in this context took a very different form, however. Very few licence-getters (1 per cent) paid direct bribes. Instead, almost all the extra-legal payments went to agents – professionals who operate as intermediaries between citizens and bureaucrats: 80 per cent of both the comparison and bonus groups and 59 per cent of the lesson group hired agents. Across the groups, individuals who used an agent paid almost double the amount to obtain a licence than those who did not (see figure 18). In return, they experienced an easier process: it took, on average, one week less to obtain a licence with an agent. Most startlingly, those who hired agents were, by and large, able to bypass the required driving exam and much more likely to fail the independently administered exam (53 per cent versus 23 per cent).

The corruption documented in this study undercuts the very rationale of the driving regulation, which is to keep bad drivers off the road. Corruption, therefore, not only raises the price of services, but also causes serious social distortions. These findings sharply contrast with a purely efficient view of corruption, and confirm that there are clear social returns to designing and implementing strong anti-corruption programmes. This study also shows that, even in a relatively simple and common process, corruption is often more complicated than a simple bribe passing from citizen to bureaucrat. In this case, most corruption appeared to operate through the agent system. In order to design anti-corruption programmes more effectively, future research should focus on these complexities.

Table 12 Obtaining a licence, by group[a]

	Comparison (1)	Bonus (2)	Lesson (3)
Obtained licence	0.48	0.71	0.6
Days to obtain permanent licence	48	32	53
Took RTO licensing exam	0.29	0.38	0.51
Failed independent exam	0.61	0.64	0.15
Total expenditures	1120	1140	964
Paid direct bribe	0.01	0.02	0.01
Hired agent	0.78	0.8	0.59

[a] Sample includes the 409 individuals who obtained a licence.

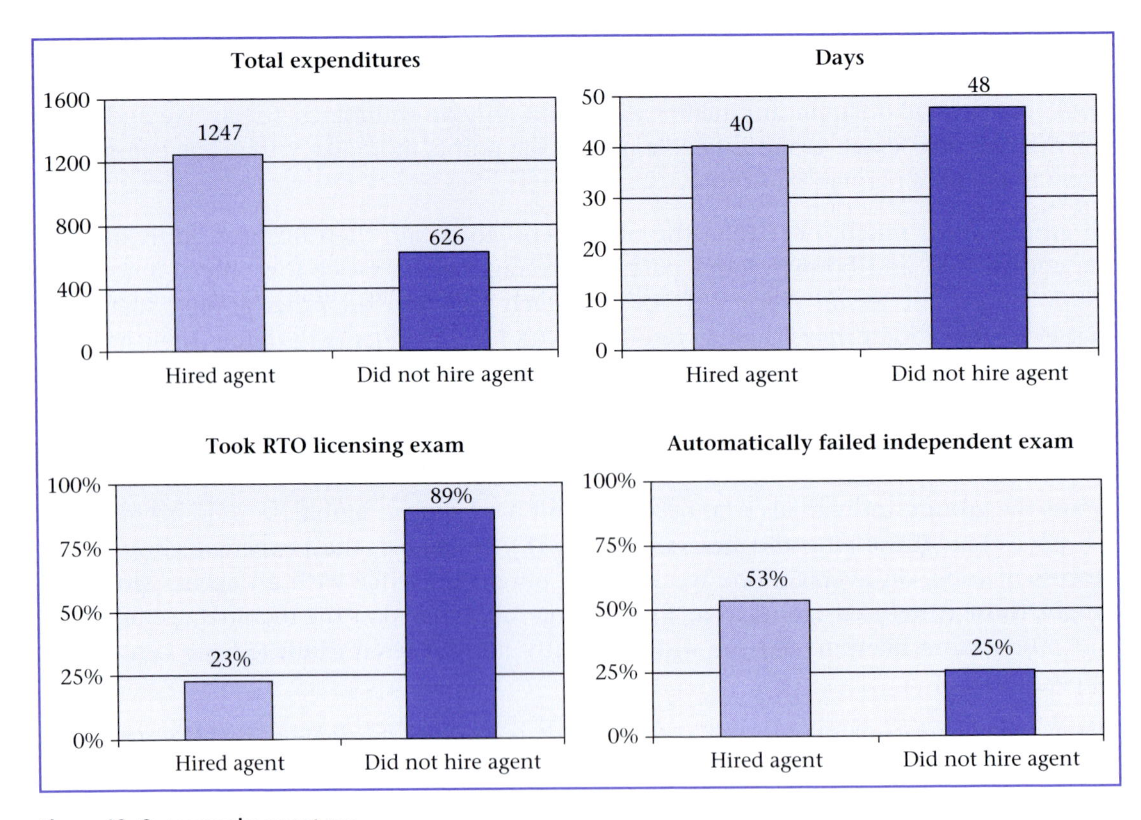

Figure 18 Outcomes by agent use

Corruption and institutional trust in sub-Saharan Africa

Emmanuelle Lavallée[1]

For many years 'efficient grease' theories prevailed in the analysis of corruption in the economic and political sciences. These theories argue that, in an environment in which levels of bureaucratic burden and delay are high, bribery is an efficient way to reduce the red tape, and therefore that corruption can improve economic and political development.[2] In political science, corruption is presented as facilitating political parties' development and the emer-

1 Emmanuelle Lavallée is an economist and research fellow at DIAL, a public institute for development research in Paris.

2 See N. Leff, 'Economic Development through Bureaucratic Corruption', *The American Behavioural Scientist,* vol. 8, no. 3 (1964), and S. Huntington, *Political Order in Changing Societies* (New Haven, CT: Yale University Press, 1968).

gence of a stable political environment. Corruption could also increase citizens' loyalty and trust towards their political institutions.[3]

This research project tests the central argument of 'efficient grease' theories. It challenges the idea that corruption may increase citizens' institutional trust, in particular the trust of those citizens who face a lot of red tape.[4]

This section's empirical basis is the Afrobarometer surveys. The Afrobarometer is an independent, non-partisan research project that measures the social and political atmosphere in Africa. The Afrobarometer surveys are conducted in more than a dozen African countries and are repeated on a regular basis. This study uses Round 2 surveys, which were conducted from May 2002 to October 2003 in fifteen countries: six austral African countries (Botswana, Lesotho, Malawi, Namibia, South Africa and Zambia), four eastern African countries (Uganda, Tanzania, Mozambique and Kenya) and five West African countries (Senegal, Mali, Cape-Verde, Ghana and Nigeria). Round 3 surveys, conducted in 2005, are used for Madagascar.

These datasets are particularly interesting for four major reasons. First, so far as can be ascertained, the corruption and trust nexus has never been explored in comprehensive empirical fashion in these countries, despite the fact that corruption is widespread in this area of the world. Second, these countries are young democracies; this makes an analysis of the consequences of corruption on these regimes' consolidation particularly important, since institutional trust and state legitimacy may be key elements to political stability. Third, the survey includes questions about both the experience with and perception of corruption. Thus the consequences of both these facets of corruption on institutional trust can be analysed. Fourth, the survey also contains information about citizens' perception of the quality of public services. Therefore the effects of corruption on institutional trust according to the level of red tape can be explored, and in this way 'efficient grease' theories can be rigorously tested.

The section relies on four key composite indicators drawn from the survey: institutional trust, experienced corruption, perceived corruption and bureaucratic quality. Institutional trust measures citizens' trust in political institutions such as the courts, the national government and political parties. Experienced corruption captures how frequently the respondents have had to pay a bribe in order to access a public service in the past year, while perceived corruption taps the popular perceptions of the overall prevalence of corruption among politicians and public officials. The quality of bureaucracy index assesses whether clients consider public services to be 'easy to use'. Every indicator uses a 0 to 10 scale, where 10 means a high degree of trust, quality of bureaucracy, experienced corruption or perceived corruption.

The following correlation coefficients show a negative and significant relationship between experienced and perceived corruption and trust in political institutions. Moreover, table 13

---

3 See D. Bayley, 'The Effects of Corruption in a Developing Nation', *Western Political Quarterly*, vol. 19, no. 4 (1967) and J. Becquart-Leclerq, 'Paradoxes of Political Corruption: A French View', in A. J. Heidenheimer *et al.* (eds.), *Political Corruption: A Handbook* (New Brunswick, NJ: Transaction, 1989).

4 An article was published on this research in *Afrique Contemporaine*, vol. 4, no. 220 (2006).

Table 13 Correlations between corruption and institutional trust according to the level of red tape

	Experienced corruption	Perceived corruption
Complete sample	−0.09 [0.00]	−0.14 [0.00]
Sample restricted to high level of red tape	−0.08 [0.00]	−0.14 [0.00]
Sample restricted to low level of red tape	−0.07 [0.00]	−0.13 [0.00]

Source: Author's computations.[5]

indicates that the strength and direction of that relationship does not change with the quality of public service delivery.

This section also uses a multivariate model[6] that controls for additional factors to analyse further the trust – corruption nexus and to test systematically the 'efficient grease' hypothesis. The calculations, performed on the complete sample of African countries, indicate that 'experienced' or 'perceived' corruption has a negative influence on institutional trust, whatever the quality of bureaucracy. Furthermore, the results show that these negative effects increase as the quality of administration increases.[7] For instance, when access to public services is seen as very difficult, a one-unit increase in perceived corruption leads to a drop in institutional trust of 2 per cent. In contrast, when access to public services is considered to be very easy, the same increase in corruption lowers institutional trust by almost 15 per cent.

In order to verify the soundness of these results the same calculation was performed for each country. Although the results may vary drastically from one country to another, they never indicate that corruption has a positive influence on institutional trust. In the best case, corruption has no impact on institutional trust, such as in Malawi, Namibia, Tanzania, Mozambique or Senegal. Every time corruption has a significant impact on institutional trust, it is negative. The central argument of 'efficient grease' theories can therefore not be substantiated. Higher levels of corruption most of the times coincide with low levels of institutional trust. Corruption does not appear to grease the working of institutions.

5 The significance level of each correlation coefficient is reported in brackets. The sample is spilt in two parts according to the quality of the bureaucracy reported by the respondent. The mean of the bureaucratic quality index is used as the splitting threshold (other thresholds have been tested, and these results appear to be quite robust).

6 Other covariates are introduced, such as demographic variables and variables denoting the economic situation of the respondent and his or her political attitudes and preferences.

7 This counterintuitive result is probably explained by the fact that corruption may be perceived as only one problem among many in a weak institutional environment.

Index

In this index:
(1) entries in **bold** denote main entries;
(2) the following abbreviations are used.

ACP	Panama Canal Authority
ADB	Asian Development Bank
AfDB	African Development Bank
BPCB	Business Principles for Countering Bribery
CDM	Clean Development Mechanism
CPI	Corruption Perceptions Index
CSOs	civil society organisations
EC	European Commission
ECAs	export credit agencies
EIA	Environmental Impact Assessments
EITI	Extractive Industries Transparency Initiative
EPFIs	Equator Principles Financial Institutions
EU	European Union
FOI	freedom of information
GDP	gross domestic product
GIPs	Governance Improvement Plans
IBA	International Bar Association
IFC	International Finance Corporation
IFIs	international financial institutions
IHA	International Hydropower Association
IMT	Irrigation Management Transfer
IPs	Integrity Pacts
IWRM	integrated water resources management
MDGs	Millennium Development Goals
NIS	National Integrity System
NRB	New Russia Barometer
ODA	Official Development Assistance
OECD	Organisation for Economic Co-operation and Development
OPIC	Overseas Private Investment Corporation
PIM	Participatory Irrigation Management
PPAs	Power Purchase Agreements
PPPs	public–private partnerships
PRT	Promoting Revenue Transparency Project
TI	Transparency International
UK	United Kingdom
UN	United Nations
UNCAC	United Nations Convention Against Corruption
UNCESCR	United Nations Committee on Economic, Social and Cultural Rights
UNCLNUIW	United Nations Convention on the Law of the Non-Navigational Uses of International Watercourses
UNESCO	United Nations Economic, Scientific and Cultural Organisation
UNMD	United Nations Millennium Declaration
US	United States
WCD	World Commission on Dams
WGA	World Governance Assessment
WIN	Water Integrity Network
WINS	Water Integrity National Survey
WRM	water resources management
WSSCC	Water Supply and Sanitation Collaborative Council
WUAs	water users associations